Oxford Series in Ecology and Evolution
Edited by Paul H. Harvey, Robert M. May, H. Charles J. Godfray, and Jennifer A. Dunne

The Comparative Method in Evolutionary Biology
Paul H. Harvey and Mark D. Pagel
The Cause of Molecular Evolution
John H. Gillespie
Dunnock Behaviour and Social Evolution
N. B. Davies
Natural Selection: Domains, Levels, and Challenges
George C. Williams
Behaviour and Social Evolution of Wasps: The Communal Aggregation Hypothesis
Yosiaki Itô
Life History Invariants: Some Explorations of Symmetry in Evolutionary Ecology
Eric L. Charnov
Quantitative Ecology and the Brown Trout
J. M. Elliott
Sexual Selection and the Barn Swallow
Anders Pape Møller
Ecology and Evolution in Anoxic Worlds
Tom Fenchel and Bland J. Finlay
Anolis Lizards of the Caribbean: Ecology, Evolution, and Plate Tectonics
Jonathan Roughgarden
From Individual Behaviour to Population Ecology
William J. Sutherland
Evolution of Social Insect Colonies: Sex Allocation and Kin Selection
Ross H. Crozier and Pekka Pamilo
Biological Invasions: Theory and Practice
Nanako Shigesada and Kohkichi Kawasaki
Cooperation Among Animals: An Evolutionary Perspective
Lee Alan Dugatkin
Natural Hybridization and Evolution
Michael L. Arnold
The Evolution of Sibling Rivalry
Douglas W. Mock and Geoffrey A. Parker
Asymmetry, Developmental Stability, and Evolution
Anders Pape Møller and John P. Swaddle
Metapopulation Ecology
Ilkka Hanski
Dynamic State Variable Models in Ecology: Methods and Applications
Colin W. Clark and Marc Mangel
The Origin, Expansion, and Demise of Plant Species
Donald A. Levin
The Spatial and Temporal Dynamics of Host-Parasitoid Interactions
Michael P. Hassell

The Ecology of Adaptive Radiation
Dolph Schluter
Parasites and the Behavior of Animals
Janice Moore
Evolutionary Ecology of Birds
Peter Bennett and Ian Owens
The Role of Chromosomal Change in Plant Evolution
Donald A. Levin
Living in Groups
Jens Krause and Graeme D. Ruxton
Stochastic Population Dynamics in Ecology and Conservation
Russell Lande, Steiner Engen, and Bernt-Erik Sæther
The Structure and Dynamics of Geographic Ranges
Kevin J. Gaston
Animal Signals
John Maynard Smith and David Harper
Evolutionary Ecology: The Trinidadian Guppy
Anne E. Magurran
Infectious Diseases in Primates: Behavior, Ecology, and Evolution
Charles L. Nunn and Sonia Altizer
Computational Molecular Evolution
Ziheng Yang
The Evolution and Emergence of RNA Viruses
Edward C. Holmes
Aboveground–Belowground Linkages: Biotic Interactions, Ecosystem Processes, and Global Change
Richard D. Bardgett and David A. Wardle
Principles of Social Evolution
Andrew F. G. Bourke
Maximum Entropy and Ecology: A Theory of Abundance, Distribution, and Energetics
John Harte
Energetic Food Webs: An Analysis of Real and Model Ecosystems
John C. Moore and Peter C. de Ruiter

Energetic Food Webs

An Analysis of Real and Model Ecosystems

JOHN C. MOORE
Natural Resource Ecology Laboratory, Colorado State University, USA

PETER C. DE RUITER
Biometris, Wageningen University, The Netherlands

With the assistance of Kim Melville-Smith

OXFORD
UNIVERSITY PRESS

Great Clarendon Street, Oxford, OX2 6DP,
United Kingdom

Oxford University Press is a department of the University of Oxford.
It furthers the University's objective of excellence in research, scholarship,
and education by publishing worldwide. Oxford is a registered trade mark of
Oxford University Press in the UK and in certain other countries

© Oxford University Press 2012

The moral rights of the authors have been asserted

First Edition published in 2012

Impression: 1

All rights reserved. No part of this publication may be reproduced, stored in
a retrieval system, or transmitted, in any form or by any means, without the
prior permission in writing of Oxford University Press, or as expressly permitted
by law, by licence or under terms agreed with the appropriate reprographics
rights organization. Enquiries concerning reproduction outside the scope of the
above should be sent to the Rights Department, Oxford University Press, at the
address above

You must not circulate this work in any other form
and you must impose this same condition on any acquirer

British Library Cataloguing in Publication Data

Data available

Library of Congress Cataloging in Publication Data

Data available

ISBN 978–0–19–856618–2 (Hbk)
     978–0–19–856619–9 (Pbk)

Printed and bound by
CPI Group (UK) Ltd, Croydon, CR0 4YY

Links to third party websites are provided by Oxford in good faith and
for information only. Oxford disclaims any responsibility for the materials
contained in any third party website referenced in this work.

Contents

Chapter 1   Approaches to studying food webs — 1

1.1   Introduction — 1
1.2   Traditions in ecology — 3
    1.2.1   The community perspective — 5
    1.2.2   The ecosystem perspective — 7
1.3   Food webs and traditions in ecology — 11
    1.3.1   Theoretically based food webs — 11
    1.3.2   Empirically based food webs: architecture — 11
    1.3.3   Empirically based food webs: information — 13
    1.3.4   How useful are these descriptions? — 15
1.4   Bridging perspectives through energetics — 18
    1.4.1   Core concepts and elements — 18
    1.4.2   Comments on our approach to studying food webs — 19
1.5   An overview of the parts and chapters — 21
1.6   Summary — 22

Part I   Modeling simple and multispecies communities

Chapter 2   Models of simple and complex systems — 27

2.1   Introduction — 27
2.2   Model structure and assumptions — 27
    2.2.1   Dimensions of mass, area, and time — 30
    2.2.2   Functional responses — 30
    2.2.3   Energetic efficiency and conversion rates — 33
    2.2.4   Intraspecific competition and self-limitation — 36
2.3   Stability — 36
    2.3.1   Local stability — 37
    2.3.2   Qualitative stability — 38
    2.3.3   Negative diagonal dominance and quasi-diagonal dominance — 39
        2.3.3.1   Diagonal dominance — 39
        2.3.3.2   Quasi-diagonal dominance — 41

vi • *Contents*

	2.3.4 Quasi-diagonal dominance and loops	42
	2.3.4.1 Global stability	44
2.4	Simple food chains	44
	2.4.1 Primary-producer-based food chains	45
	2.4.2 Detritus-based food chains, internal cycling, and donor control	46
	2.4.2.1 Models of donor-control dynamics	47
	2.4.2.2 The stability of donor control and detritus	48
2.5	The dynamics of primary-producer-based and detritus-based models	50
2.6	Summary and conclusions	51

Chapter 3 Connectedness food webs   54

3.1	Introduction	54
3.2	Soil food webs	54
3.3	The CPER soil food web	55
	3.3.1 Food web components	57
	3.3.1.1 Basal resources	57
	3.3.1.2 Inorganic nitrogen and carbon	59
	3.3.1.3 Consumers	59
	3.3.2 Aspects of food web structure	61
	3.3.2.1 Diversity and food chain length	61
	3.3.2.2 Diversity and complexity	64
	3.3.2.3 Omnivory	68
	3.3.3 Spatial and temporal averaging	70
3.4	Summary and conclusions	71

Chapter 4 Energy flux food webs   72

4.1	Introduction	72
4.2	Biomass and physiological parameters	72
4.3	Feeding rates and mineralization rates	75
	4.3.1 Feeding rates, mass balance, and energy budgets	75
	4.3.2 Mineralization rates	78
4.4	Energy flux descriptions	79
4.5	Summary and conclusions	92

Chapter 5 Functional webs   94

5.1	Introduction	94
5.2	Interaction strengths	94
	5.2.1 Population models	97
	5.2.2 The Jacobian matrix	98
	5.2.3 Estimating interaction strength from energy flux	99
	5.2.4 Stability	102

5.3	A functional food web for the CPER	103
	5.3.1 Energy flux and interaction strengths	103
	5.3.2 Mortalities from predation	111
	5.3.3 Functional groups and food web stability	113
5.4	Summary and conclusions	113

Part II The dynamics and stability of simple and complex communities

Chapter 6 Energetic organization and food web stability 127

6.1	Introduction	127
6.2	Energetic organization and stability	128
6.3	Distribution of interaction strengths: trophic-level-dependent interaction strengths	131
	6.3.1 Distribution of biomass and flux rates: the role of energetics	132
	6.3.2 Distribution of interaction strengths: the role of trophic interaction loops	135
	6.3.3 Food web stability: loops and energetics	137
6.4	Summary and conclusions	140

Chapter 7 Enrichment, trophic structure, and stability 141

7.1	Introduction	141
7.2	Simple primary-producer-based and detritus-based models	143
	7.2.1 Feasibility	144
	7.2.2 Stability and resilience	146
7.3	Trophic structure and dynamics along a productivity gradient	147
	7.3.1 Transitions in trophic structure	149
	7.3.2 Transitions in dynamic states	150
	7.3.3 Energetic efficiencies	152
	7.3.4 Body size and home range	155
7.4	More complex models	157
	7.4.1 Attack rates and dynamics	157
	7.4.2 Paradox of enrichment	159
	7.4.3 Enrichment and donor control	160
	7.4.4 Chaos and complex dynamics	161
7.5	Connections to real-world productivity	165
7.6	Summary and conclusions	169

Chapter 8 Modeling compartments 172

8.1	Introduction	172
8.2	Complexity, diversity, compartments, and stability	173

8.3	Defining compartments	176
8.4	Approaches to studying compartments	177
	8.4.1 Evidence of compartments—binary data I	178
	8.4.2 Evidence of compartments—binary data II	179
	8.4.3 Evidence of compartments—carbon and nitrogen flux	180
	8.4.4 Evidence of compartments—carbon flux and interaction strengths	184
8.5	The energy channel	187
	8.5.1 Time—flux rates and turnover	187
	8.5.2 Habitats—spatial arrangement	192
	8.5.3 Temporal arrangements and spatial interactions	192
	8.5.4 Structural and dynamic properties	195
	8.5.5 Quasi-independence of dynamics	196
8.6	Energy channels—structure and stability	200
	8.6.1 Coupled pathways and weak links	201
	8.6.2 Resistance and resilience	204
	8.6.3 Enrichment, predators, and energetic bottlenecks	205
	8.6.4 Donor-controlled dynamics	206
8.7	Summary and conclusions	207

Chapter 9 Productivity, dynamic stability, and species richness — 208

9.1	Introduction	208
9.2	Trophic structure, dynamics, and productivity	213
9.3	Feasibility revisited	214
9.4	Feasibility and the hump-shaped curve	218
	9.4.1 Life history strategies and adaptations	220
	9.4.2 Oscillations, instabilities, and population turnover rates	220
	9.4.3 Energetic properties of populations	222
9.5	Trophic structure and the diversity of production	224
9.6	A review of hypotheses	226
	9.6.1 Time	227
	9.6.2 Dynamic stability and disturbance	227
	9.6.3 Environmental heterogeneity	227
	9.6.4 Area	228
9.7	Summary and conclusions	229

Part III Dynamic food web architectures

Chapter 10 Species-based versus biomass-based food web descriptions — 235

10.1	Introduction	235
10.2	Dynamic food webs—playing Jenga	236

10.3	Two case studies	239
	10.3.1 Island biogeography	239
	10.3.2 Tuesday Lake	243
	10.3.3 Summary of the case studies	243
10.4	Stability, disturbance, and transition	244
	10.4.1 Stability	245
	10.4.2 Agents of change	248
10.5	Summary and conclusions	250

Chapter 11 Dynamic architectures and stability of complex systems along productivity gradients — 252

11.1	Introduction	252
11.2	Food web structure in a cave ecosystem	253
11.3	Food web structure and stability along the primary succession gradient at the Wadden island of Schiermonnikoog, The Netherlands	260
11.4	Food web structure in a changing Arctic	268
11.5	General framework	280
11.6	Summary and conclusions	282

Chapter 12 Food web dynamics beyond asymptotic behavior — 284

12.1	Introduction	284
12.2	Variability, equilibrium states, and asymptotic stability	285
	12.2.1 Variability, steady state, and the calculation of energy flow and nutrient cycling	285
	12.2.2 Variability, trophic interaction strengths, and asymptotic stability	287
12.3	Transient dynamics	289
12.4	Spatial systems	296
	12.4.1 Spatial networks, structure, and resilience	298
	12.4.2 Dynamical implications of space	301
12.5	Asymptotically ambiguous states	303
	12.5.1 Permanence	303
	12.5.2 Apparent complexity	305
12.6	Reconciling asymptotic stability, spatial structure, and transient dynamics	306
12.7	Summary and conclusions	307

References — 310

Index — 329

1

Approaches to studying food webs

"I think you should be more explicit here in step two."
Sydney Harris (1977)

1.1 Introduction

Human activity has altered basic functions within ecosystems and accelerated the decline in biodiversity. Dealing with these issues is complex and will require a synthesis of theories related to species diversity, how communities are structured, how key nutrients flow through ecosystems, and how ecosystems respond to disturbances. Yet, current paradigms about how species interact and the factors that limit their distributions are based largely on studies and observations of organisms that are important to us economically, for health reasons or ones that we can build emotional attachments to. We do not have a complete picture at this stage.

Recently we published a paper (Moore et al., 2005) in which we asked readers to consider how ecosystems and biogeochemical cycles are depicted in many college-level introductory textbooks to illustrate the need for a broader synthesis. We take some liberties here in borrowing much of the flow of the argument and some text to make our point. For terrestrial systems the diagrams typically include a tree with roots extending into the soil, a charismatic herbivore (often a deer, rabbit, or rodent) feeding on the understory, followed by arrows leading to a predator (typically a mountain lion, fox, or hawk) lurking in the background. This portion of the diagram is familiar. The soil is labeled with process-arrows (decomposition, nutrient transformations) leading nutrients back to the roots of the plant. Some texts might discuss the processes and incredible diversity of the organisms within soils, but the connections between what is happening above ground to the processes below ground are unclear. The same types of diagrams are found for freshwater and marine systems. We noted that in many respects these diagrams remind us of the cartoon by Sydney Harris depicting two mathematicians at the blackboard, one pondering the works of the other (Figure 1.1). The writing on the blackboard contains two sets of equations interconnected by the text "... *then a miracle occurs.*" The one pondering the work says to his colleague: "*I think you should be more explicit here in step two.*" To be fair, the diagrams found in textbooks are incomplete, as they are attempting to simplify complex phenomena. But what is included and is not included tells us something about the state of ecosystem science (O'Neill, 2001).

2 • Energetic Food Webs

"I THINK YOU SHOULD BE MORE EXPLICIT HERE IN STEP TWO."

Figure I.1 Cartoon by Sydney Harris (1977), ScienceCartoonsPlus.com. In many ways the cartoon captures the sentiment expressed by O'Neill (2001) on the state of ecosystem science. Can energetics provide the bridge between the community and ecosystems approaches to studying food webs?

This brings us to a second observation. During our education we learn very early on that life begins with photosynthesis, and that plant life is consumed by herbivores, which in turn are controlled by predators. This simple concept is introduced at multiple scales. This is a fundamental equation of life at the molecular level view:

$$6CO_2 + 6H_2O \leftrightarrow C_6H_{12}O_6 + 6O_2 \qquad (1.1)$$

It manifests as consumer and resource interactions and food chains at the organism and population levels. At the community level we speak of food webs. Finally, we come full circle and introduce biogeochemical cycles involving molecules and elements at the ecosystem level.

Common themes in each of the examples are the transfer of matter from one organism or pool to another, and the interaction between the living and nonliving components of the ecosystem. These ideas have been around for some time.

Embedded in these ideas is the notion of regulation: more to the point, the controls that resources have on consumers (i.e., bottom-up regulation) and the controls that consumers have on their resources (top-down regulation). One of the earliest written examples can be found in the Rhind Papyrus (Newman, 1956), an Egyptian document from approximately 1700 BC (Figure 1.2):

"In each of seven houses are seven cats; each cat kills seven mice; each mouse would have eaten seven ears of spelt (wheat); each ear of spelt would have produced seven hekat (half a peck) of grain. Query: How much grain is saved by the seven houses' cats?"

At the time we asked how far modern ecology had progressed from this. If we parse the riddle we see several familiar themes: (1) a simple food chain based on living organisms, (2) the notions of bottom-up and top-down control, (3) trophic cascades, (4) direct and indirect effects of organisms on the densities of others, and (5) the interdependence of organisms in terms of dynamics and stability. This is a comprehensive list, one that summarizes a body of theory that is based on linear food chains originating from plant life through vertebrate herbivores to vertebrate top carnivores. Yet this pathway represents but a small fraction of the fate of plant life on Earth. Most plant life is neither consumed by a herbivore, nor by a vertebrate, but rather it dies and is decomposed by microorganisms and a myriad of invertebrates.

Pictures and anecdotes are representations of our theories based on our observations. Yet often our theories and observations seem out of synchronization with one another. We observe complex and diverse systems that seem to persist, yet we still grapple with the relationships between diversity, complexity, and stability. Elements and nutrients are fundamental to every trophic interaction, yet nutrients are often viewed as not relevant to system stability. Our intuition and observations suggest connections between nutrient dynamics, species interactions, and stability, but our theories have failed to provide adequately for them.

We offer a view of ecological systems that is based on the premise that the interactions among species constitute the energetic organization of a community that forms the basis of ecosystem stability. To paraphrase Gould (1982), we address the question of whether food webs are mere incarnations of the interactions of the moment, or whether a food web represents a primary constraining stable structure, with change as a "difficult" phenomenon, usually accomplished rapidly when a stable structure is stressed beyond its buffering capacity to resist and absorb.

1.2 Traditions in ecology

There are two different approaches within ecology, with their own epistemologies developed in the twentieth century: an individualistic community-based approach and a holistic ecosystem-based approach. What follows is certainly an oversimplification of the history of the traditions in ecology and will be extremely food-web centric, but it revolves around the following questions. Are ecological communities simply a collection of species shaped by the resource needs and tolerances of

4 • *Energetic Food Webs*

Figure I.2 This papyrus, written around 1700 BC, describes mathematical problems, one of which is the problem of the *Seven Cats*, and their solutions using ancient Egyptian mathematics and problem-solving techniques. The scroll was found in some ruins near the city of Thebes and was purchased at Luxor in 1858 by A. Henry Rhind. It currently resides in the British Museum. The concepts presented in the problem of the *Seven Cats* include those of a linear food chain, trophic cascades, linear functional responses, and top-down and bottom-up control, pre-dating modern ecology by more than 3000 years. This image is in the public domain.

individual species and by their interactions with other species? Or are ecological communities (i.e., ecosystems) entities in-and-of themselves, possessing properties that reinforce their persistence? The two perspectives are not as mutually exclusive as the phrasing of the questions suggests, but rather offer different approaches to how we view communities and the factors that regulate them. Both perspectives invoke information theory and systems theory but use them in different ways. Our aim here is not to advocate one or the other, but rather to highlight aspects of the perspectives that we have found useful in our study of food webs and advocate an integrative approach that emerges from both.

1.2.1 The community perspective

The community side of the ledger has a rich history steeped in the natural history of species, their distributions, and the nature of their interactions. Community ecology is based on the study of the ways in which individual species interact and sort themselves within communities. The perspective focuses on the tolerances and responses of species to different factors within the environment, which include climate, the availability of limiting resources, and interactions with other organisms. The field balances the views that communities represent assemblages of species that have coevolved with one another with views that communities are the winnowed products of the chance encounters of preadapted species. Regardless, it is the adaptations of species to the environment that is central to the community perspective.

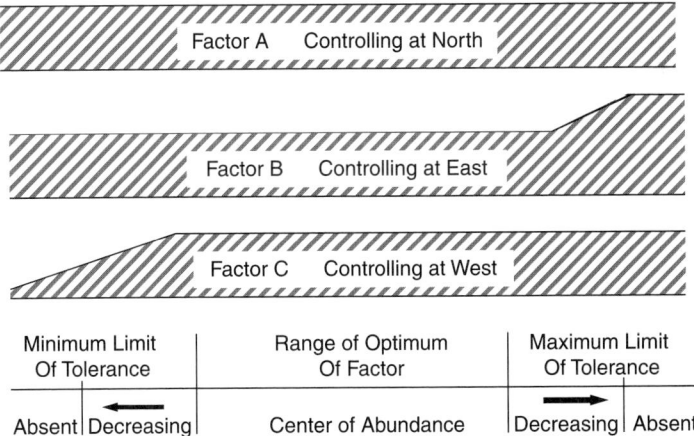

Figure 1.3 An adaptation of the illustration of Shelford's *Law of Tolerance* (modified from Shelford, 1935). Communities are compilations of species that result from their attributes and adaptations to one another and abiotic factors.

6 • Energetic Food Webs

Shelford (1931) is credited with the *law of tolerance*, which proposes that the abundances and distribution of organisms were shaped by their ability to survive and reproduce in the face of biotic and abiotic factors (Figure 1.3). This view is an extension of the individualistic concept of plant associations (Gleason, 1926, 1939) and holds that communities were by and large a coincidence. A classic illustration of this principle can be found in the studies conducted by Connell (1961a, 1961b), aimed at determining the causation of the stratification of the barnacles *Balanus balanoides* and *Chthamulus stellatus* in the rocky intertidal (Figure 1.4). Through observation and manipulative experimentation in the field, Connell (1961a, 1961b) demonstrated how random colonization followed by the species adaptive responses and tolerances to physical factors in the environment, coupled with the biotic interactions of predation and competition, could explain the patterning of barnacles within the littoral zone.

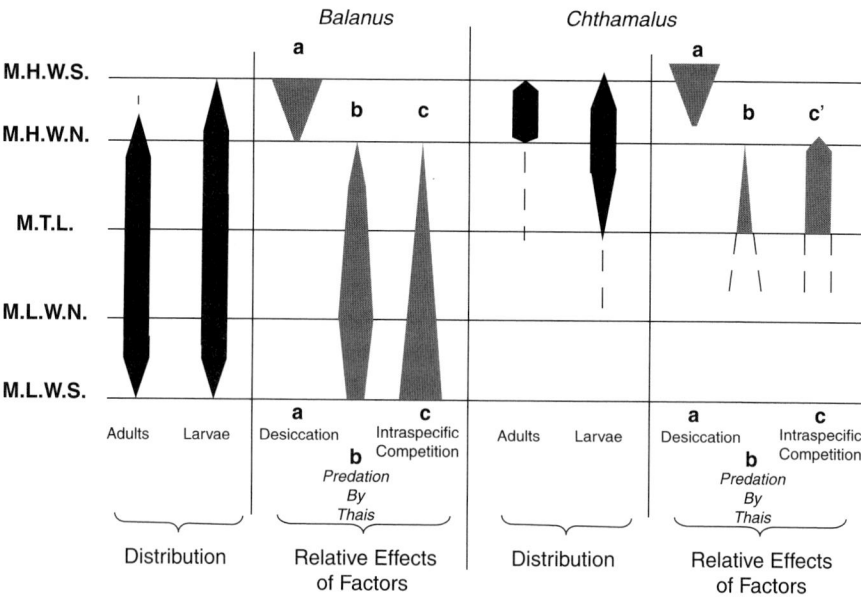

Figure I.4 The summary of the results of the observations and manipulative studies conducted by Connell (1961b) at Millport, which illustrates the distributions of settled larvae and adults of *Balanus balanoides* and *Chthamulus stellatus* and the relative effects of: (a) dessication; (b) predation by *Thais*; (c) intraspecific competition; and (c') interspecific competition in shaping the distributions. Tide levels: M.H.W.S., mean high water spring; M.H.W.N., mean high water neap; M.T.L., mean tide level; M.L.W.N., mean low water neap; M.L.W.S., mean low water spring (modified from Connell, 1961b). Community structure is viewed as the additive effects of species interactions (predation and competition in this case) and the adaptation of species to abiotic conditions.

The community perspective holds species interactions as being preeminent in the structure and maintenance of communities. In his address to the Society of American Naturalists, Hutchinson (1959) asked "Why are there so many kinds of animals?" and in doing so laid out a framework heavily dependent on trophic interactions. The work led to an important series of papers that addressed aspects of community development, species diversity, and the factors governing community structure (Hairston et al., 1960, MacArthur and Wilson, 1963, Paine, 1966, MacArthur and Wilson, 1967, MacArthur, 1972).

Hairston et al. (1960) asked "Why is the world green?" and "Why isn't coal being produced?" The answers they provided to both questions relied less on trophic dynamics and the importance of resource limitations, and more on the role of predation, and particularly on Lotka–Volterra notions of density dependence, the donor-controlled nature of detritus not withstanding. In answer to the former, they concluded that herbivores were controlled by predators, thereby releasing green plants from predation pressure. They concluded that coal was not being formed as a result of decomposers being resource-limited. From a more complex food web perspective, Paine (1966) demonstrated the importance of certain predators, termed "keystone species," in maintaining the integrity of ecological communities. Through predation pressure on organisms at lower trophic positions, they maintain these populations at levels that mollified their competitive and predatory interactions, avoiding species loss through competitive exclusion and overconsumption. These ideas of trophic cascades and the complementary influences of top-down and bottom-up controls were not accepted without criticism (Murdoch, 1966, Ehrlich and Birch, 1967), but spawned a generation of debate and studies that still resonate (Estes and Palmisano, 1974, Carpenter et al., 1987, Strong, 1992, Hairston and Hairston, 1993, Preisser, 2003, de Ruiter et al., 2005).

The *theory of island biogeography*, as presented by MacArthur and Wilson (1963, 1967), possesses many of the assumptions behind the Law of Tolerance and embraces the view of the winnowing of chance encounters of preadapted species. We will discuss the details of this theory in Part III of the book. Central to the theory are the importance of colonization and extinction of individual species, both assumed to operate at random, without regard for coevolution or preadaptation, or any underlying mechanism for the success that either generates. Yet, the theory is qualitatively accurate in its predictions on the diversity of communities at climax following either primary or secondary successional development.

1.2.2 The ecosystem perspective

Ecosystem ecology is based on the study of the ways in which species interact among themselves and their environment (Golley, 1993). The ecosystem concept can be traced back to Möbius (1877) and the idea of biocoenosis of communities, and the work of Clements (1916, 1920, 1936) on plant succession. In the extreme, Clements viewed mature communities as types of superorganisms with the capacity

8 • Energetic Food Webs

to reproduce, and the ecological succession leading to their development as a form of ontogeny akin to the developmental stages of an individual species. The concept was further developed by Elton (1927) to include the role of energetics in shaping animal communities. His work led to the modern concept of a food chain and food web, defining them not only in terms of the species that were present but of the matter and energy contained within, as illustrated in his concepts of the pyramids of numbers and biomass, respectively. The peakedness of the ecological pyramids was due to limitations in the availability of resources and the metabolic requirements of species.

Lindeman (1942) took the approach a step further by representing the ecosystems in terms of species interactions in the context of the abiotic environment and by quantifying not only the biomasses of the species but the flows among species (Figure 1.5). His description of the Cedar Creek Bog food web defined groups of species interactions in energetic terms, and included detritus (ooze), microorganisms, and nutrients, as direct inflows and outflows.

If trophic dynamics and density dependence represent the core of community ecology, hierarchy theory and cybernetics govern ecosystem thinking (Tansley, 1935,

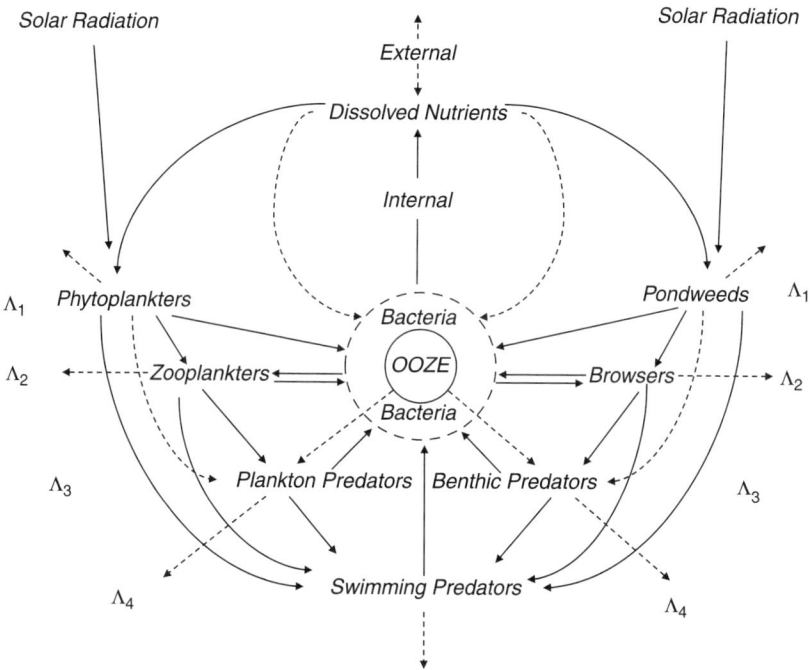

Figure I.5 Description of the food web diagram of the trophic interactions and energy flows through the four trophic levels (Λn) for the Cedar Creek Bog (modified from Lindeman, 1942). The solid lines represent direct trophic interactions. The dashed lines represent the transfer and loss of inorganic nutrients.

Margalef, 1961, Simon, 1962, Margalef, 1968, Odum, 1969). Tansley (1935) coined the term "ecosystem," defining it as the fundamental unit of nature formed by the coupling of the interactions among organisms and the physical factors within the environment. Tansley rejected the extreme nature of the superorganism concept suggested by Clements, but embraced the hierarchical systems thinking that it embodied. In fact, in defining the ecosystem, Tansley (1935) referred to it as forming "... *one category of the multitudinous physical systems of the universe, which range from the universe as a whole down to the atom."*

Odum (1969) presented the strategy of ecosystem development, in what now appears as the ecosystem ecology alternative perspective to the theory of island biogeography. The paper did not present a theory per se, but rather juxtaposed characteristics of ecosystems during their developmental and mature stages (Table 1.1). The ideas are more phenomenological than explanatory, with "strategy" being the operative word. The ecosystem, rather than the species that comprise it, is at the center. Similarly to the theory of island biogeography, the mechanisms that lead to the changes underlying the contrasts between the developing and mature communities are not clearly spelled out, but unlike that theory it does not assume that processes operate in an equal manner at random on species or other components of the system. Odum (1969) embraces the coevolution between species and feedbacks between organisms and abiotic factors (inorganic nutrients), to bring an ecosystem and its components and processes into balance. The ecosystem possessed cybernetic properties governed by the first and second laws of thermodynamics and the conservation of matter.

The ideas presented by Odum (1969) generated intense debate over the entire concept of the ecosystem and its purported cybernetic properties (Engleberg and Boyarski, 1979, Knight and Swaney, 1981, McNaughton and Coughenour, 1981). Wiener (1948) defined cybernetics as the science of control and communication. The term "cybernetics" was coined from the Greek word *kubernetes*, meaning "steersman" or "governor," from the radical *kuberman*, meaning "to steer" or "to govern." The concept relies on the idea of the flow of information among constituents to control processes within an organized system. McNaughton and Coughenour (1981) made the case that an ecosystem is a cybernetic system in that it possesses coordination, regulation, communication, and control through the information flow embedded in trophic interactions (sensu Lindeman 1942, Hairston et al. 1960) among species and their interactions with the abiotic environment.

Although elements of cybernetic organization are apparent, O'Neill et al. (1986) make a strong case that the concept is incomplete, in that it relies on homeostasis and does a poor job of explaining the reaction of systems to stress and instability. Ecosystems might possess cybernetic properties, but the concept does not represent the fundamental organizing principle that shapes and guides them. In its place, O'Neill et al. (1986) apply the concept of hierarchical organization to ecosystems. O'Neill (2001) later argued that even taking a hierarchical view of ecosystems, the use of the "machine analogy" at the base of ecosystem thinking has limited its ability to address many questions. Two conclusions emerged from this work. First, the

Table I.I. The tabular model of ecological succession presented by Odum (1969). The table presents trends that might be expected in the development of ecosystems during the developmental stages to the mature, or climax, community.

Ecosystem Attributes	Developmental Stages	Mature Stages
Community Energetics		
Gross production/community respiration (P/R ratio)	Greater or less than 1	Approaches 1
Gross production/standing crop biomass (P/B ratio)	High	Low
Biomass supported/unit energy flow (B/E ratio)	Low	High
Net community production	High	Low
Food chains	Linear, predominantly grazing	Weblike, predominantly detritus
Community Structure		
Total organic matter	Small	Large
Inorganic nutrients	Extrabiotic	Intrabiotic
Species diversity: richness	Low	High
Species diversity: evenness	Low	High
Biochemical diversity	Low	High
Stratification and spatial diversity (pattern diversity)	Poorly organized	Well organized
Life History		
Niche specialization	Broad	Narrow
Size of organisms	Small	Large
Life cycles	Short, simple	Long, complex
Nutrient Cycling		
Mineral cycles	Open	Closed
Nutrient exchange rate, between organisms and environment	Rapid	Slow
Role of detritus in nutrient regeneration	Unimportant	Important
Selective Pressure		
Growth form	For rapid growth ("r-selection")	For feedback control ("K-selection")
Production	Quantity	Quality
Overall Homeostasis		
Internal symbiosis	Undeveloped	Developed
Nutrient conservation	Poor	Good
Stability (resistance to external perturbations)	Poor	Good
Entropy	High	Low
Information	Low	High

interaction of natural selection on individual species and the internal feedbacks afforded by the interactions within the ecosystem are what drive dynamics. Adaptation of species to the environment is crucial, but the adaptations of species to the system of interactions within an ecosystem are important too. Second, ecosystem stability is governed by processes and constraints that occur outside of the boundaries of the local ecological system in question.

1.3 Food webs and traditions in ecology

A food web describes the feeding relationships among species or groupings of species. Several different strategies have been developed and describe food webs for different purposes. May (1973) differentiated the tactical model from the strategic model in his treatment studying the stability and complexity of ecosystems, which highlighted the two traditions that had developed in ecology: community ecology and ecosystem ecology. We have opted to categorize food web descriptions as being either theoretically based or empirically based. Theoretically based food webs are generally developed as the core for analyses to study general properties of model systems. Empirically based food web models capture observed trophic interactions and are parameterized with field data.

1.3.1 Theoretically based food webs

The use of mathematical equations to model food webs has been around for some time (see the example found on the Rhind Papyrus, Figure 1.2). Lotka (1922) and Volterra (1927) presented mathematical representations of the interactions between species to study their dynamics. What have come to be known as the Lotka–Volterra equations have, in a strict sense, been criticized for their lack of biological realism. The original formulations did not include terms for intraspecific competition or self-regulation, and the predation terms and the resulting functional responses of the predators to prey populations were far too simplistic. The dynamic properties of the early models did not capture the range of dynamic states observed in nature. For example, the purely oscillatory dynamics of the Lotka–Volterra equations that were defined as neutral stability might easily have been confused with the types of limit-cyclic behavior or chaotic dynamics generated by more complex models. This said, the basic premise of modeling systems using mathematical equations has held, even though more sophisticated or realistic equations have been developed to address their deficiencies.

1.3.2 Empirically based food webs: architecture

Food webs are described for different reasons. Cohen (1978) categorized the descriptions of food web as community food webs, source food webs, or sink food

12 • *Energetic Food Webs*

(a)
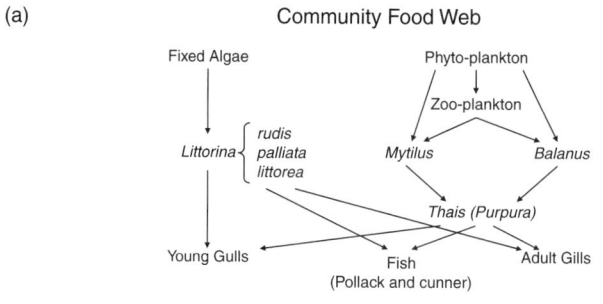

(b)
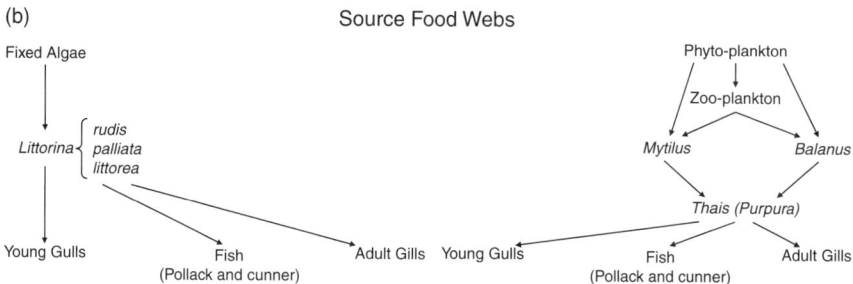

(c)
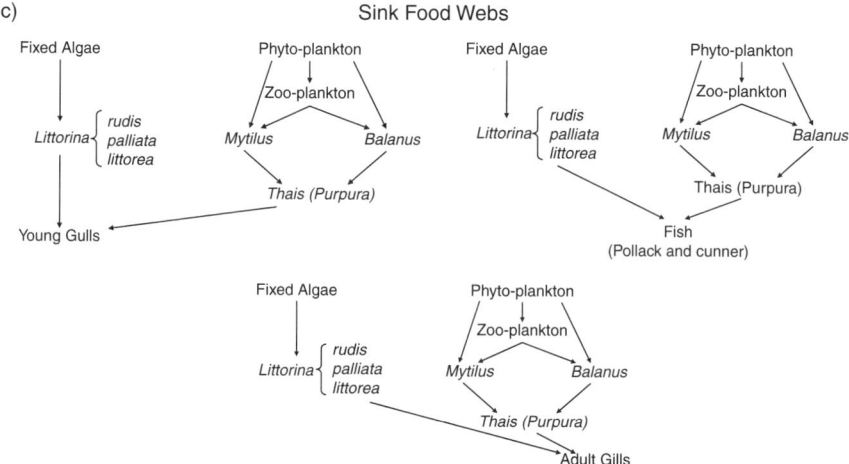

Figure I.6 The three categories of food web descriptions proposed by Cohen (1978) based on the frame of reference of the observer, using an adaptation of Colton's (1916) description of the Mount Desert Island, Maine, rocky intertidal. (a) The community food web description. (b) The source food web descriptions. (c) The sink food web descriptions.

webs, based on the frame of reference of the observer. Differences in perspectives affect which species are included or not included in the descriptions (Figure 1.6). One perspective is not necessarily superior to another, but it is important that we understand their limitations.

The *community food web* is a description based on a type of habitat and all the species and interactions among the species that reside within the habitat. Community food web descriptions are the most comprehensive and inclusive of the three descriptions, but nonetheless they are incomplete. These descriptions are affected by the boundaries that define the habitat and are influenced by the natural histories of the individual species. Typically species that live at the boundaries of the habitat, that frequent the habitat but do not reside in the habitat for extended periods of time, or that do not reproduce in the habitat are excluded. For example, shore birds or predatory mammals that forage in the rocky intertidal were often excluded from the descriptions. Similarly, surface-active arthropods that frequent the soil litter layer and that may have edaphic life stages are often excluded from the descriptions of soil food webs (see Chapter 3).

A *source food web* identifies resources at the base (i.e., basal resources) and includes all consumer species and species that interact with them, up to top predators. The basal resources included in a source web are a subset of the basal resources included in a community food web. The types of basal resources used in the descriptions are not necessarily restricted to primary producers (plants, algae, or certain microorganisms) or detritus (nonliving organic material) but might include consumers. Source food webs give an incomplete picture of the number of trophic interactions and energy flow, as consumers often feed on multiple resource types, some of which obtain energy that may have originated from a basal resource not included in the description.

A *sink food web* identifies a consumer and works backward to include its prey, and its prey's prey, down to the basal resources. The choice of the sink is defined by the objectives of the study and may not necessarily be a top predator. Like the source food webs, sink food webs may exclude linkages and energy pathways in instances where the system has multiple top predators and basal resources. We will make use of this concept in our analyses of energy flow (see Chapter 4) and community architecture (see Chapter 8) throughout this book.

1.3.3 Empirically based food webs: information

Paine (1980) took a slightly different approach to categorizing empirical food webs. Rather than focus on the perspective of the observer, Paine based his distinctions between *connectedness* descriptions, *energy flow* descriptions, and *functional* descriptions on the information on which the observer based the linkages. These three different types of food web descriptions use the same basic structure but depict different aspects of the trophic relationships with different types of information (Figure 1.7).

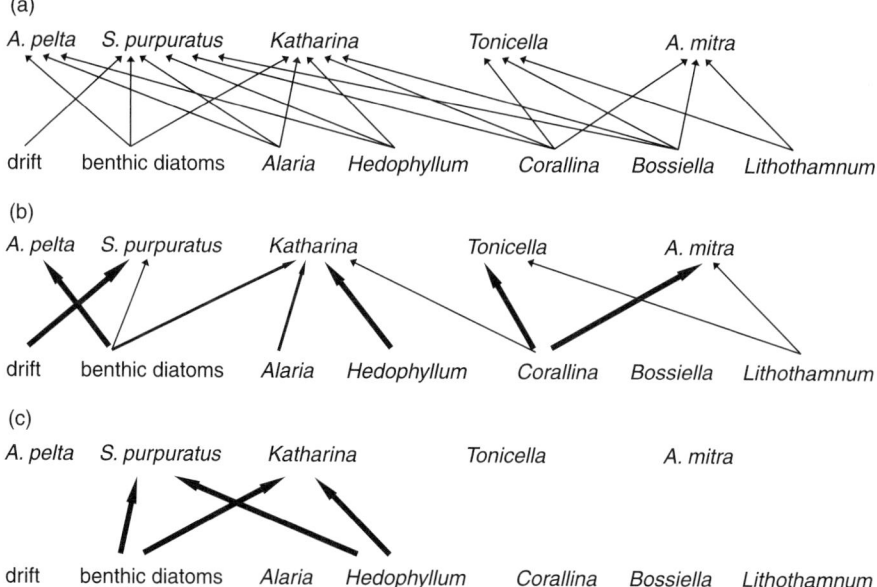

Figure 1.7 The three categories of food web descriptions using different depictions of a subset of trophic interactions from an intertidal food web as proposed by Paine (1980). (a) The connectedness web description depicts the feeding relationships between organisms. The information is obtained from field and laboratory observations. (b) The energy flow web description depicts the transfer of energy or matter from resource to consumer. The thickness of the arrows indicates the flow rate. The information is obtained from measurements, models, and literature values. We will refer to these as energy flux webs. (c) The functional web description depicts the strength of interactions among species. The information is obtained from controlled manipulations (Redrawn from Paine, 1980).

The *connectedness web* description includes a collection of interacting species with vectors indicating the trophic interactions among the species. Current convention has the vectors pointing in the direction of material or energy transfer. The descriptions are based on field and laboratory observations. Connectedness descriptions vary in terms of which components to include or not to include, as we discussed above, depending on whether they are community, source, or sink food web descriptions.

The *energy flow web* description weights the vectors that were included in the connectedness description with estimates of the amount of energy or material that is transferred from one species to another over a specified time period. When presenting this concept, Paine (1980) gave a nod to the approach but was clearly not enamored by it or by its utility, viewing it as static and as a mere accounting exercise. He points out that the relationship between the magnitudes of particular flows of energy does not necessarily equate to its importance to community organization. Recent work on the importance of weak interactions would bear this out (McCann et al., 1998). However, we will demonstrate in chapters 4 and 5 that the information contained in the energy flows is important to our understanding of community organization, dynamics, persistence, and stability.

The *functional web* described the influence of the removal of a species on the population sizes of the remaining species (MacArthur, 1972, Paine, 1992, see Laska and Wooton, 1998). The strength of the trophic links that a species possesses are determined by the responses of other species in the food web following controlled manipulations—usually the removal of the species. In Chapter 5 we present our take on functional web descriptions. We estimate the interaction strengths directly from the differential equations used to describe the dynamics of the system, estimates of steady-state densities obtained from field observations, and estimates of life history and physiological attributes of the species in the description.

1.3.4 How useful are these descriptions?

The standards used to describe food webs and to parameterize them, whether they are theoretically based or empirically based, have been too broad. The criticisms are usually directed at all aspects of the endeavor, from the underlying assumptions in the construction of the description of the food web and if a model were constructed, the structure and parameterization of the model (Paine, 1988, Moore et al., 1989, Polis and Strong, 1996).

Theoretically based food web models have been criticized for being too simple, being too abstract, and lacking biological realism (Roberts, 1974). The models often set the number of species involved and then randomly construct networks of interactions among species with varying degrees of complexity (Rosenzweig, 1971, May, 1972). The analyses focus on the eigenvalues of the interaction matrix of the system of equations used to describe the system, and their dynamic properties.

Empirically based food web descriptions have been criticized at several levels. One criticism focuses on the completeness of the descriptions in terms of which species are included. We have already discussed this above. For example, food web descriptions often include only those species with significant resident times in the communities. Descriptions of many intertidal food webs generally do not include littoral and terrestrial birds, mammals, and fish that forage during high and low tides, but rather focus on those species that adhere to rocks or live within the tide pools. Our own example of the soil food web described in Chapter 3, which we will use as a guide throughout the book, does not include predators that are in low abundance (e.g., ciliates) or predators that venture only briefly below the soil surface (e.g., carabid beetles), but nonetheless may be important to the system. Along these same lines, many food webs are simply incomplete. An assessment of webs compiled in the literature (Cohen, 1978, Moore et al., 1989, Cohen, 1990) revealed several instances in which basal species were in fact intermediate consumers.

A second complicating factor is that the descriptions of food webs rarely include the interactions among species per se, but rather the interactions among groups of species, or collections of taxa, which may have been grouped for convenience rather than on ecological criteria. Ideally the food web would be described at the species level, but this is rarely feasible given the high diversity of systems, and in the case of some taxa the diversity is not known. Instead, ecologists have sought to aggregate groups of species into units that are internally homogeneous relative to other units that are based on a set of ecologically relevant criteria (Table 1.2). For example, Root (1967) introduced the notion of feeding guilds based on the major food sources and feeding modes of insects and birds. In our own treatment of microorganisms and invertebrates within soil food webs, we adopted the notion of functional groups (Hunt et al., 1987, Moore et al., 1988). Ehrlich and Holm (1962) articulated the rationale for this approach in their discussion of different approaches to studying patterns in populations. They argue that to obtain the *"...highest information content and predictive value, specialized taxonomies designed for optimal usefulness under restricted conditions could be created...In a taxonomy based on ecological requirements, whales will be more closely related to sharks than to bears. Such a relationship is no more or less 'true' or 'natural' than the classical one; it is merely based on different attributes...What sort of taxonomy is desirable depends on what one wishes to use it for."*

Third, the taxonomic resolution and groupings of organisms within the descriptions are often irregular, with a tendency for the resolution to increase as one moves up the trophic ladder (Pimm, 1982, Paine, 1988). One of the earliest food web descriptions of a rocky intertidal illustrates this point (Colton, 1922, Fisher, 2005). Assessments of subsequent descriptions revealed a tendency for top predators to be described at the species level, whereas intermediate and basal species are placed in functional or phylogenetic groupings (Pimm, 1982).

A fourth issue revolves around how to deal with detritus. Many food web descriptions include detritus links, but the theoretical treatments ignore them. This is not so much a problem with the treatment of empirical descriptions. Problems do arise when the results of theoretical treatments are adapted or applied to these descriptions. Most of the theoretical studies were based on variations of Lotka–Volterra producer and consumer equations. May (1973), Pimm (1982), and Cohen (1978) all acknowledge the importance of detritus and the donor-controlled mathematical representations describing their dynamics, but were clear in their tomes that they focused on consumer resource models of living systems. Similarly, in our own treatment of trophic loops and loop weight (Neutel et al., 2002), which we discuss in Chapter 5, we do not focus on loops that are generated by the cycling of matter from each species back to the detritus pool. Having said this, there are at least two areas involving detritus that we should consider. First, the dynamic and stability of donor-controlled systems share many similarities with the producer and consumer-based systems but, as we will point out later, possess key differences. Second, studies that have focused on the topology of food webs, i.e., the number of species and linkages among species, generally do not consider the flows from each of the species back to

Table 1.2. Candidate functional groups of taxa found in the soils of the shortgrass steppe at the USDA Central Plains Experimental Range in Colorado (adapted from Moore et al., 1988).

Functional group	Description	Examples (genera)
Arthropods		
General predators	Attack most soil invertebrates small enough to overcome	*Hypoaspis, Asca, Amblyseius, Rhodacarellus, Rhodacarus, Gamasellodes, Macrocheles, Symphylellina, Arebius*
Predators of arthropods	Attack only arthropods	*Japyx, Campodea, Bembidion, Tachys, Spinibdella, Cyta, Pulaeus, Stigmaeus, Cocorhagidia*
Predators of nematodes	Attack only nematodes	*Alliphis, Eviphis, Alycus, Alicorhagia*
Fluid-feeding fungivores	Pierce and suck fluids of fungi and/or nematodes	*Tydeus, Eupodes, Tarsonemus, Bakerdania, Pedicularer, Scutacarus, Speleorchestes*
Engulfing fungivores: mites	Ingest particles of fungal hyphae, spores, algae, and/or nematodes	*Oribatula, Zygoribatula, Pilogalumna, Tectocepheus, Oppiella, Joshuella, Ceratozetes, Tyrophagus*
Engulfing fungivores: Collembola	Ingest particles of fungal hyphae, spores, algae, and/or nematodes	*Folsomia, Isotoma, Isotomodes, Hypogastrura, Tullbergia, Anotylus*
Root engulfers	Ingest roots	*Phyllophaga, Trichiorhyssemus*
Root-fluid feeders	Pierce roots and suck sap	*Margarodes, Orthezia, Eriococcus, Amonostherium, Cryptoripersia, Radicoccus, Tuckerella, Linotetranus*
Detritivores	Ingest dead plant matter	*Reticulotermes*
Nematodes		
Predators	Attack primarily nematodes, but ingest protozoa and bacteria	*Discolaimium, Monochus*
Omnivores	Ingest bacteria and protozoa	*Mesodiplogaster*
Bacterial feeders	Ingest bacteria and soil	*Acrobeloides, Pelodera, Rhabditis*
Fungal feeders	Pierce fungal hyphal walls and suck fluids	*Aphelenchus*, some *Aphelenchoides* spp.
Root-fluid feeders	Pierce roots and suck sap	*Helicotylenchus, Tylenchorhynchus, Xiphinema*
Protozoa		
Amoebae	Engulf primarily bacteria, some flagellates, soil detritus and fungal spores	*Acanthamoeba, Hartmannella, Trichamoeba, Mayorella, Vannella*
Flagellates	Ingest primarily bacteria, some soil and detritus	*Pleuromonas, Bodo, Mastigamoeba*
Ciliates	Ingest bacteria, amoebae, flagellates and soil	*Colpoda*

the detritus pool. These feedbacks to the detritus pool can be important drivers of the system's productivity and dynamics.

Another issue arises with the accuracy of the feeding relationships that are depicted. The diets of many species are diverse, yet most descriptions focus on the dominant food sources obtained from observations, feeding studies, and gut content analyses. Paine (1988) noted that many species change their diets with change in size, age, or ontogenetic life history stages.

Given the objections raised above, one might ask why to bother modeling food webs at all. We find ourselves echoing the cautions that were raised by May (1973) and shared by others, that the food webs and their models "... *are at best caricatures of reality, and thus have both the truth and the falsity of caricature.*"

1.4 Bridging perspectives through energetics

Our thesis represents a bridging of the community ecology and ecosystem ecology perspectives through energetics. It ultimately comes down to the basic conundrum faced by any study of a system. From an ecological standpoint we need to balance or determine the extent to which the contributions that individual species, through their interactions, bring to the understanding of what constitutes an ecological system and its dynamics, and how these collective activities of species combine to generate system-level properties that feed back to affect the individual species and their interactions. We will employ both the theoretical underpinnings and the methodological approaches used by community and ecosystem ecologists. We will attempt to explain the observed regularities in ecosystem development, structure, function, and dynamic responses to disturbance in terms of species adaptations to abiotic factors and interactions with other species.

We have approached the challenge by following the tenets of Kuhn (1970) by first addressing the following questions:

1. What are the fundamental entities of which the universe is concerned?
2. How do they interact?
3. What are legitimate questions about such entities?
4. What techniques can be used in seeking solutions?

1.4.1 Core concepts and elements

We base our approach on two important underpinnings. First, ecological systems are thermodynamically open. Energy and matter enter and leave the system and must be accounted for. Second, ecological systems are complex adaptive systems (Levin, 1999, Levin, 2002) that hinge on the diversity and individuality of components, localized interactions among the components, and an autonomous process that uses the outcomes of those localized interactions to select for a subset of components for replication or enhancement.

Two additional properties that emerge from these underpinnings are that ecological systems are hierarchically organized and that they possess cybernetic properties. The concepts of niche, adaptations, and natural selection are prominent as we view ecosystems as complex adaptive systems (Holland, 1995, Levin, 1998, 1999). The larger-scale ecosystem properties and patterns (Odum, 1969) emerge from the properties of the constituent species and the interactions among them. Two such properties are that the ecosystems are hierarchically organized (Allen and Starr, 1982, O'Neill et al., 1986) and that they are cybernetic (Odum, 1969, McNaughton and Coughenour, 1981). These properties—hierarchical organization and cybernetic behavior—lead us to posit that the ecosystem properties that emerge from the interactions among the constituent species in turn operate as selective forces that select for subsets for replication or enhancement (Levin, 1999). The self-organizing tendencies are by-products of evolutionary processes of the organizing units, i.e., individuals and species, while the evolutionary processes that affect the units are influenced by the by-products of the self-organizing tendencies (e.g., nitrogen availability).

In short, we take a seamless view of the community and ecosystem approaches and will use the terms "community" and "ecosystem" interchangeably in the discussions of our models. To do so would require that we relax our definition of a species or other component of a system, as conveyed in the following quote:

"Our language and culture include a prejudice for applying the concept of individual only to bodies, but any coherent entity that has a unique origin, sufficient temporal stability, and a capacity for reproduction with change can serve as an evolutionary agent. The actual hierarchy of our world is a contingent fact of history, not a heuristic device or a logical necessity." (Unattributed)

1.4.2 Comments on our approach to studying food webs

The ecosystem will serve as the universe from which we will develop our food webs. Ecosystems represent the community of organisms and the physical environment within a particular region bound by biotic interactions and biogeochemical cycles.

The food webs and models we present are designed to capture the biotic interactions, functional attributes, and biochemical cycles of ecosystems. Our models will be based on aggregations of species, or functional groups. Species within a functional group share similar: (1) food sources, (2) feeding modes, (3) life history characteristics, and (4) habitats (Moore et al., 1988). The choice of criteria to form the functional groups was based on the prevailing ideas of the ecological niche and the principal dimensions niche space (Schoener, 1974)—food (food source, feeding mode), habitat, and time (life history characteristics). The criteria used to form these groups all but assures that functional groups will be comprised of closely related taxa given the similarity in the grouping criterion to the those used in phylogeny (Norberg et al., 2001). In fact, more often than not, single broad taxonomic groups are parsed into finer functional groups (nematodes into separate functional groups) rather than groupings of several unrelated taxa into a single functional group (nematodes and arthropods into the same functional group). From a practical standpoint, aggregating species in this manner into dynamic models has been shown to have little effect on the overall model performance if the taxa share similar food and predators, and have similar turnover rates, as gauged by birth, death, and consumption rates (Gardner et al., 1982, Hunt et al., 1987).

The densities of the functional groups and sizes of biochemical pools, and all other functional attributes of ecosystems will be expressed in terms of matter (e.g., carbon or nitrogen) on a dry weight basis per unit area (e.g., g C m^{-2} or kg C ha^{-1}). Estimates of the population densities of organisms are obtained from field studies and constitute either point estimates of the population size at a given time or averages over several points in time. The densities are then converted to biomass, and using known elemental ratios of the biomass (e.g., C:N ratio) into the desired units.

Estimates of the physiologies, life history attributes, and behaviors (e.g., prey preferences, feeding rates, and functional responses) of species within functional groups are essential to our models. These estimates are obtained from laboratory and field studies and from estimates gleaned from the literature. The level of detail used is arguably coarse, but given the nature of the models these estimates can form the bases of sensitivity analyses.

Our models employ predator–prey interactions and mutualistic interactions that involve the direct transfer(s) of matter between functional groups of species, and transfers of matter between functional groups and detritus, and functional groups and pools of inorganic nutrients. Intraspecific competition will be incorporated into our models. Interspecific competition and other types of interactions will not be described directly in our equations, but are inferred through the other interactions. The approach does not preclude their inclusion.

Three general types of models will be used throughout the book—empirically based simulation models, theoretically based abstract models, and empirically based abstract models. The empirically based simulation models are akin to the tactical models described by Holling (1966, 1968) and May (1973), designed to study species interactions and biogeochemical cycles within ecosystems. The theoretically based abstract models are of the strategic types (May 1973), designed to explore general features and trends in responses to species interactions. The empirically based simulation models are a hybrid of the other two, employing the details of the physiologies and life history traits of organisms, and details on the rate-limiting processes on flow of matter. The hybrids are our attempt to use the tactical and strategic forms in the complementary manner proposed by May (1973).

Throughout this book we will present general ideas that emerge from the models but back them up with the results of empirical studies in the form of observations, trends, and manipulative studies. We will make extensive use of case studies.

1.5 An overview of the parts and chapters

We have organized the book into three parts, which correspond to the means of describing a food web (Part I), assessing dynamics and stability and their implications for diversity and food web architecture (Part II), and addressing the "dynamics of dynamics," wherein we focus on the development of food web architecture and the dynamic nature of structure over space and time (Part III).

Part I includes four chapters that detail our approach to describing and modeling food webs using elements of the community perspective and elements of the ecosystem perspective. Chapter 2 presents the framework for developing the equations to describe the dynamics among organisms and resources. We define stability, develop the means to assess it, and apply it to models of simple primary producer-based and detritus-based food chains. As an organizational tool we use the belowground food web description for a shortgrass steppe ecosystem at the Central Plains Experimental Range (CPER) in northeastern Colorado. Chapter 3 presents the connectedness description of the CPER food web. The chapter works through the rationale behind the selection of species included in the description, how they were aggregated into functional groups, and how the trophic interactions were determined. We go beyond the strict box and arrow description of a connectedness description and present estimates of the steady-state biomasses of the functional groups and the physiological and life history parameters associated with each group. Chapter 4 presents the energy flux description (*aka* energy flow description sensu Paine 1980) of the CPER food web. Using the biomass and parameter estimates we present a means of estimating the flux rates of carbon and nitrogen that would be required to maintain the assumed steady state. Chapter 5 presents a functional description of the CPER food web. This is where we take the information embodied in the connectedness description and the energy flux description, and map that to the terms of a series of differential equations that describe the dynamics of the community. This exercise provides a means to link energy flux to interaction strength.

Part II includes four chapters that explore stability and how energy flux impacts it. In Chapter 6 we explore the importance of structural and dynamic patterns within the CPER food web revealed in Part I. The analyses reveal pyramidal distributions of biomass and energy flux, and distinct pathways of energy flux. These pyramidal distributions equate to an asymmetric patterning in the paired positive negative and interaction strengths associated with each trophic interaction that is important to stability. The asymmetric patterning repeats itself within each of the distinct pathways. In the remaining chapters in Part II we parse different aspects of the observed patterns individually, and then bring them together in an integrative fashion. In Chapter 7 we explore how enrichment affects the structure, dynamics, and stability of the primary producer-based and detritus-based food chains. We expand the analysis to include more complicated models capable of generating a wider array of dynamic responses. In Chapter 8 we expand this theme and study of how enrichment might affect the distribution of biomass, trophic pyramids, and stability. Chapter 9 focuses on the distinct pathways if energy flux and the degree to which food webs possess hierarchical organization. We use the energy flux description and the functional description of the CPER food web to identify compartments based on energy source, energy flux and the patterning of interaction strengths.

In parts I and II we assumed that the connectedness structure of the food web was static. Part III presents three chapters that explore the dynamics of food web architecture during development and at steady state. Chapter 10 uses the principles discussed in chapters 6–9 to develop a model to explain the observed patterns of community development and diversity along gradients of productivity. Chapter 11 introduces the concept of a dynamic architecture and its implications on how we study food webs. We compare the notion of the keystone species that emerged from Paine's (1966) seminal work in the intertidal as a controlling factor in community stability to that of a flexible and dynamic architecture controlling the placement of the keystone species in the community. Chapter 11 addresses the changes in architecture that communities undergo during development along a gradient of productivity. We invoke the *theory of island biogeography* and link it to the concepts presented in Part II. Rather than focus entirely on species richness, we ask how the trophic structure and its stability in terms of the distribution of biomass, the patterning of energy flux, and the patterning of interaction strengths interact along a gradient of productivity. In Chapter 12 we study the spatial and temporal heterogeneity in the architecture of food webs at their presumed steady state. We demonstrate that there are parallels between species diversity and trophic structure along a productivity gradient and in the degree of temporal and spatial heterogeneity. We conclude by exploring the importance of transient dynamics, ambiguous dynamic states, and the concept of apparent complexity.

1.6 Summary

We propose to study food webs as a melding of the community ecology perspective and the ecosystem ecology perspective. These perspectives present different worldviews of how ecological systems are structured and persist, but we argue that they are not mutually exclusive. Individual species interact to form assemblages that we recognize as communities. Interactions within an assemblage interact with the abiotic components of the environment to cycle matter in a way that we collectively identify as an ecosystem. Species, matter, and energy are common features of all ecological systems. Ultimately the dynamics and stability of the first two are governed by the third. In short, in the context of food webs, we will explore the extent to which this inextricable relationship between species, matter, and energy can explain how systems are structured and how they persist in real and model systems.

Part I

Modeling simple and multispecies communities

Part I develops modeling approaches for studying simple and multispecies food webs. In Chapter 2 we will start with a presentation of how we structured our equations to describe the dynamics of individual and interacting populations, and the approaches that we will use to assess their stability. We have organized the remaining three chapters of this section following the categorizations of food webs presented by Paine (1980; see Figure I.7). We will use the soil food web of the North American shortgrass prairie described by Hunt et al. (1987) from the Central Plains Experimental Range (CPER) in Colorado, USA as the tutorial model to construct the different web types. Chapter 3 presents the connectedness description of the soil food web. Chapter 4 presents the energy flow description of the soil food web. Chapter 5 presents the functional web description of the soil food web. Each chapter builds on the previous one, providing background and observations for the topics in Parts II and III.

Chapter 2 is a toolbox of sorts, combining aspects of the community-based and ecosystem-based modeling approaches. We present two basic model types: ones based on primary producers and ones based on detritus. The primary-producer-based models should be familiar to students of ecology. The detritus-based models, although present in the literature for some time are often overlooked or treated as a special case. We start by introducing the primary-producer-based and detritus-based models separately to compare their structures and dynamic properties. In the remaining chapters we present integrated forms that include both primary producers and detritus.

Although both classes of models follow the Lotka–Volterra form, we present more complex alternatives to suit different needs. Approaches to modeling intraspecific competition or self-limitation that were missing in the original Lotka–Volterra models are discussed. More complex functional responses to describe the dynamics of predator–prey interactions than the linear responses in the original Lotka–Volterra models are also discussed. We move on to discuss the concept of stability and how to assess it.

Most importantly, we have structured the models to accommodate our emphasis on energetics. Traditional food web models developed from the community ecological perspective include the transfer of matter and energy from resource to consumer and from predator to prey. Missing from the traditional trophic interactions are the fates of matter that result from the death of all living organisms, and the unassimilated and mineralized matter that results from a consumer feeding on a resource or prey. For carbon, this translates as the corpses and discarded parts of microorganisms, plants, and animals, carcasses of prey and feces, and CO_2 respiration, respectively. For nitrogen this takes nitrogen-bearing organic (e.g., nucleic acids and proteins) and inorganic (e.g., NH_4^+ or NO_3^-) forms of the above, respectively. The models we develop account for these unassimilated and mineralized fractions of matter. The fractions are integral to the cycling of matter as allochthonous and autochthonous sources, and important resources to consumers that are often overlooked.

The organic fractions discussed above constitute detritus (dead organic matter). The detritus-based models include both allochthonous and autochthonous forms of dead organic material as a resource at the base of the system and then consumption in the same fashion as the more familiar primary-producer-based models. Initially we will present the detritus-based models as separate analogs to their primary-producer-based counterparts. By juxtaposing the two models in this way we can highlight differences but also point out important similarities in their structure and dynamics. Arguably, an ecology based on the structural and dynamic properties of detritus, or at least one that originated from primary production that is integrally linked to detritus, makes more sense than the traditional approaches that focus on plants, herbivores, and predators. The vast majority of primary production is not consumed by herbivores, but rather is discarded or dies, forming the resource base for microorganisms and detritivores. As we will see, this arrangement sets up interesting parallels with the more traditional modeling approach that focuses on primary production alone, but there are some key differences as well.

In Chapter 3 we present the *connectedness web* description based on species composition and functional grouping. These descriptions are quintessentially community ecology based, but are crafted in a way that allows us to use them as the foundation of the energy flux and functional food web descriptions. Given our focus on energetics, and short of basing the description on individual species, we emphasize the importance of establishing criteria for the descriptions that take into consideration aspects of a species physiology, life history, home range, and behavior. Each of these attributes of species is measurable with one or more of the dimensions of mass, space, and time that are used in the energy flux and functional descriptions.

In Chapter 4 we incorporate quantitative measures of observed biomasses and modeled feeding rates in order to construct the ecosystem-based *energy flux web* that emerges from the trophic interactions presented in the connectedness descriptions. The energy flux descriptions and underlying models conform to the laws of thermodynamics and mass balance. These descriptions address the immobilization of inorganic matter to organic matter as captured by primary producers, the transformation of one organic form to another form embodied in the trophic interactions and resource uptakes among organisms, the mineralization of organic matter to inorganic matter that results from the metabolism of all organisms, and the cycling of carbon and nitrogen into and within the food web. We present a modeling approach to estimating fluxes that when parameterized with field and laboratory data provides remarkable accuracy in estimating processes related to trophic interactions.

Chapter 5 focuses on the *functional web* description. This description represents a melding of the community and ecosystem perspectives. We use the tools presented in Chapter 2, the connectedness structure presented in Chapter 3, and the energy flux estimates in Chapter 4 to estimate the interaction strengths associated with the trophic interactions among organisms. From these estimates we can evaluate stability and other dynamic properties of the system.

As mentioned above, the description of the soil food web for the CPER in Colorado will serve as our tutorial. To explore patterns, we will apply the approaches discussed above to six additional soil food webs: two webs from the Kjettslinge Experimental Farm (Uppsala, Sweden) from barley practices with two different fertilizer rates (Andrén et al., 1990), two webs from the Horseshoe Bend Research Site (Athens, Georgia, USA) from a no-tillage and a conventional-tillage treatment (Hendrix et al., 1986), and two webs from the Lovinkhoeve Experimental Farm (Marknesse, The Netherlands) from winter wheat fields under integrated and under conventional farming (de Ruiter et al., 1993a). We will present observations and analyses of all seven soil food webs, but in some cases we will use a particular food web as a representative example. Although the food webs we deal with in detail are all soil food webs, we will show in Parts II and III that most patterns we describe also apply to food webs in aboveground ecosystems and aquatic ecosystems.

2

Models of simple and complex systems

2.1 Introduction

We have four objectives in this chapter. First, we present general approaches to developing the models that we will use throughout the book. The models will be deterministic models in nature, although they can be used and adapted to study the uncertainty of random fluctuations in the environment and in the parameters used to characterize the populations. Second, we discuss stability, the means that have been used to assess stability and other dynamic properties of the models, and special configurations of model systems that are inherently stable. Third, we will develop and assess the dynamics and stability of models that are based on primary producers and models that are based on detritus. Fourth, we will then compare the structure and dynamics of the two classes of models. We will revisit the approaches, concepts, and tools in the chapters that follow.

2.2 Model structure and assumptions

We describe two classes of models: one based on primary production and one based on detritus (Figure 2.1). Primary-producer-based models, as their name implies, start with populations of one or more primary producers that possess the ability to transform inorganic matter into an organic form. Detritus-based models start with one or more forms of nonliving organic matter (*aka* detritus) that originate from outside the system (an allochthonous source) and from within the system (an autochthonous source). The models will be based on variations of the familiar Lotka–Volterra form, by accommodating intraspecific competition and, in the case of nonliving materials, incorporating detritus. Furthermore, the models we will use possess a common currency of biomass (living or nonliving) bound by the conservation of matter and the first and second laws of thermodynamics. As such, whether viewed as open or closed to matter, a full accounting of inputs, outputs, and internal cycling of matter will be possible.

At this stage we will start with descriptions of the basal species and resources of the primary-producer-based and detritus-based models, and then follow with descriptions of the growth equations for the consumers of these resources. These descriptions will include a brief rationale for the structure and definitions of the parameters. The sections that follow provide a dimensional analysis of the growth

28 • Energetic Food Webs

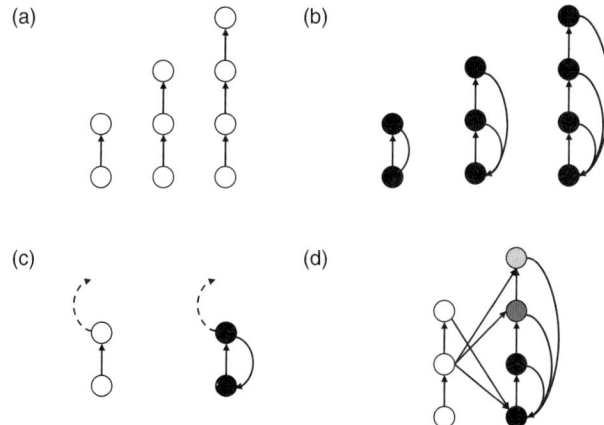

Figure 2.1 Simple food chains and food webs. Circles represent organisms or resources. Arrows represent flows of matter. Open circles are primary producers and their consumers. Black circles are detritus and their consumers. Shaded circles are consumers that consume or obtain energy from both types of food chains. (a) Primary-producer-based food chains of length 2, 3, and 4 include an autotroph at the base followed by consumers. (b) Detritus-based food chains of length 2, 3, and 4 include inputs of detritus from an outside (allochthonous) source, and cycled unassimilated organic matter and corpses from internal (autochthonous) sources. (c) Consumers in both types of food chains mineralize organic matter (dashed arrows, not depicted in the chains in a and b for clarity) and return unassimilated organic matter to the environment. For the detritus models, the unassimilated organic matter from consumption and the corpses of consumers is returned to the detritus pool as a basal resource. For the primary-producer-based models the unassimilated organic matter and corpses are accounted for but not included as an additional resource (i.e., they exit the system). (d) Models that link primary producers and detritus may cycle the unassimilated organic material and corpses through the detritus pool in an autochthonous manner or have a portion exit the system. The choices of the degrees to which organic material enters, cycles within, or leaves the system is entirely up to the modeler.

equations, and discussions on the functional responses, intraspecific competition coefficients, and the energetic efficiencies of fate of consumed matter.

The primary-producer-based models with n species will possess the following form:

$$\frac{dX_i}{dt} = r_i X_i - \sum_{i=1}^{n} f(X_i) X_j \qquad (2.1)$$

where X_i and X_j represent the population densities of the primary producer and consumers, respectively, r_i is the specific growth rate of the primary producer, and $f(X_i)$

represents the functional response of the interaction between (species $i \neq j$) or within the populations (species $i = j$) (more of which we will discuss below).

The detritus-based models that include detritus and n species possess the following form presented by Moore et al. (1993):

$$\frac{dX_D}{dt} = R_D + \sum_{i=1}^{n}\sum_{j=1}^{n}(1-a_j)f(X_i)X_j + \sum_{i=1}^{n}d_iX_i - \sum_{j=1}^{n}f(X_D)X_j \quad (2.2)$$

Here we use the subscript D to simply highlight and keep track of where detritus arises in the equations. In subsequent chapters the subscript d may be replaced by the subscript 1. For the detritus based models, X_i and X_j represent the n living species and X_D is the nonliving detritus. Detritus enters the system from allochthonous and autochthonous sources. The allochthonous source, R_D, includes detritus that enters from outside the system. Examples might include kelp washing up onto a beach system, or leaves entering a stream. In this representation we include a single allochthonous source, but there is no reason that multiple sources could not be included. The model identifies two allochthonous sources, one from the unassimilated fractions of prey that are killed by consumers, $\sum_{i=1}^{n}\sum_{j=1}^{n}(1-a_j)f(X_i)X_j$, which includes feces, orts, and leavings, and a second source from the corpses that die from causes other than predation, $\sum_{i=1}^{n}d_iX_i$. The functional response of the interactions between prey (living or nonliving) and consumers is presented by $f(X_i)$. Finally, the formulation for the direct consumption of detritus, $-\sum_{j=1}^{n}f(X_D)X_i$, resembles the consumption within the primary-producer-based equation (Equation 2.1).

We assume that detritus and living organisms within primary-producer-based systems and detritus-based systems are consumed in a density-dependent manner in the same way. For both classes of models the growth equations for consumers, X_i, can be described as follows:

$$\frac{dX_i}{dt} = -dX_i + \sum_{j=1}^{n}a_ip_if(X_j)X_i - \sum_{j=1}^{n}f(X_j)X_i \quad (2.3)$$

Consumer growth is offset by natural deaths represented by a specific death rate, d_i, and as a function of being consumed by other consumers, $-\sum_{j=1}^{n}f(X_j)X_i$, or is offset as a result of intraspecific competition (when $i = j$). In the case of intraspecific competition, the functional response, $f(X_i)$, is structured to model the impact of individuals of the same species (species i = species j) on their growth and dynamics, as above in Equation 2.1. The consumer populations grow in a density-dependent manner as a function of the living or nonliving prey that is consumed. This involves the summation of the consumption of individual prey types: $\sum_{j=1}^{n}a_ip_if(X_j)X_i$. The consumption term incorporates the functional response, $f(X_i)$, found in Equations 2.1 and 2.2, but accounts for the assimilation efficiency, a_i, and consumption efficiency, p_i, of the consumer.

2.2.1 Dimensions of mass, area, and time

The dimensions of the models are expressed in units of biomass per area per time. The choice of units is arbitrary, but when scaled to the system under study they provide greater insight into the system. The population densities (X_i) and detritus densities (X_D) are expressed in terms of biomass per area. Biomass can be represented as the total live weight (e.g., g or kg), carbon (e.g., g C), nitrogen (e.g., g N), or another element, but on a dry weight basis. Area can be scaled to any set of units but generally is expressed in meters (m^2) or hectares (ha^2). Time is generally scaled to the birth and death rates of the primary producers at the base of the food web. For plant populations, we traditionally use an annual time step. Throughout the book the growth equations will be expressed as $g\ C\ m^{-2}\ yr^{-1}$ or $kg\ C\ ha^{-1}\ yr^{-1}$.

The models possess two important rate-determining parameters: the specific birth (r_i) and specific death (d_i) rates. As explained below, these specific rates possess units of "per time." The specific birth rate (r_i), sometimes referred to as the intrinsic rate of increase, represents births minus deaths unrelated to consumption that is possible given the organism's life history and physiology within the chosen time step. The specific death rate (d_i) represents death not related to consumption. As a specific rate, it expresses the likelihood of the death of a unit of biomass per individual unit of biomass within the designated time step, having units of biomass per biomass per time, or simply per time. For the specific birth rate, we need to take into consideration the number (biomass) of progeny produced by an individual (biomass) within the time step. For the specific death rate, the inverse of the organism's life span is considered as a first approximation. We can also include deaths not related to consumption that are not explicitly included in the model (e.g., diseases, and deaths unaccounted for by predators).

2.2.2 Functional responses

The functional response, $f(X_i)$, characterizes how a predator adapts or adjusts its attack rate to changes in prey density. Many forms of functional responses have been proposed and their dynamic properties have been discussed at length elsewhere. Holling (1959) described three basic forms of functional responses, which he dubbed type I, type II, and type III (Figure 2.2). We will focus on these now, as each has important implications on how systems are structured and on their dynamic properties.

The type I, or linear functional, response is the simplest form of functional response, represented as follows:

$$f(X_i) = c_{ij}X_i \qquad (2.4)$$

where c_{ij} represents the consumption coefficient with unit per biomass per time, and X_i represents the density in biomass of the prey species. The type I functional response depicts a relationship wherein the attack rate is constant, regardless of prey density. This approach has been criticized as being unrealistic, as consumers are likely to adapt their feeding rates with changes in prey density.

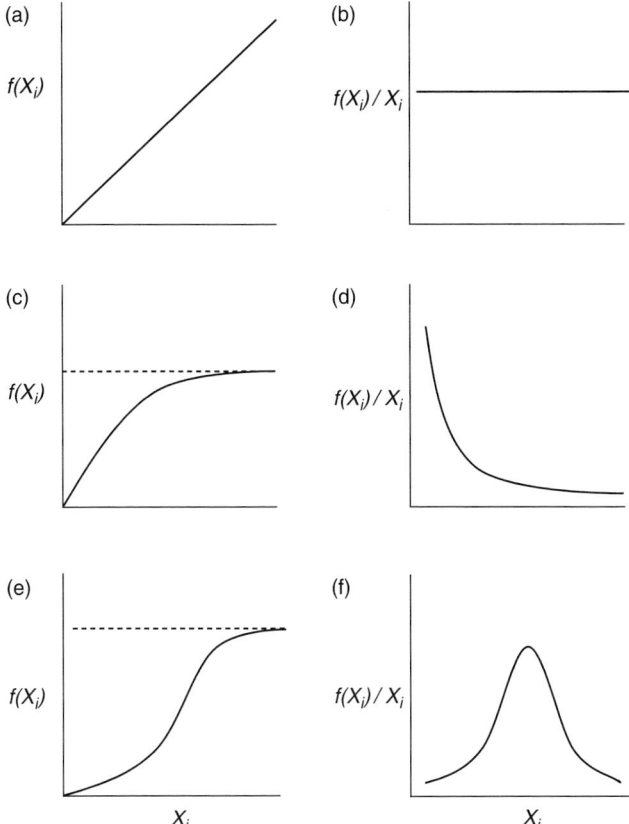

Figure 2.2 Functional response types described by Holling (1959). (a) Type 1 linear response: the feeding rate, $f(X_i)$, of the consumer increases linearly with prey density, X_i, and then becomes a constant value at the consumer's saturation point. (b) Relative mortality due to predation ($f(X_i)/X_i$) in the case of a type 1 linear response. (c) Hyperbolic type 2 functional response: the feeding rate increases with prey density but this increases continuously decreases until it becomes constant at saturation. (d) Relative mortality due to predation in the case of a type 2 hyperbolic response. (e) Sigmoid type 3 functional response: the feeding rate increases with prey density, first accelerating and then decelerating until the constant value is reached at the saturation level. (f) Relative mortality due to predation in the case of a type 3 sigmoid response.

The type II, or saturating functional, response is more realistic, and was originally represented by Holling (1959) as follows:

$$f(X_i) = \frac{cX_i}{1 + chX_i} \tag{2.5}$$

where c is the consumption or attack rate and h is the handling time. The type II functional response asserts that the attack rate increases with prey density to a point and then approaches a constant as the predator becomes satiated or is limited in its ability to process food. For this reason the form has all the properties of the Michaelis–Menton equations used to define enzyme kinetics. In fact with a little algebra, Equation 2.3 can be described by the Michaelis–Menton equation:

$$f(X_i) = \frac{\alpha X_i}{\beta + X_i} \tag{2.6}$$

where α is the maximum feeding rate when the consumer is satiated, and β is the prey density that generates half the maximum feeding rate, or the value of X where $\beta = \alpha/2$ defines the half-saturation point.

The type III, or switching functional, response can be derived directly from the generalized Michaelis–Menton equations as follows (Real, 1977):

$$f(X_i) = \frac{\alpha X_i^x}{\beta + X_i^x} \tag{2.7}$$

where x is the encounter rate where the predator is most efficient. When $x = 1$ the functional response is a Holling type II. When $x > 1$ the functional response is represented by a type III sigmoidal curve whose sill, like that of the type II functional response, is defined by handling time or satiation. The principal way it differs from the type II functional response is that its shape depicts an attack rate that is disproportionately low at low prey density, disproportionately high at high prey density, and constant when the predator is satiated. This type of response may represent a learning curve for consumers on prey types or switching behavior in cases where multiple prey are involved. The disproportionately low attack rate at low prey densities effectively serves as a refuge for the prey.

At first glance the graphical depictions of functional responses presented in Figure 2.2 might suggest that predators have a controlling effect on prey dynamics with increasing prey numbers, as the predation rate increases. From the standpoint of control or regulation, it is more important that we focus on how the *relative* mortality in the prey population, due to predation ($f(X_i)/X_i$), responds to changes in prey numbers (X_i). To illustrate this point, we have paired the traditional graphical representations of the different functional responses in Figure 2.2 with representations that show the relative contributions. For example, comparing the predation according to a type 1 linear response (Figure 2.2a) causes mortality that is neutral (Figure 2.2b), i.e., constant percentage-wise. The type 2 hyperbolic response (Figure 2.2c) means that mortality decreases with prey numbers, creating a destabilizing positive feedback (Figure 2.2d), i.e., the more prey the less mortality. The type 3 sigmoid response (Figure 2.2e), however, causes negative feedback at the relatively low prey numbers, i.e., the more prey (over that range) the higher the mortality (Figure 2.2f). This shows that a functional response has only a controlling effect on the prey numbers when the response has an accelerating shape and that an increase in prey numbers is followed by a disproportional increase in predation rate.

Functional responses have been established for many types of predator–prey systems and have been used to evaluate the consequences of the shape of the functional response for population dynamics and stability (e.g., Holling, 1965, Hassell, 1979). In most cases functional responses are established for single predator–prey interactions, but some include alternative prey from competitive predators (e.g., Hassell, 1979, Koen-Alonso, 2007). Measuring feeding rates through functional responses is a very laborious approach conducted under controlled field and laboratory conditions. Feeding rates have been based on direct observations of kills or consumption, and by indirect means including the number of corpses or amount of fecal deposition.

2.2.3 Energetic efficiency and conversion rates

The models take into consideration the energetic efficiencies of the trophic interactions. For our models we have defined the *energetic efficiency, e,* of a trophic interaction as the product of two components: the assimilation efficiency (a) and the production efficiency (p). Put another way, the energetic efficiency represents the ratio of immobilized matter or energy that forms new biomass in the form of individual growth or reproduction to the amount of matter or energy consumed, where

$$e = \frac{\text{Growth and Reproduction}}{\text{Consumed Biomass}} = ap \qquad (2.8)$$

Equation 2.8 acknowledges that the energetic efficiencies of organisms are not perfect, yet must be greater than zero otherwise the organism dies, e.g., $0 < e < 1$. The fraction, $(1-e)$, represents the fraction of consumed prey or resource that remains in an organic form that is unassimilated (e.g., leavings and feces) by the consumer and the fraction that is assimilated by the consumer and mineralized to an inorganic form, i.e., the maintenance fraction.

$$(1 - e) = \frac{\text{Unassimilated Consumption and Maintenance}}{\text{Consumed Biomass}} \qquad (2.9)$$

Most models account for the energetic efficiencies of organisms. For our purposes, we need to refine the fractions further to account for fates of organic and inorganic matter. From Figure 2.3 we see that the matter that is consumed is separated into an assimilated fraction and an unassimilated fraction. *Consumption* of matter involves ingestion into a food vacuole or oral cavity, and then into a blind sac or gut. Not all the material that is consumed by an organism necessarily makes its way into the organism. For example, protozoa engulf materials into vacuoles or, in the case of amoebae, surround materials with pseudopodia. Once engulfed by the protozoan, materials are digested by enzymes secreted into the vacuole and move by diffusion

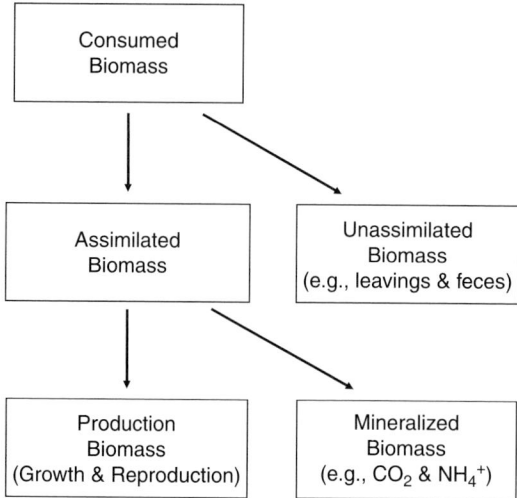

Figure 2.3 Scheme for the fate of consumed matter for given trophic interactions. The consumed biomass is divided into compartments determined by two energy conversion efficiencies. The assimilation efficiency defines the fraction of biomass of the prey or resource that is assimilated by the consumer. The unassimilated fraction is left behind, is egested, or passes through the consumer. The production efficiency is the fraction of assimilated that remains in the consumer and used for growth and reproduction. The remaining fraction is mineralized from organic to an inorganic form and returned to the environment. Estimates of these efficiencies for various taxa are readily available. If they are not available, they can be obtained through experimentation or approximated based on the body size, morphology, and physiology of the organisms.

across the cell membrane into the cytoplasm, with undigested materials remaining in the vacuole and then *egested*, or returned to the surroundings. Likewise, metazoans capture and kill prey, leaving unconsumed prey behind or returning the undigested prey to the environment as feces.

Of the material that has been *ingested*, a portion is used for growth, reproduction, and maintenance. *Assimilation* refers to the intake of molecules by an organism across cellular membranes so that they can then be used for these three processes. The *assimilation efficiency, a,* of an organism is the fraction of ingested or consumed matter that is assimilated.

$$a = \frac{\text{Assimilated Consumption}}{\text{Consumed Biomass}} \quad (2.10)$$

The unassimilated fraction, $(1 - a)$, represents the fraction that remains or is returned to the environment in either organic or inorganic form.

$$(1 - a) = \frac{\text{Egested Consumption}}{\text{Consumed Biomass}} \quad (2.11)$$

It is important to remember that the unassimilated material is organic material and is not lost to the environment, but rather serves as an energy source for other organisms. The unassimilated faction of consumed matter is left behind (leavings and orts), is evacuated from vacuoles (e.g., Protozoa) or gastric caecae (e.g., Cnidaria and Platyhelminthes), or is passed through the consumer as fecal material (e.g., protostomes and deuterostomes) and returned to the environment. These different forms of unassimilated organic compounds constitute either the autochthonous inputs of detritus that are returned to the labile or resistant detritus pool depending on their respective C:N ratios, or outputs (allochthonous inputs for another system) that leave the system. We have opted to assign a single value for the assimilation efficiency of a given taxon in our models. Clearly the assimilation efficiency will vary depending on the quality of the resource the taxon is consuming.

The second fate of consumed matter applies to the fraction that was assimilated. The assimilated fraction can be transformed into new biomass, i.e., *production*, as added biomass or reproduction, or can be used for maintenance. *Production* refers to the creation of new biomass as growth and reproduction from assimilated molecules. The *production efficiency, p,* of an organism is defined as the fraction of assimilated matter that is transformed into new biomass as growth and reproduction.

$$p = \frac{\text{Growth and Reproduction}}{\text{Assimilated Consumption}} \quad (2.12)$$

The maintenance fraction of the assimilated matter, $(1 - p)$, is the fraction that is returned to the environment in an inorganic form. For carbon, the maintenance fraction is the CO_2 respired; for nitrogen the fraction is nitrogenous waste (e.g., urea, ammonium).

$$(1 - p) = \frac{\text{Maintenance}}{\text{Assimilated Consumption}} \quad (2.13)$$

Coming full circle, we arrive at our definition for the *energetic efficiency, e,* of a trophic interaction presented in Equation 2.8 by taking into account both the assimilation efficiency, a, and the production efficiency, p.

The energy budget scheme presented above can be expanded on. We could define many more compartments than the four broad compartments used in this scheme. For example, we could further partition the unassimilated fractions into different forms, each serving as a separate basal resource. We could also expand the way in which the scheme is parameterized. Here we have opted to treat the assimilation and production efficiencies as constants. These were based on the physiology of the predator and the average quality of the prey. Prey types that are more recalcitrant than others might be assimilated differently and require greater metabolic energy for breakdown than more labile materials. The simplifications we used are arguably departures from reality, but for the purpose of the analyses, our assessment through sensitivity analyses leads us to believe that the impact of these simplifications on the outcome of the model is relatively small.

2.2.4 Intraspecific competition and self-limitation

Intraspecific competition, or self-limitation, represents the negative effects that individuals of a species within a population have on the growth of other individuals of the same species within the population. The process is modeled as being density dependent and tied to the acquisition or utilization for a resource including, but not limited to, a prey item, mates, space, essential nutrients, light, or water.

The original Lotka–Volterra models did not include intraspecific competition. Arguably, most, if not all, populations are subject to intraspecific competition and self-regulation to some degree. A simple representation starts with a model of a single species following Equation 2.1, with intraspecific competition modeled using a type I functional response (Equation 2.4):

$$\frac{dX_1}{dt} = r_1 X_1 - c_{11} X_1 X_1 \tag{2.14}$$

where c_{11} is the coefficient defining the degree of intraspecific competition. Equation 2.14 is a different representation of the equation of logistic population growth:

$$\frac{dX_1}{dt} = rX_1 \left(1 - \frac{X_1}{K}\right) \tag{2.15}$$

where population equilibrium $X_1^* = r_1/c_{11}$, which is the carrying capacity K. What becomes clear from our discussion of stability, in the sections that follow, is that the functional response associated with intraspecific competition, $c_{11} X_1$, is central to the dynamics and stability of Equation 2.14. Below we will, in fact, demonstrate the importance of the functional responses for intraspecific competition, $c_{ii} X_i$, by extension of the systems of equations defining the interactions of multiple species.

2.3 Stability

Several definitions of stability have been developed and associated with ecological systems (McCann, 2000). Most meanings center on the ability of the system to maintain or return to a steady state in species composition, population size, and function in the face of perturbations. How we observe and study real ecosystems aligns well with our mathematical representations in models, but the fit is not perfect. Mathematical definitions of stability are less forgiving than those that we might base on observations given the assumptions on which they are based, and the contexts in which they are viewed. For example, many mathematical models of food webs often assume that the system operates at equilibrium or near equilibrium and are closed to immigration and emigration. In either case, if a species were to become locally extinct following a disturbance, the system would be deemed unstable in a mathematical sense. For real systems we might question the notion of a strict equilibrium and recognize that species often become locally extinct, only to return after some time. Other models do not take into consideration spatial variation or temporal variation in species composition, densities, or dynamics, but instead rely on spatial and temporal averaging. In these cases, local extinctions in space and time, and the ability of species to migrate from one locale to another may be important aspects of a systems dynamics and ability to persist, yet are missed when averages are employed.

Having said this, models require assumptions from which we can assess hypotheses and generate predictions. The challenge for us will be to select a framework or perspective to work with that captures enough biological realism and complexity and yet is tractable, both empirically and mathematically. As May (1973) pointed out, the factors that contribute to the stability at one level of resolution (locally) may not be the same as those that apply to the system as a whole (globally). That is, when a component of a system is viewed in isolation it may be stable, yet when coupled with others may not, and vice versa. We will review these notions of global and local stability, and properties of models that exhibit these types of dynamics.

2.3.1 Local stability

Local stability focuses on the dynamics of a system near equilibrium, regardless of the value of the equilibrium. For our purposes, we will focus on the subset of systems that have a feasible equilibrium, X_i^*, wherein all species have positive equilibrium densities ($X_i^* > 0$ for all i species). With local stability, if the system were disturbed causing any species to deviate, x_i, a small distance from its equilibrium ($x_i = X_i - X_i^*$) and in time the system returns to its original equilibrium, then the system is locally stable, and otherwise unstable. To model this we need to rewrite the system of equations using Taylor's expansion to represent the deviation (x_i) of a population (X_i) from its equilibrium (X_i^*), i.e., $x_i = X_i - X_i^*$. To understand the dynamics of the deviation it is useful to express these new equations in terms of their eigenvalues (λ_j) as follows:

$$\lambda x_i(t) = \sum_{j=1}^{m} \alpha_{ij} x_j(t). \qquad (2.16)$$

Equation 2.16 represents an nth order polynomial equation in λ of the Jacobian matrix A, whose elements are α_{ij}.

We can assess the local stability of a set of equations by analyzing the eigenvalues of the Jacobian matrix A as proposed by May (1972):

$$\frac{dx}{dt} = Ax \qquad (2.17)$$

where x is the vector describing the state of the system in terms of the departures of the population sizes ($X_i - X_i^*$) from the equilibrium state (X_i^*), and A is the Jacobian matrix, in which the elements, α_{ij} describe the per capita effects of the populations upon one another under equilibrium conditions. The elements, α_{ij}, of A are defined as follows:

$$\alpha_{ij} = \left(\frac{\partial \frac{dX_i}{dt}}{\partial X_j}\right)^* \quad (2.18)$$

The stability of the system is governed by the eigenvalues (λ) of the matrix A. The eigenvalues are determined as follows:

$$\det|A - \lambda I| = 0 \quad (2.19)$$

where I is the unit matrix with the same dimensions as A.

The eigenvalues may be complex numbers with real (ζ) and imaginary (ξ) parts, i.e., $\lambda = \zeta + i\xi$, each of which describes a component of the dynamics of the deviations x_i. The real part ζ influences the degree of exponential growth or decay in the deviation, while the imaginary part ξ introduces sinusoidal oscillations. If the real part ζ of all the eigenvalues is less than zero, the disturbance x_i decays and the system returns to its original equilibrium. The system is stable in the region of the equilibrium if the real parts of the eigenvalues, λ_i, of matrix A are negative. If one or more of the eigenvalues possess positive real parts, then the system is not stable, and if perturbed, will deviate from its original equilibrium. In this case we simply do not know what the configuration of the system will take over time, other than it will not return to its original state.

2.3.2 Qualitative stability

As expected, as the eigenvalues are defined in terms of the elements of A, there are conditions under which the patterns of the distribution and signs of the elements of A ensure stability. The Routh–Hurwitz criteria are based on one such patterning that can be used to gauge the stability of a system by knowing the signs of the interactions within the matrix A, without knowing the magnitudes of the elements of the matrix A. This can be useful as many of the published descriptions in the literature are connectedness descriptions devoid of any quantitative measures. For an $n \times n$ interaction matrix A with elements α_{ij} the conditions for local stability are as follows:

1. $\alpha_{ii} \leq 0$ for all species i.
2. $\alpha_{ii} \neq 0$ for at least one species
3. $\alpha_{ij}\alpha_{ji} \leq 0$ for all (i,j); $(i \neq j)$.
4. All sequences of three or more subscripts have cyclical products equal to zero, i.e., u, $v, w, \ldots, y, z$, such that $\alpha_{uv}\alpha_{vw}\ldots\alpha_{yz}\alpha_{zu} = 0$.
5. $\det A \neq 0$.

The biological meanings of the conditions are fairly straightforward. Condition 1 focuses on the diagonal elements of matrix A and requires that no species have a positive self-feedback term. If a positive feedback term were present, the deviation from equilibrium would simply grow in time. Condition 2 also focuses on the diagonal elements of matrix A, and further adds that at least one of the species is self-regulating, possessing a damping term. Without this condition the system could at best have positive terms, thereby violating Condition 1, or have all the terms equal to zero, leading to neutral stability wherein the population once disturbed cycles around the equilibrium, but never returns. Taken together Conditions 1 and 2 require that at least one of the species is constrained in some fashion by a limiting resource. Condition 3 requires that no two species that interact with one another have similar signs. This condition rules out direct inter-specific competition ($\alpha_{ij} < 0$ and $\alpha_{ji} < 0$) and mutualism ($\alpha_{ij} > 0$ and $\alpha_{ji} > 0$). Condition 4 eliminates the possibility of closed cycles of three or more species. This condition rules out most omnivorous interactions where a species preys on more than one species that occupy different trophic positions, and all such interactions within an isolated linear food chain or food chain embedded in a more complex web. Condition 5 insures that all species are either self-regulating and/or regulated by other species.

Qualitative stability is a useful tool for identifying stable configurations when the magnitudes of the elements of the community matrix are not known. However, we must remember that a system may be stable even though it is not considered to be qualitatively stable. In many cases, stability may depend on the magnitude of the elements and/or the patterning of elements of the community matrix.

2.3.3 Negative diagonal dominance and quasi-diagonal dominance

For descriptions that have estimates of the interactions' strengths the means to gauge stability are clear; evaluate the eigenvalues. There are cases where the configuration of the system ensures stability. For example, if the interaction matrix A possesses either a negative diagonal dominance configuration or a quasi-diagonal dominance configuration the equilibrium is stable. Both of these configurations involve assessing the relationships between the magnitudes of the diagonal elements to those of the off-diagonal elements of matrix A.

2.3.3.1 Diagonal dominance

A matrix A is diagonally dominant if the absolute value of each diagonal element is greater than the sum of the absolute values of the elements within its row:

$$|\alpha_{ii}| > \sum_{j \neq i} |\alpha_{ij}| \qquad (2.20)$$

McKenzie (1960) proposed the theorem that if a matrix A is dominant diagonally and the diagonal elements are all negative ($\alpha_{ii} < 0$ for all i), i.e., A is negative diagonally dominant, then the real parts of the eigenvalues of A are negative ($\lambda_i < 0$ for all i), hence A is stable.

This result has important implications when assessing the stability of ecological communities. At first glance the utility of McKenzie's theorem for ecological communities might appear tenuous, as the condition that all the diagonal elements are negative is rarely met. Traditional formulations of multispecies models include a negative self-limitation or intraspecific competition term along the diagonal for living basal species and consumption terms of detritivores on detritus that result in negative elements, $\alpha_{ii} < 0$, along the corresponding diagonal positions. The equations that describe the dynamics of the consumers in these descriptions often do not possess terms that lead to self-limitation or negative feedback along the diagonal, but typically result in $\alpha_{ii} = 0$.

The reason for the omissions appears grounded in part on a belief that not all species self-regulate at equilibrium and in part on a desire to simplify the equations. Regardless of the reason for the omissions, we argue that all living organisms exhibit some form of self-limitation due to crowding, the overexploitation of food resources or limiting nutrients, or a shortage of mates. If we add a density-dependent self-limitation term in the form of $-c_{ii}X_i^2$ to the equations for each living organism in the community, include feedback loops for their death, $-d_iX_i$, and unassimilated consumption to detritus, $(1-a_j)c_{ij}X_iX_j$, each of the diagonal elements of the Jacobian matrix A will be negative ($\alpha_{ii} < 0$ for all i). We will illustrate this point below for living species within the following three species omnivorous food chains:

$$\frac{dX_1}{dt} = r_1X_1 - c_{11}X_1^2 - c_{12}X_1X_2 - c_{13}X_1X_3 \tag{2.21a}$$

$$\frac{dX_2}{dt} = -d_2X_2 - c_{22}X_2^2 - c_{23}X_2X_3 + a_2p_2c_{12}X_1X_2 \tag{2.21b}$$

$$\frac{dX_3}{dt} = -d_3X_3 - c_{33}X_3^2 + a_3p_3c_{13}X_1X_3 + a_3p_3c_{23}X_2X_3 \tag{2.21c}$$

The Jacobian matrix A for this system of equations is:

$$A = \begin{bmatrix} -c_{11}X_1^* & -c_{12}X_1^* & -c_{13}X_1^* \\ a_2p_2c_{12}X_2^* & -c_{22}X_2^* & -c_{23}X_2^* \\ a_3p_3c_{13}X_3^* & a_3p_3c_{23}X_3^* & -c_{33}X_3^* \end{bmatrix} \tag{2.22}$$

The Jacobian matrix A presented in Equation 2.22 does not satisfy Condition 4 of the Routh–Hurwitz criteria for a qualitatively stable matrix and is, therefore, not qualitatively stable. Nonetheless the system may be stable depending on the magnitudes of the elements of matrix A, particularly since its diagonal elements are all negative. Matrix A would be negative diagonally dominant and therefore stable at equilibrium if, after some algebra, the following conditions were met:

1. $c_{11} > c_{12} + c_{13}$
2. $c_{22} > a_2 p_2 c_{12} + c_{23}$ (2.23)
3. $c_{33} > a_3 p_3 c_{13} + a_3 p_3 c_{23}$

Most of the information required to evaluate the inequalities presented in Equation 2.23 can be readily obtained, as we will demonstrate in Chapter 4. The more difficult parameters to estimate are the coefficients for the self-limitation terms (c_{ii}). Knowing c_{ii} or having a convenient way to estimate it would simplify our analyses, but as we demonstrate below that all is not lost.

2.3.3.2 Quasi-diagonal dominance

An $n \times n$ matrix A representing community interactions with elements α_{ij} is quasi-diagonally dominant if there exists a set of n positive numbers $\pi_1, \pi_2, \ldots \pi_n$, such that the condition

$$\pi_i \alpha_{ii} + \sum_{j \neq i}^{n} \pi_j |\alpha_{ij}| < 0 \quad (2.24)$$

holds for every $i = 1, 2, \ldots, n$. Moreover, if a matrix A is a negative diagonal matrix (i.e., all $\alpha_{ii} < 0$) and quasi-diagonally dominant, it is also stable (McKenzie, 1960).

DeAngelis (1975) conducted a series of analyses on models of simple food webs that provided plausible biological conditions (see inequalities presented in Equation 2.23) that would increase the likelihood of creating a system with a quasi-diagonal dominant matrix and meeting the conditions of Equation 2.24. These conditions included that: (1) consumers are inefficient either in terms of their energetic efficiencies or in terms of their capture rates, (2) consumers have high levels of self-limitations, and (3) the systems possessed donor-controlled dynamics. The first two conditions are apparent in the inequalities presented in Equation 2.23. The left-hand side of the inequalities contains the diagonal elements represented by the coefficients for self-limitation, c_{ii}. The right-hand side of the inequalities contains the off-diagonal elements composed of the consumption coefficients, c_{ij} and the predator energetic efficiencies, a_j and p_j. The first two conditions clearly tilt the system so that the diagonal elements of matrix A are large relative to the off-diagonal elements. The third condition deals with the impact of donor-controlled systems similar to those that involve detritus. We will discuss these types of systems in more detail below, but for now will leave the reader with the counterintuitive result that donor-controlled systems produce strong negative diagonal terms.

We will rely on the properties of negative diagonal dominance and quasi-diagonal dominance and employ variations of the procedures described above, offering a means for assessing and comparing the stability of complex multispecies models.

2.3.4 Quasi-diagonal dominance and loops

To understand the stability of complex communities, Neutel et al. (2002) studied food web structure in terms of trophic interaction loops. A trophic interaction loop is a closed chain of trophic links defined as a pathway of interactions that originate from a species through the web back to the same species without passing other species more than once (Figure 2.4). Being a pathway of interactions, an important distinction is that each step on the path going from one species to another refers to the interaction strength between the two species and not to the energy flux rate or to the feeding rate. For predator–prey interactions, the interaction loop includes the positive effects of prey on predators and the negative effects of predators on prey. Two important metrics of the trophic interaction loop are its length and weight. The length of a loop is the number of groups or species in the loop. Loop weight is a measure of the strengths of the interactions in the loop calculated as the geometric mean of the absolute values of the interaction strengths, (i.e., elements of matrix A) in the loop.

To illustrate these points, we turn to the example of a three-species omnivorous interaction within the food web by Colton (1916) that we presented in Chapter 1. The interactions among phytoplankton, zooplankton, and *Balanus* form two loops of length 3 (Figure 2.4). One of the loops consists of two top-down effects and one bottom-up effect; the other consists of two bottom-up effects and one top-down effect. The two loop weights for this omnivorous subsystem of length 3 can be calculated as:

$$W_1 = \sqrt[n]{(\alpha_{32}\alpha_{63}|\alpha_{26}|)} \tag{2.25}$$

$$W_2 = \sqrt[n]{(|\alpha_{23}||\alpha_{36}|\alpha_{62})} \tag{2.26}$$

The weights of the trophic interaction loops in food webs are an interesting property, as the maximum of all loop weights is an indicator for matrix stability (Neutel et al., 2002). By assuming $\alpha_{ii} = \alpha < 0$ for all i, and defining a positive matrix A with $\alpha_{ij} = |\alpha_{ij}|$ and $\alpha_{ii} = 0$, Neutel et al. (2002) argued that the dominant eigenvalue $\lambda_d(A)$ of A is the smallest value for intraspecific interaction necessary for A to be quasi-diagonally dominant and hence a sufficient condition for stability. In this way, the maximum loop weight provides an approximation of the level of intraspecific interaction, α_{ii}, sufficient for stability, and hence a means of estimating the α_{ii} elements of the A matrix at equilibrium. We will argue that the diagonal elements of the interaction matrices of our complex food webs can be expressed as $\alpha_{ii} = s_i d_i$, where d_i refers to all forms of nonpredatory death, both density-dependent and density-independent death, and s_i is the fraction that is density dependent. Hence for any food web with negative diagonal elements, we can adjust the value of s_i to find the critical values of s_i required for matrix stability.

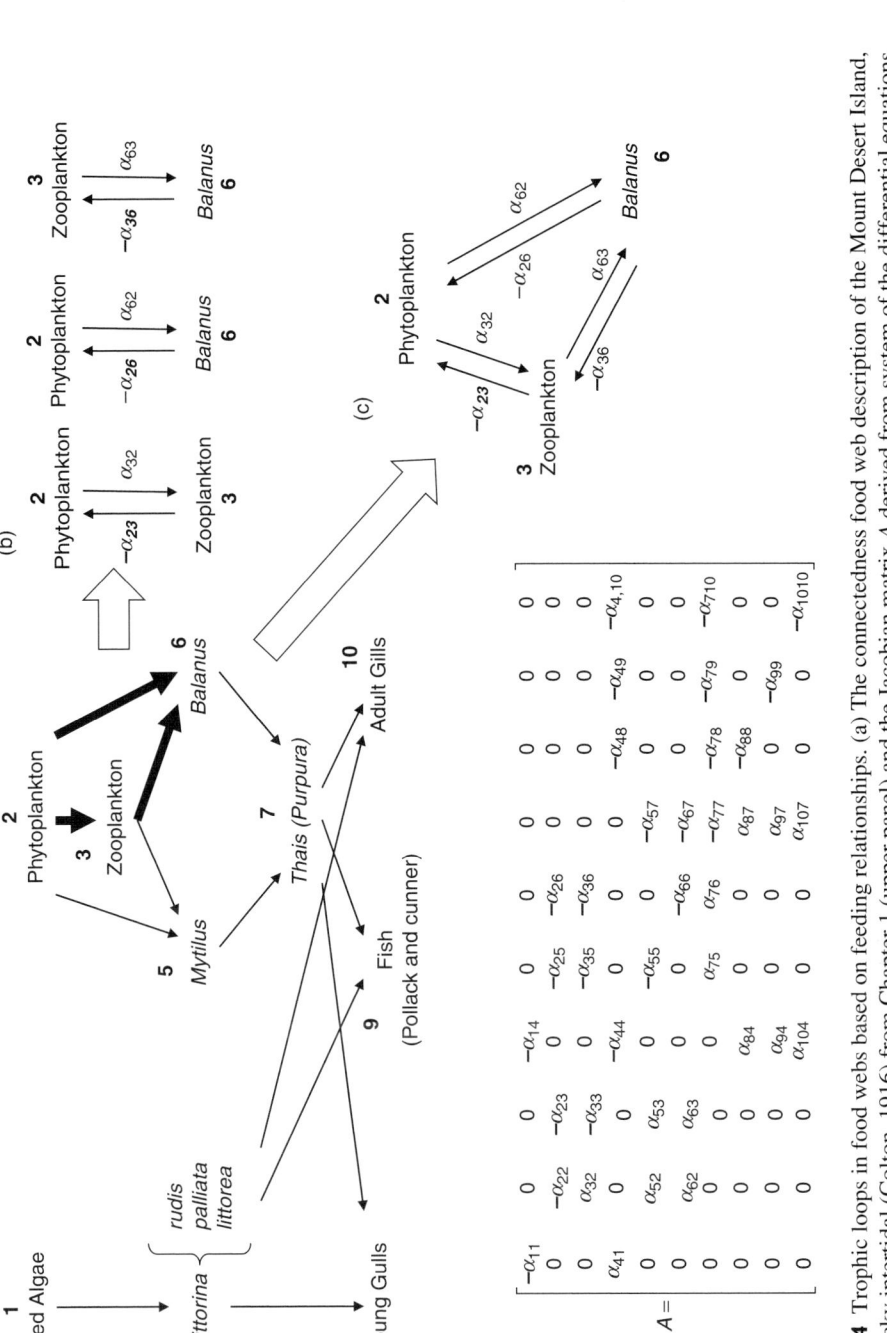

Figure 2.4 Trophic loops in food webs based on feeding relationships. (a) The connectedness food web description of the Mount Desert Island, Maine, rocky intertidal (Colton, 1916) from Chapter 1 (upper panel) and the Jacobian matrix A derived from system of the differential equations that describe the trophic dynamics (lower panel). The black arrows isolate a three-species omnivorous trophic loop that includes phytoplankton, zooplankton, and *Balanus*. (b) Given that every two-species trophic interaction creates a simple trophic loop of length 2, the three-species omnivorous loop for three loops of length 2: $(\alpha_{3,2}, -\alpha_{2,3})$, $(\alpha_{3,2}, -\alpha_{2,3})$, and $(\alpha_{3,2}, -\alpha_{2,3})$. (c) The three-species omnivorous loop possess two loops of length 3: $(\alpha_{6,2}, -\alpha_{3,6}, -\alpha_{2,3})$ and $(-\alpha_{2,6}, \alpha_{6,3}, \alpha_{3,2})$.

2.3.4.1 Global stability

Global stability analyses study the system over a wider landscape, encompassing larger perturbations. A system that is globally stable will return to equilibrium from any initial condition, i.e., deviation from equilibrium, and not just those that are close to equilibrium. Complications arise in assessing global stability because the trajectory that the system takes following the disturbance may involve moving away from the equilibrium.

Physical analogies of landscapes are often invoked to convey the distinction between local and global stability (May, 1973, Pimm, 1982). Picture a hilly, sloping landscape pocked with craters. The analogy for a local stability analysis would be as follows. Assume our system were a round object nestled at the bottom of a crater (at equilibrium) that was pushed up the wall of the crater but not over its lip (a short distance from the equilibrium). If released, our system would return to its original position at the basin (locally stable). To be globally stable, the analysis would have to take into account what would happen to our system if it were pushed over the lip of the crater. In this case the trajectory is less clear, but if it were globally stable it would eventually return to the basin of the crater.

Global stability can be assessed by identifying a Lyapunov function, V, and determining whether the time derivative of V is less than or equal to zero, i.e, $dV/dt \leq 0$, provided that all the solutions are bounded for $t \geq 0$. We will demonstrate below that for simple food chains described as primary producer-based and detritus-based food chains Lyapunov functions have been identified that meet these conditions, indicating that both types of models are globally stable.

2.4 Simple food chains

Here we will explore the structural and dynamic properties of simple food chains based on primary producers and detritus (Figure 2.1). We will focus more attention on the discussion for the detritus-based models as the literature is replete with detailed descriptions and analyses of the primary producer-based models. More complex forms will be presented in Chapter 6.

The connectedness descriptions of the food chains are presented in Figure 2.1. The number of trophic levels refers to the number of energy states (sensu Odum, 1953) that the system possesses. Energy at the base of the system, whether from a primary producer or detritus, resides in the first trophic level. Energy from the first trophic level that is consumed and incorporated into the biomass of the consumer resides in the second trophic level. Energy from the second trophic level that is consumed and incorporated into the biomass of the consumer resides in the third trophic level, and so on. The trophic position of an organism is defined by the relative contribution of each trophic level plus 1 to its steady state biomass. For our simple linear food chain, the energy that resides in each trophic level corresponds to the organism's trophic position within the food chain. For systems that include omnivores (organisms that obtain energy from more than one trophic level) the concordance between trophic levels and the trophic positions of organisms abates.

Before delving much further into the discussion we first define the notion of feasibility as it is related to the densities of species at steady state (Roberts, 1974). A system is feasible if all the species possess positive densities at steady-state, i.e., $X_i^* > 0$ for all i species. This may seem like a trivial point, but it is important if we are to model systems that are biologically realistic. It does not make sense to model the dynamics of a system in which one or more of the populations possess negative densities, much less to try to define such circumstances biologically.

2.4.1 Primary-producer-based food chains

A simple two-species primary-producer-based food chain depicting a plant, X_1, (Equation 2.12a), and a herbivore, X_2, (Equation 2.12b) is given by:

$$\frac{dX_1}{dt} = r_1 X_1 - c_{11} X_1^2 - c_{12} X_1 X_2 \tag{2.27a}$$

$$\frac{dX_2}{dt} = -d_2 X_2 + a_2 p_2 c_{12} X_1 X_2 \tag{2.27b}$$

The equilibrium values are as follows:

$$X_1^* = \frac{d_2}{a_2 p_2 c_{12}} \tag{2.28a}$$

$$X_2^* = [(r_1 a_2 p_2 c_{12}) - (c_{11} d_2)]/(a_2 p_2 c_{12}^2) \tag{2.28b}$$

The Jacobian Matrix, A, is given by:

$$A = \begin{bmatrix} -c_{11} X_1^* & -c_{12} X_1^* \\ a_2 p_2 c_{12} X_2^* & 0 \end{bmatrix} \tag{2.29}$$

The critical eigenvalue, λ_{max}, that dominants the dynamics of this model is as follows:

$$\lambda_{max} = \frac{\alpha_{11} + \sqrt{\alpha_{11}^2 + 4\alpha_{12}\alpha_{21}}}{2} \tag{2.30}$$

The equilibrium of this two-species system is locally stable as both the eigenvalues are negative. Goh (1977) demonstrated that if food chains described using generalized Lotka–Volterra models of the form as those described above, are feasible (all $X_i^* > 0$), then the equilibrium is globally stable as well. If a Lotka–Volterra model for n interacting species has the following form:

$$\frac{dX_i}{dt} = r_i X_i + \sum_{i,j=2}^{k} c_{ij} X_i X_j \quad i = 1, \ldots, n \qquad (2.31)$$

and has a nontrivial equilibrium (i.e., the system is feasible as all $X_i^* > 0$), then the Lyapunov function

$$V(X) = \sum_{i=1}^{n} l_i \left[X_i - X_i^* - X_i^* \ln \frac{X_i}{X_i^*} \right] \qquad (2.32)$$

has the following time derivative expressed only in terms of the intraspecific parameters (Harrison 1979):

$$\frac{dV}{dt} = \sum_{i=1}^{n} l_i c_{ii} (X_i - X_i^*)^2 \qquad (2.33)$$

As for the intraspecific competition coefficient, $c_{ii} \leq 0$, $dV/dt \leq 0$, the system is globally stable. Harrison (1979) went on to prove that simple food chains described in this manner with an intraspecific competition term at the first level assures global stability.

2.4.2 Detritus-based food chains, internal cycling, and donor control

Here we will discuss models that include detritus at the base, followed by detritivores and predators. While there are similarities in the ways in which detritus-based and primary-producer-based models behave, there are key differences that may have important implications.

We know that not all the energy that is captured by primary producers is consumed in the manner described above and not all the energy that sustains a food web arises directly from living primary producers within the ecosystem. Most primary production is not consumed while living, but rather is released into the environment as nonliving detritus to be consumed by a host of organisms. In many instances, detritus leaves the system from which it originated, only to enter another system. In fact, few ecosystems are based entirely on homegrown energy, but instead are a composite of internally (autochthonous) and externally (allochthonous) derived energy, a point we will discuss further below. Autochthonous inputs include feces, exuvia, fallen leaves, dead roots, orts, and corpses that remain in the system. Allochthonous inputs include the same materials imported from another system.

The balance between the levels of autochthonous and allochthonous inputs of energy into a food web creates an interesting array of dynamics. DeAngelis (1980) defined food chains and webs that are based on detritus inputs as donor-controlled systems. Under donor control, the population density and rate of input of the donor has an effect on the consumer population density and dynamics, whereas the population density and dynamics of the consumer has no direct effect on the dynamics of the donor population or resource. Odum (1969) noted that the balance of internal and external levels of detritus and nutrient inputs played a more important role in the maintenance—and, by implication, stability—of mature ecosystems. Although the empirical evidence supports the notion that the internal origins of detritus and nutrients become more prevalent in mature systems, the basis for the proposition that it is important to the stability and persistence of the system is not obvious from observation alone.

2.4.2.1 Models of donor-control dynamics

We will elaborate on the equations we introduced in Section 2.2 that included both allochthonous and autochthonous inputs, the conceptual model of which is presented in Figure 2.1. The equations for dynamics of the detritus pool and its consumers were defined as follows:

$$\frac{dX_1}{dt} = R_D - a_2 c_{12} X_1 X_2 + d_2 X_2 \quad (2.34a)$$

$$\frac{dX_2}{dt} = -d_2 X_2 + a_2 p_2 c_{12} X_1 X_2 \quad (2.34b)$$

where X_1 and X_2 are the densities of detritus and the consumer, respectively, R_D is the allochthonous input of detritus, d_2 is the specific death rate of the consumer, a_2 is the assimilation efficiency, p_2 is the production efficiency, and c_{12} is the consumption coefficient of the consumer on the detritus.

For a two-species linear detritus food chain based on Equations 2.34a and 2.34b, the equilibrium values for detritus (X_1) and the consumer (X_2) are as follows:

$$X_1^* = \frac{d_2}{a_2 p_2 c_{12}} \quad (2.35a)$$

$$X_2^* = \frac{R_D p_2}{d_2 (1 - p_2)} \quad (2.35b)$$

The equilibrium for the detritus using this formulation is somewhat fixed, as it is solely dependent on the death rate, energetic efficiencies, and feeding behavior of the consumer. The equilibrium of the consumer is feasible for all R_D and d_2, as the production efficiency is defined as being $0 < p_2 < 1$, and increases as R_D increases.

The Jacobian matrix (A) for our two-species detritus-based system is as follows:

$$A = \begin{bmatrix} -a_2 c_{12} X_2^* & d_2 - a_2 c_{12} X_1^* \\ a_2 p_2 c_{12} X_2^* & 0 \end{bmatrix} \quad (2.36)$$

An interesting feature of the model is that the diagonal term of the interaction matrix corresponding to detritus (α_{11}) is negative even though the model does not possess a term for intraspecific competition (*aka* self-limitation). Instead, the negative diagonal term arises from the internal, or autochthonous, cycling of detritus.

Three important points that merit clarification and qualification emerge when discussing allochthonous inputs and donor-controlled systems. First, the definition of an allochthonous input is scale dependent and, more specifically, boundary dependent. The movement of resources across habitat boundaries is a common phenomenon. Leaf litter entering a soil system might be considered allochthonous if we consider the aboveground realm separate from the belowground realm. However, this is hardly the case, as consumption and utilization of leaf litter by detritivores and microorganisms provide crucial nutrients that impact plant growth and development (Coleman et al., 1983, Moore et al., 2003, Wardle et al., 2004). The corpses of microorganisms and invertebrates, and the unassimilated fraction of each trophic interaction, have the potential to feed back within the system, affecting plant growth.

Second, few ecosystems are devoid of allochthonous inputs and few are based entirely on them. Interestingly, the examples of systems that represent either end of the spectrum involve isolated aphotic systems such as deep-sea hydrothermal vents and caves. Those devoid of allochthonous inputs rely on chemolithotrophic microorganisms to obtain their energy from inorganic compounds. Sarbu et al. (1996) found that the Movile Cave ecosystem in Romania was based entirely on in situ chemoautotrophic energy sources with no allochthonous sources derived from photosynthetic or detritus sources. Sealed from the surface for over 5 million years, the chemoautotrophic bacteria use hydrogen sulfide as an energy source to fix inorganic carbon, resulting in thick mats of bacteria and fungi that float on the water surface. The mats support an assemblage of 48 species of invertebrates that include microbial grazers and predators. Cave ecosystems also provide examples of systems that rely entirely on allochthonous inputs from surface communities (Culver, 1982). Moore et al. (1996) found that the Wind Cave ecosystem in South Dakota was based largely on allochthonous energy sources in the form of windblown plant debris, hair, skin, and feces from rodents, and clothing fibers, hair, and skin from tourists. The allochthonous energy sources supported a diverse array of bacteria, fungi, and invertebrates (Jesser, 1998, Horton, 2005).

Finally, although not restricted to detritus, the majority of the cited examples of donor-controlled systems include detritus as the allochthonous energy source (Polis and Strong, 1996, Huxel and McCann, 1998, Moore et al., 2004). So it is not surprising that detritus-based systems are thought to be the quintessential model of donor-controlled systems. The original formulations of donor-controlled systems differed from our presentations in an important manner, as they often did not include the internal cycling of detritus. For this reason, we could argue that the detritus-based models presented here contain elements of a donor-controlled model, but are not strictly donor controlled.

2.4.2.2 The stability of donor control and detritus

The two-species system of detritus and decomposers described above in Equations 2.34a and 2.34b are not only locally stable, but globally stable as well (Neutel et al.,

1994). A critical feature of the results hinged on the inclusion of the allochthonous input, R_D, and the internal cycling of materials due to natural death of the decomposers, d_2X_2. In their analysis, Neutel et al. (1994) noted that for the energy function V as defined by Goh (1977) to serve as a Lyapunov function, as in Equation 2.32, would require some substitutions, as the direct manner employed by Harrison (1979) for the Lotka–Volterra analogs could not be applied. Neutel et al. (1994) demonstrated that the two-species detritus-detritivore model is globally stable, as V is indeed a Lyapunov function applicable to this system with a time derivative as follows:

$$\frac{dV}{dt} = -l_1 \frac{(X_1 - X_1^*)^2}{X_1 X_1^*} (R_D + d_2 X_2) \leq 0. \tag{2.37}$$

Huxel and McCann (1998) add a different dimension to the influence of allochthonous resources, and detritus in particular, on the dynamics of a system. They studied simple tri-trophic models that coupled an allochthonous resource with an autochthonous resource through a common consumer, where each consumer–resource interaction used a type II functional response. The dynamics and stability of the system was governed by two factors: 1) the level of allochthonous input, and 2) the degree of the consumer's preference for the autochthonous source. If the flow of energy shifted too strongly from the autochthonous source towards the allochthonous source either through feeding preferences or increased rates of input of the allochthonous source, the food web was prone to collapse.

The rationale and discussion provided above gives us a place to start our assessment of Odum's proposition about the increasing importance of detritus and internal nutrient cycling to the maintenance and persistence of ecosystems during the early stages of development to their mature stage (Odum, 1969). Assume that a mature ecosystem is at a steady state in terms of primary production, the inputs that it receives from outside sources, and respiration. At a steady state the primary production of the system is in balance with respiration. If outside sources of detritus enter the system, and the system is at a steady state, then the composition of the decomposer community would adapt to compensate for the internally generated materials and the externally generated materials. In either case, net primary production equals respiration.

Recall that the stability of the system is governed by the eigenvalues (λ) of the matrix A. For our simple model of detritus and a detritivore (Equation 2.34) the eigenvalues are determined as follows:

$$\det|A - \lambda I| = 0 \tag{2.38}$$

where,

$$\lambda^2 + (a_2 c_{12} X_2^*)\lambda - (a_2 p_2 c_{12} X_2^*)(d_2 - a_2 c_{12} X_1^*) = 0 \tag{2.39}$$

The solution to the quadratic equation in λ, yields the following for the critical eigenvalue λ_{max}:

$$\lambda_{max} = \frac{-a_2 c_{12} X_2^* + \sqrt{(-a_2 c_{12} X_2^*)^2 + 4(a_2 p_2 c_{12} X_2^*)(d_j - a_2 c_{12} X_1^*)}}{2} \quad (2.40)$$

As we discussed above, the equilibrium of the system is both locally stable and globally stable. The discriminant, μ, of Equation 2.40 is less than $-a_2 c_{12} X_2^*$ as the constant of the quadratic is negative, and $X_2^* > 0$, hence λ_{max} is always negative. The rate that the deviation from steady state decays is governed by the linear coefficient $-a_2 c_{12} X_2^*$, which at steady state (from Equation 2.35b) can be expressed as follows:

$$-c_{12} R_D \frac{e_2}{d_2 \left(1 - \frac{e_2}{a_2}\right)} \quad (2.41)$$

where $e_j = a_j p_j$ represents the energetic efficiency of the detritivore. Leading up to the steady state, the dynamics are dominated by the influence of the increased input of detritus, as seen for both the steady-state density of the detritivores and the linear coefficient of the solution to the quadratic for λ_{max}. If we were to view the system at a steady state with a fixed rate of consumption of detritus, $c_{12} R_1$, the dynamics are controlled internally by the ecological efficiency, e_2, and the specific death rate, d_2, of the detritivore, both of which reflects the rate of turnover of the of the detritivore and detritus. As $e_2 \to 1$ and $d_2 \to \infty$, the linear coefficient $\alpha_{11} \to 0$. Although our two-species detritus-based system is globally stable, as $\alpha_{11} \to 0$, the capacity of the system to stabilize more complex arrangements as proposed by Huxel and McCann (1998) at a given rate of input diminishes. This is a topic we will revisit.

2.5 The dynamics of primary-producer-based and detritus-based models

How different are the dynamics of primary-producer-based and detritus-based models? We will evaluate factors that affect the feasibility and resilience of our simple two-species, three-species, and four-species primary-producer-based and detritus-based food chains. We will follow a variation of the approach used by Pimm and Lawton (1977), wherein the feasibility and resilience of our simple two-species, three-species, and four-species primary-producer-based and detritus-based food chains were compared. For this exercise we constructed 10,000 food chains of each type using equations 2 and 3, where the consumer–resource interactions used type I functional responses. All parameters were sampled from a uniform distribution, I (0,1). We estimated resilience as the return time (RT) defined as RT = –1/real (λ_{max}), where real(λ_{max}) is the real part of the dominant (least negative) eigenvalue (Pimm and Lawton 1977). Under this definition, RT represents the time it takes the deviation x_i to decay to $1/e$ ($\approx 37\%$) of its original value.

At first glance the primary-producer-based and detritus-based food chains behaved similarly (Figure 2.5). In both cases, feasibility decreased and return times increased with increased food chain length. However, direct comparisons of the two types of food chains of the same length revealed that if a higher proportion of the detritus were feasible, the return times were shorter than for their primary-producer counterparts. The decline in feasibility for both types of food chains was precipitous, to the point that fewer than 5% of the detritus-based and 1% of the primary-producer-based food chains could persist under the parameter selection, and subsequent level of productivity, they represented. One surprising result is that all of the two-species detritus-based food chains were feasible over the range of productivity that we chose. As we will demonstrate in a subsequent analysis in Part II, this result recurs over a wide range of productivities and suggests that the lower limit of production (detritus input) needed to support a detritivore may indeed be a limit defined by minimum cell size and physiology.

We can draw three conclusions from this exercise. The first conclusion is that both the primary-producer-based and detritus-based food chains behave in a qualitatively similar manner for the measures that we employed. The second conclusion is that the decline in the proportion of feasible food chains with increased food chain length are consistent with the hypothesis that food chains are limited by the rate of production (Hutchinson, 1959). In this exercise the terms used to define the rate of production at the base of the food chain (r_i and R_D) were on average the same within each type of food chain, but the realized rates were higher on average for the detritus-based equations owing to the internal cycling of detritus and the way in which inputs were defined ($r_1 X_1^*$ for the primary-producer-based food chains and R_D for the detritus-based food chains). Nonetheless, in both cases increasing the length of the food chains from two species to four species places a greater exploitative pressure on a fixed rate of production. The third conclusion is that the increase in the return times for both types of food chains with increased length suggests that dynamics cannot be ruled out as a limiting factor. To understand this we need to invoke the concepts of persistence and variability for the simple formulations that we used here. As Pimm (1982) pointed out, systems with long return times are neither more nor less vulnerable to disturbances than those with short return times, as both are stable. However, in a stochastic environment, systems with long return times are likely to experience the type of catastrophic perturbations that could occur in any system.

2.6 Summary and conclusions

We have developed primary-producer-based and detritus-based models that possess familiar forms in ways that meld the traditional approaches of theoreticians and empiricists in the fields of population, community, and ecosystem ecology. The models track the dynamics of living and nonliving biomass. Both classes of models are flexible in nature and could be adapted and scaled to a number of community and ecosystem types. In developing the equations and assessing their dynamics and

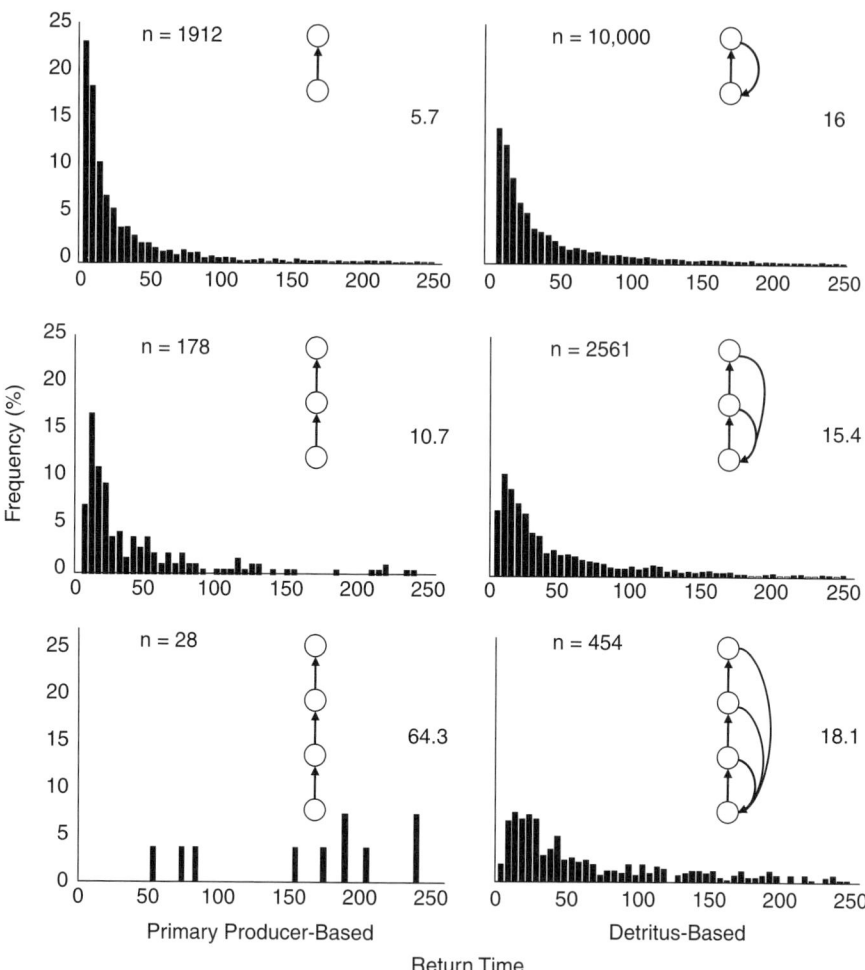

Figure 2.5 Frequency distributions of return times for the primary-producer-based and detritus-based food chains of length 2, 3, and 4 depicted in Figure 2.1 using Equations 2.1–2.3 with type I functional responses. All parameters in the models were assumed to possess a uniform distribution over the interval (0,1). The analysis included 10,000 trials for each food chain. The number of trails that produced feasible food chains (n) wherein each species possessed a positive equilibrium is presented in the upper left corner of each graph. The number to the right of the food chains represent the percentage of return times that exceeded 250 time units (the selection of 250 as a stopping point was arbitrary).

stability we are left with the unmistakable conclusion that trophic structure, dynamics, energy flow, and stability are interrelated.

We discussed the concept of stability, drawing on themes and approaches that have been treated at length elsewhere (May, 1973, Pimm, 1982, DeAngelis, 1992, Cushing et al., 2003). We assessed the stability of simple two-species primary-producer-based and an analogous simple two-species detritus-based model. The models were found to possess similar Jacobian matrices, and hence similar dynamic properties. The detritus-based models included elements that were donor controlled and elements that were density dependent. These attributes led to an interesting condition wherein the growth and dynamic of nonliving detritus is constrained in a self-limiting manner, much like living organisms.

At this stage we have treated primary producers and detritus separately, in part for pragmatic reasons to provide a historical perspective and a familiar basis for comparison. We will continue along these lines in later chapters, but will link the two in our discussions of complex multispecies food webs.

3

Connectedness food webs

'*Life in the earth is a kind of troglodyte existence...*'
(Schaller, 1968)

3.1 Introduction

A connectedness food web description is a simple box-and-arrow diagram that depicts the trophic interactions among species in an ecosystem. The boxes represent species or groups of species, and the vectors (arrows) represent trophic interactions between the species, or resource uptake in cases where detritus and nutrient pools are included. The descriptions can tell us something about the dynamics too, as the number of species that are present represent the dimension of the Jacobian matrix, the vectors indicate the positions of the nonzero elements of the matrix, and the direction of the vectors relative to the species indicate the signs of the elements.

A well-crafted connectedness description provides a snapshot of the natural history of a portion of the ecosystem, and the patterns of material flows within the ecosystem. When constructing a connectedness food web, decisions need to be made about the boundaries of the ecosystem, which species to include or not include, and the level of resolution in presenting the species. In this chapter we present an approach to constructing a connectedness description that optimizes the information about species which captures the natural history in a way that will allow us to seamlessly connect trophic interactions, energy flow, and stability. We will use the well-studied soil food web of the Central Plains Experimental Range (CPER) developed by Hunt et al. (1988) as a model to walk through this approach. The CPER food web description and the model behind it were initially designed to study the carbon and nitrogen dynamics among plants and the dominant functional groups of soil microorganisms and invertebrates. Since its introduction, the approach used to describe and model the CPER food web has since been used to estimate the contributions of soil organisms to the process described above in a number of systems across the globe (de Ruiter et al., 1993a, de Ruiter et al., 1993b, Berg and Matzner, 1997, Doles, 2000, Schröter et al., 2003, Rooney et al., 2006).

3.2 Soil food webs

Soil harbors a large part of the world's biodiversity organized in highly complex community food webs (e.g., Wolters, 1997, Griffiths et al., 2000). The primary energy sources of soil food webs are formed by various types of organic matter and root-derived materials. Within soil webs we find organisms from many of the dominant taxa at various trophic levels, representing a wide array of morphological and physiological types, and life history strategies, that span several orders of magnitudes in mass, movement, and processing rates. The lower trophic levels consists of microorganisms, such as bacteria and fungi, that utilize detritus, soil organic matter, and plant products, and herbivorous nematodes and insects that feed on roots. The microorganisms are by far the most dominant groups of soil organisms, in terms of numbers as well as biomass (Andrén et al., 1990, Bloem et al., 1994). At the higher trophic levels, soil food webs generally contain a large variety of fauna such as protozoa (amoebae, flagellates, ciliates), nematodes (bacterivores, fungivores, omnivores, herbivores, and predators), micro-arthropods such as mites (bacterivores, fungivores, predators), Collembola (fungivores and predators), and enchytraeids and oligochetes. Furthermore, there are also larger organisms that feed on almost all trophic levels, such as earthworms. Soil food webs have been the subject of several large multidisciplinary research programs, as soil processes are controlled by the dynamics and interactions in the soil community food web (de Ruiter et al., 1993a, de Ruiter et al., 1993b, Moore and Hunt, 1988). Moreover, because of the large amounts of materials that are decomposed and processed, soil food webs are thought to govern major components in the global cycling of materials, energy, and nutrients (Wolters, 1997, Griffiths et al., 2000).

3.3 The CPER soil food web

A connectedness food web diagram of the CPER food web is presented in Figure 3.1. The CPER is located in northeastern Colorado. The site is a characteristic shortgrass steppe ecosystem of the North American Great Plains dominated by the C_4 grass, *Bouteloua gracilis*, and the C_3 grass, *Pascopyrum smithii*. The ratio of shoot production to root production is roughly 1:1 (Milchunas and Lauenroth, 1992, Milchunas and Lauenroth, 2001), although some estimates are closer to 1:2 (Clark, 1977). The aboveground plant parts provide 20–40% cover, with exposed soil between them (Lauenroth and Milchunas, 1991). There is little accumulation of plant litter, as much of the aboveground production remains in place as standing dead, rather than falling to the soil surface. This aspect of the shortgrass steppe makes it like many grassland ecosystems, as plant roots provide the major input of carbon to soil. As such, plant roots are the focal point of biological activity in soils and form the base of the belowground food web (Coleman et al., 1983).

The connectedness description of the CPER food web possesses 17 groups of species (functional groups) and resources. We can break down the web based on the interactions among 15 functional groups of organisms (Hunt et al., 1987, Moore et al., 1988) that obtain energy from plant roots and/or detritus (labile substrates and resistant substrates), and in the process form and utilize inorganic nitrogen (Figure 3.1). The functional groups were formed using the criteria discussed earlier—

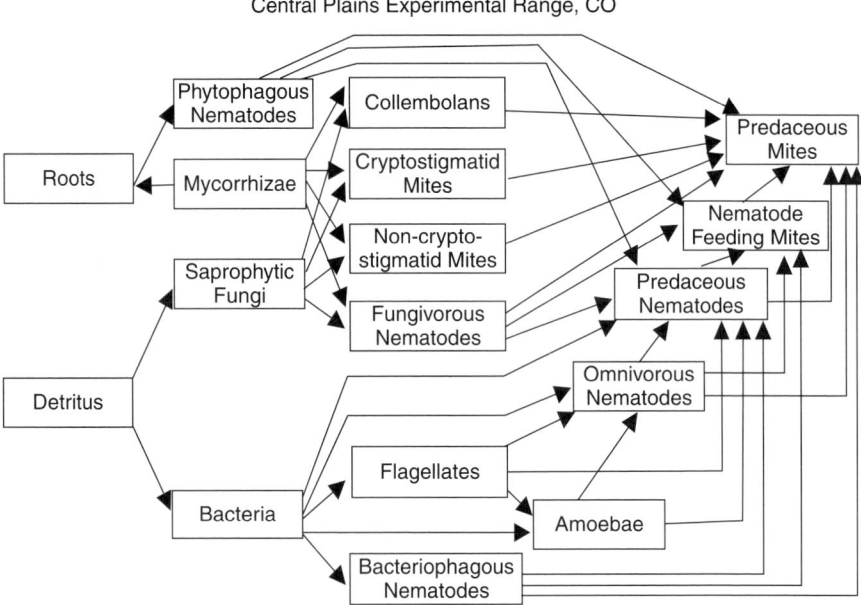

Figure 3.1 The connectedness description of the CPER food web (after Hunt et al., 1987). The diagram includes the trophic interactions among resources, plants, and functional groups of microorganisms and invertebrates. Although not depicted in the diagram, the model accounts for the respiration of CO_2 by all organisms and the contributions of materials from all living organisms to the inorganic nitrogen pool and the detritus pools (labile and resistant substrates). The energy flux description and details of how the carbon and nitrogen fluxes are estimated are provided in Chapter 4. The functional description and details of how the interactions strengths were estimated are presented in Chapter 5.

primary food source, feeding mode, habitat selection, and life history attributes (Table 3.1). These attributes are aligned with the principal niche axes of food (source and mode), habitat, and time (life history) (sensu Schoener, 1974). Food sources include the basal resources and other functional groups. Groups that are separated by feeding mode are groups that feed on similar sources but by different means and strategies. For the CPER web, habitat selection includes aspects of the size of the home range and scale of movement, dependency on water, and the need for pore space (Table 3.2). Bacteria and protozoa are essentially aquatic organisms living within water in pores and on particle surfaces. Nematodes live on water films. Fungi and arthropods live in and among particles and within air-filled pore spaces. Life history attributes capture the differences in the growth rates, death rates, and metabolic rates among the functional groups.

The CPER food web description and the underlying model were designed to track carbon and nitrogen pools and flows (Hunt et al., 1987, Moore et al., 1988). For clarity, a number of interactions that we discussed in Chapter 2 are omitted from the

Table 3.1. Examples of functional groups of belowground organisms in the North American shortgrass steppe based on principal food sources, feeding mode, life history, and habitat (Moore et al., 1988).

Functional Group	Genera
Protozoa	
Amoebae	*Acanthamoeba, Hartmannella, Trichamoeba, Mayorella, Varella*
Flagellates	*Pleuromonas, Bodo, Mastigamoeba*
Nematodes	
Bacterial feeders	*Acrobeloides, Pelodera, Rhabditis*
Fungal feeders	*Aphelenchus, Aphelenchoides*
Root feeders	*Helicotylenchus, Tylenchorhynchus, Xiphinema*
Omnivores	*Mesodiplogaster*
Predators	*Discolaimium, Mononchus*
Micro-arthropods	
Collembola—fungal feeders	*Entomobrya, Lepidocyrtus, Isotoma, Tullbergia, Brachystomella, Xenylla, Folsomia, Neelus, Dueterosminthura, Sminthurus*
Cryptostigmatid mites—fungal feeders	*Haplozetes, Zygoribatula, Passaozetes, Oppiella*
Non-cryptostigmatid—fungal feeders	*Tyrophagous, Neoraphignathus, Bakerdania, Scutacarus, Uropoda*
Nematophagous mites	*Ololaelaps, Veigaia, Gamaselloides*
Predatory mites—general predators	*Spinibdella, Cunaxa, Cunaxoides, Tydeus, Rhagidia*

description but are included in the models as described in chapters 4 and 5 (Figure 3.1). First, there are key feedbacks (vectors) from each functional group to the substrate pools that were omitted. All living organisms contribute to the labile and resistant substrate pools through natural death, and all animals (protozoans and metazoans) through evacuation of vacuoles or defecation, and excretion. Organisms that consume other organisms usually kill their prey and leave behind unconsumed portions (leavings or orts) that become part of the substrate pools. Second, all living organisms metabolize nitrogen-containing molecules (e.g., proteins and nucleic acids) and excrete them in organic and inorganic nitrogen forms that ultimately work their way into the inorganic nitrogen pool. Finally, all living organisms respire (mineralize) carbon in the form of CO_2.

3.3.1 Food web components

3.3.1.1 Basal resources

The basal resources are separated into plants (shoots and roots) and detritus (labile and resistant substrates). The plants are modeled after the dominant C_4 grass species of the shortgrass steppe, *B. gracilis*, commonly known as blue grama grass. The plants can be separated into shoots and roots with bidirectional flows of matter

Table 3.2 Summary of the different spatial, temporal, and size scales for the dominant taxa of the CPER soil food web. These parameters were used in part when assigning species to the different functional groups of the CPER food webs.

	Bacteria	Fungi	Protozoa	Nematodes	Collembola	Mites
Habitat	Water/Surfaces	Free Air/Surfaces	Water/Surfaces	Water films/Surfaces	Free Air Spaces	Free Air Spaces
Minimum generation time (h)	0.5	4–8	2–4	120	720	720
Turnover time (season − 1)	2–3	0.75	10	2–4	2–3	2–3
Assimilation efficiency (%)	100[†]	100[†]	0.95	0.38–0.60	0.5	0.3–0.9
Production efficiency (%)	0.4–0.5	0.4–0.5	0.4	0.37	0.35	0.35–0.40
Body width	1–2 μm	3–100 μm	15–100 μm	5–120 μm	0.150–2 mm	0.08–2 mm
Biomass (g C indiv.$^{-1}$)	NA	NA				
Functional response (type I–III)			I, II	I, II	I, II	I, II

[†]The assimilation efficiencies of bacteria and fungi are 100%, given that microorganisms absorb materials across their membranes as opposed to ingesting or engulfing prey or materials.

between them. This feature allows for the diversion of photosynthate to roots from shoots and nitrogen from roots to shoots, and vice versa, depending on conditions. The detritus pool can be separated into two constituents, a labile (C:N < 30) substrate pool and a resistant (C:N > 30) substrate pool, based on the C:N ratio of the detritus inputs and their susceptibility to decomposition. The resistant substrates are largely structural plant materials, feces, and the exoskeletons of arthropods, whereas the labile substrates include plant root exudates and mucigels, sloughed cells, and cytoplasm.

3.3.1.2 Inorganic nitrogen and carbon

The food web couples the interplay of organic and inorganic matter to the structure and dynamics of the functional groups. The inorganic nitrogen pool includes both ammonium and nitrate. This pool is included in the model because of the importance of nitrogen to plant growth and because of the significant contributions of organisms within the food web to the pool. The inorganic carbon pool is not depicted in the diagram, but fluxes into and out of the carbon pool can be estimated. These estimates of inorganic nitrogen and carbon can provide important indices into the rate of decomposition of organic matter and the contribution of individual functional groups to the decomposition of organic matter.

3.3.1.3 Consumers

Consumers are categorized as primary consumers (including plant mutualists), secondary consumers, and predators. The primary consumers include bacteria, saprophytic fungi, mycorrhizal fungi, and root-feeding nematodes. Bacteria and saprophytic fungi live at different spatial and temporal scales and in different soil microhabitats, but utilize similar substrate types to different degrees and at different rates (Table 3.2). Bacteria live in water-filled pore spaces and within water films, whereas fungi tend to reside in air-filled pore spaces. The bacteria are assumed to use the labile substrates at twice the rate per gram of biomass as fungi, whereas fungi are assumed to use the resistant substrates at twice the rate per gram of biomass as bacteria (Gyllenberg and Eklund, 1974, McGill et al., 1981, Hunt et al., 1987).

The description includes the symbiotic relationship between arbuscular mycorrhizal (AM) fungi and blue grama. The AM fungi are consumers of plant-derived carbon but providers of inorganic nitrogen; hence the bidirectional vectors between AM fungi and plant roots. The model does not include other forms of mycorrhizae whose fungal component might be more facultative in the symbiosis and that might derive carbon directly from one of the detritus pools.

Nematodes are the dominant belowground herbivores of blue grama (Ingham et al., 1985). Although both endoparasitic nematodes and ectoparasitic nematodes can be found on the shortgrass steppe, only those feeding on the external portions of plant roots (ectoparasitic forms) were included in the description. Root-feeding insects and pathogens are not included in the description, although both could be added using the approach described in Chapter 2.

The secondary consumers include flagellates, nematodes (fungal feeders and bacterial feeders), and arthropods (Collembola, cryptostigmatid mites, and non-cryptostigmatid mites). Food sources of Collembola and mites include fungal hyphae and spores, algae, pollen, detritus, plant roots, and nematodes, depending on the species. Although they share similar food sources, these groups show a wide variation in growth rates, habitat selection, and feeding modes (Tables 3.1 and 3.2). For example, Collembola possess chewing mandibles that rip, shred, and pulverize prey, which are then engulfed. Mites have claw-like chelicerae and palps that are used to grasp and manipulate prey. The morphology of their mouthparts differs among species. In some instances these mouthparts slice and tear prey, whereas in others they have been modified to pierce their prey and suck fluids. Once in their grasp, mites cover their prey with enzymes secreted through their mouth and imbibe the liquefied prey (Tables 3.1 and 3.2). We restricted the diets of these secondary consumers to fungi, in large part to the high percentage of preserved specimens with fungal hyphae and spores in the digestive tracks and from the propensity of these organisms to consume fungi in the laboratory.

The CPER food web terminates with a suite of predators that includes amoebae, omnivorous and predaceous nematodes, nematode-feeding mites, and predaceous mites. Amoebae and omnivorous nematodes obtain the bulk of their energy from bacteria, whereas the remaining predators tend to feed on invertebrates. The predaceous mites include predators of arthropods and general predators that consume small invertebrates that they can overcome.

The process of forming the functional groups had a qualitative and subjective component to it, and was by no means an exact science, even though quantifiable criteria were used. Decisions were made based on what was known at the time. The description could have included more basal resources, functional groups of consumers, and vectors among the groups. At the base, the resistant and labile pools of carbon could have been subdivided into any number of categories based on their molecular structure and decomposability. Likewise, the bacteria and fungi could have been placed into functional groups based on the enzymes they produce, the categories of substrates that they utilize, and the processes that they engage in. The system could be extended to the soil surface and aboveground food web in a couple ways. The plant root component links the system to aboveground plant components that are consumed by vertebrates and insects. Many of the arthropods are of sufficient body size and venture to the soil surface to become prey items for active predators, such as spiders and predaceous beetles, which in turn may be consumed by a number of vertebrates.

As the functional groups are based on food sources, adding or deleting functional groups would necessarily change the vectors of interaction within the food web. Changes of this type can alter the outcomes of both quantitative and qualitative analyses of diversity, patterns of interactions within webs, ratios of functional types, importance of key interactions and species, and the resilience and stability of food webs.

3.3.2 Aspects of food web structure

Food web research catalogs multiple descriptive statistics from the diagrams for comparative purposes, to seek patterns among systems, and to study their relationships against theoretical limits and constraints. We will present several descriptive metrics and comparisons in this chapter. In subsequent chapters we will explore the importance of these metrics to the dynamic stability of food webs. When put in context, the metrics used indicate that the descriptions of our soil food webs are structured similarly to other webs in the literature, as they are well within the ranges of those of published webs.

The connectedness descriptions of the seven additional food webs are presented in Figure 3.2. The fact that these descriptions look similar should not be surprising given that similar criteria were used to construct them. Each of the authors took some liberties in adapting the description to suit their sites by including additional functional groups and interactions. All included plants, plant roots, and detritus at the base, but most did not differentiate labile detritus from resistant detritus. Most did not include the symbiotic relationship between mycorrhizal fungi and plants.

3.3.2.1 Diversity and food chain length

The diversity or richness (S) as measured by number of species and resources, mean food chain length (FCL_{Mean}), maximum food chain length (FCL_{Max}), and connectance (C) are commonly used metrics to describe food webs. Table 3.3 presents these metrics from a sampling of published food webs compiled by Briand (1983) and our food webs.

Richness (S) is simply the number of species, or groups of species, and resources that were used in the description. The published descriptions in our survey have included as few as five species to as many as 45, although more specious descriptions are in the literature (see Yodzis, 1988). Much of this depends on whether individual species or aggregates of species were used in the description. The richness of the soil food webs varied between 17 and 22 groups.

Food chains are the individual pathways that originate from a basal resource and terminate at a top predator. Food webs consist of a minimum of a single food chain (i.e., the food chain is the food web), to multiple food chains. The length of food chains can be described as either the number of links (trophic transfers) within the chain, or as the number of species within the chains. We have adopted the latter. The FCL_{Mean} is the numerical average of the lengths of all the food chains in a food web. FLC_{Max} is the longest food chain in the food web. Studies of food web descriptions have found FCL_{Mean} length to vary between three and four, while FCL_{Max} have been reported to up to eight. Reasons for the apparent limits to FCL_{Mean} and FCL_{Max} include the availability of energy inputs (Hutchinson, 1959, Moore et al., 1993) and the increased likelihood of instabilities with increased length (Rosenzweig, 1971, Pimm and Lawton, 1977). We will discuss these topics in more detail in Part II. For

62 • *Energetic Food Webs*

(a) Lovinkhoeve Integrated Farming, NL

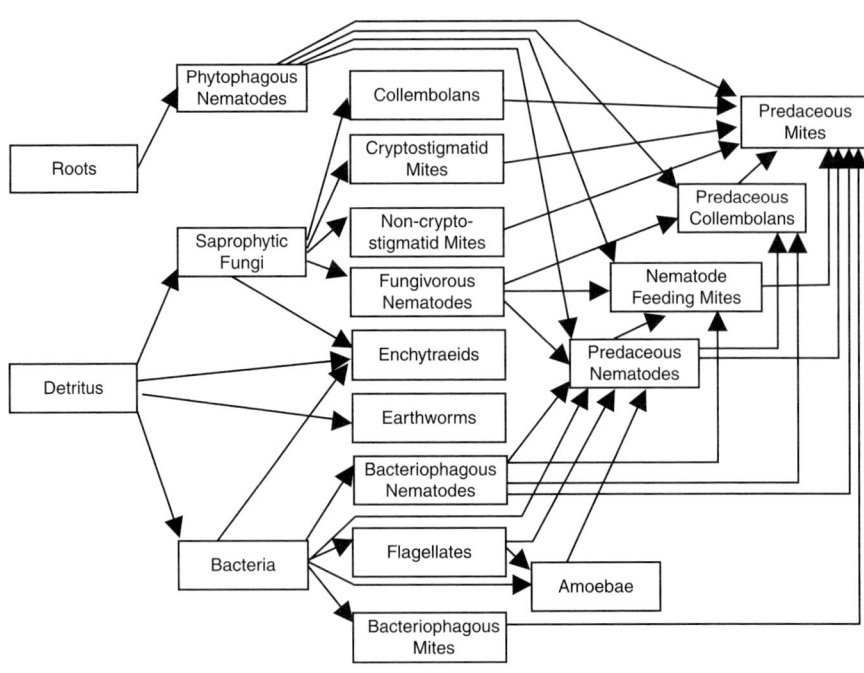

(b) Lovinkhoeve Conventional Farming, NL

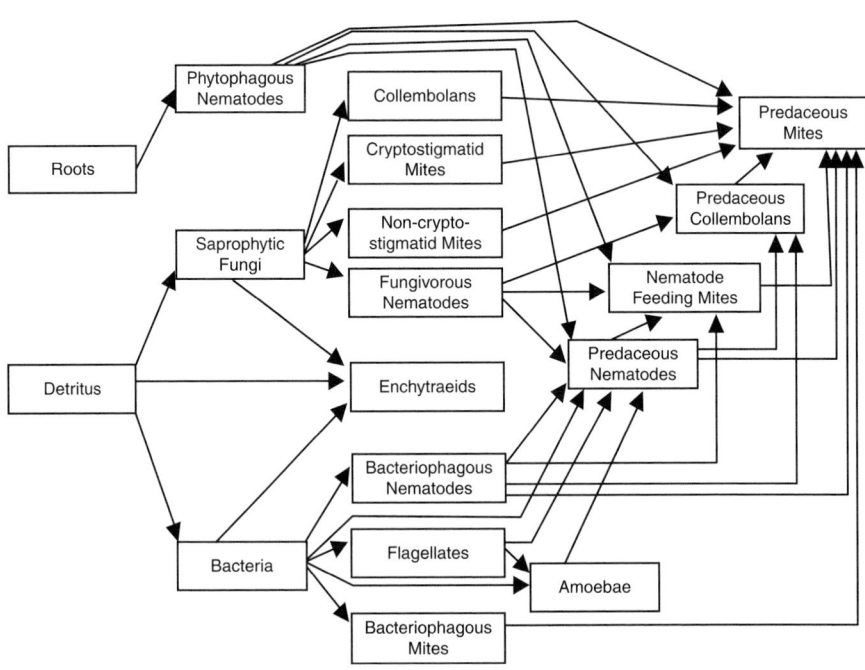

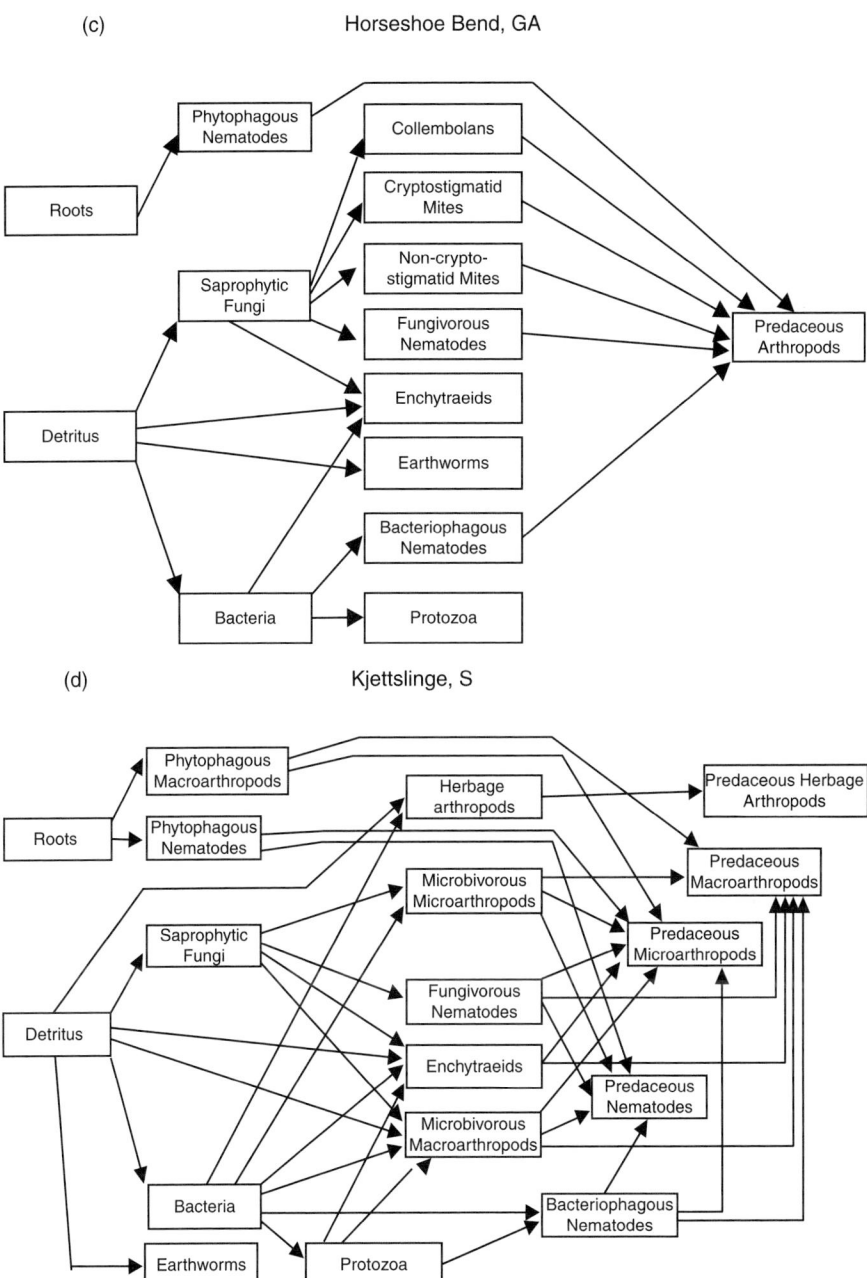

Figure 3.2 *Continues overleaf.*

(e)   Moist Acidic Arctic Tundra, AK

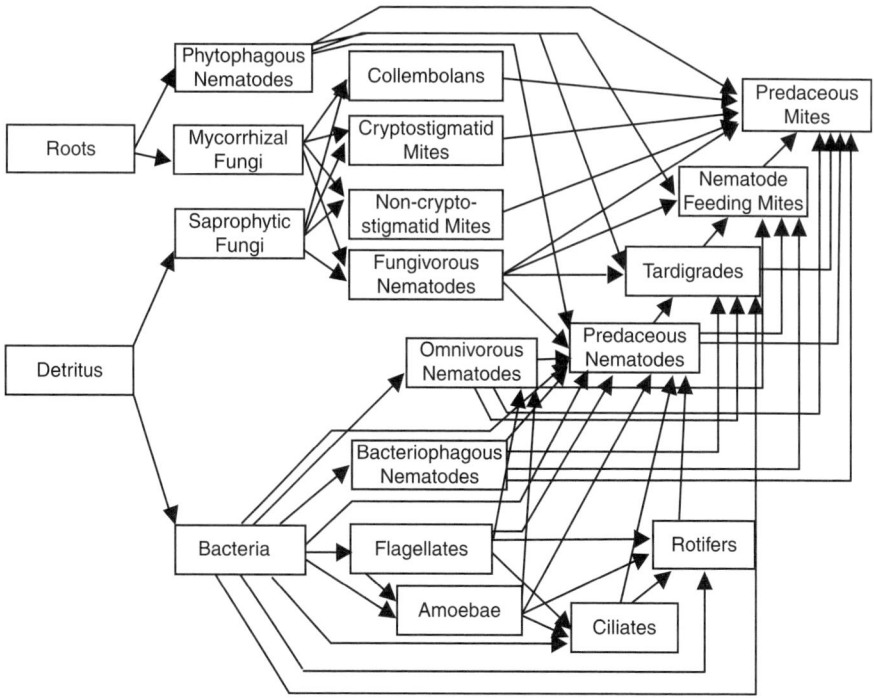

Figure 3.2 The connectedness descriptions of seven additional soil food webs. (a) Lovinkhoeve Experimental Farm (Marknesse, The Netherlands), integrated management winter wheat; (b) Lovinkhoeve Experimental Farm, conventional management winter wheat; (c) Horseshoe Bend Research Site (Athens, Georgia, USA); (d) Kjettslinge Experimental Farm (Uppsala, Sweden); (e) Toolik Field Station (Toolik Lake, Alaska, USA), moist acidic arctic tundra.

our food webs the FCL_{Mean} and FLC_{max} are between 3.41 ± 0.99 and 5.00 ± 2.33, respectively.

3.3.2.2 Diversity and complexity

Connectance (C), defined as the proportion of possible interactions within a food web among species (excluding self-interactions) that are realized, is a simple measure of food web complexity. In terms of the community or Jacobian matrix A, connectance is the proportion of off-diagonal elements in matrix A. To calculate connectance one needs to know only the richness (S) of the community, and hence the dimension of matrix A, and the number of trophic links (L):

Table 3.3 The diversity (S), the number of basal resources (S_R), mean food chain length (FCL_{Mean}), maximum food chain length (FCL_{Max}), and connectance of the soil food webs presented in Figure 3.1 and Figure 3.2 and the 40 terrestrial food webs compiled by Briand (1983).

Briand Food Webs	S	S_R	C	SC	FCL_{Max}	FCL_{Mean}	Reference
Cochin estuary	9	2	0.694	6.25	4	3.13	Qazim (1970)
Knysna estuary	15	3	0.476	7.14	3	2.71	Day (1967)
Long Island estuary	24	4	0.214	5.14	3	2.30	Woodewell (1967)
California salt marsh	13	2	0.564	7.33	4	2.74	Johnston (1956)
Georgia salt marsh	7	3	0.333	2.33	2	2.00	Teal (1962)
California tidal flat	25	2	0.303	7.58	6	3.82	MacGinitie (1935)
Narragansett Bay	20	3	0.332	6.64	4	2.79	Kremer and Nixon (1978)
Bissel Cove marsh	15	4	0.419	6.29	4	2.44	Nixon and Oviatt (1973)
Lough Ine rapids	10	2	0.511	5.11	3	2.86	Kitching and Ebling (1967)
Exposed intertidal (New England)	5	2	0.700	3.50	2	2.00	Menge and Sutherland (1976)
Protected intertidal (New England)	8	3	0.429	3.43	2	2.00	Menge and Sutherland (1976)
Exposed intertidal (Washington)	13	3	0.462	6.01	3	2.25	Menge and Sutherland (1976)
Protected intertidal (Washington)	13	3	0.487	6.33	3	2.50	Menge and Sutherland (1976)
Mangrove swamp (station 1)	8	1	0.571	4.57	3	2.40	Walsh (1967)
Mangrove swamp (station 3)	9	1	0.583	5.28	3	2.33	Walsh (1967)
Pamlico River	14	4	0.363	5.08	3	2.14	Copeland et al. (1974)
Marshallese reefs	14	3	0.286	4.00	5	3.56	Hiatt and Strasburg (1960)
Kapingamarangi atoll	27	8	0.208	5.62	4	2.00	Niering (1963)
Moosehead Lake	17	2	0.426	7.24	4	3.00	Brooks and Deevey (1963)
Antarctic pack ice zone	19	3	0.298	5.66	5	3.26	Knox (1970)
Ross Sea	10	3	0.556	5.56	7	4.61	Patten and Finn (1979)
Bear Island	28	6	0.275	7.70	7	3.69	Summerhayes and Elton (1923)
Canadian prairie	15	1	0.591	8.87	4	2.40	Bird (1930)
Canadian willow forest	12	4	0.364	4.37	4	2.70	Bird (1930)
Canadian aspen communities	25	3	0.307	7.68	4	2.16	Bird (1930)
Aspen parkland	34	9	0.200	6.80	6	2.93	Bird (1930)
Wytham Wood	22	4	0.273	6.01	4	2.89	Varley (1970)

(continued)

Table 3.3 Continued

Briand Food Webs	S	S_R	C	SC	FCL_{Max}	FCL_{Mean}	Reference
New Zealand salt-meadow	45	7	0.135	6.08	3	1.96	Paviour-smith (1956)
Arctic seas	22	2	0.312	6.86	5	3.14	Dunbar (1954)
Anarctic seas	14	1	0.527	7.38	7	5.02	Mackintosh (1964)
Black Sea epiplankton	14	2	0.835	11.69	6	3.90	Petipa et al. (1970)
Black Sea bathyplankton	14	2	0.846	11.84	6	3.86	Petipa et al. (1970)
Crocodile Creek	33	5	0.390	12.87	4	1.93	Fryer (1959)
River Clydach	12	4	0.561	6.73	4	2.56	Jones (1949)
Morgan's Creek	13	2	0.744	9.67	4	2.72	Minshall (1967)
Mangrove swamp (station 6)	22	8	0.348	7.66	4	2.07	Walsh (1967)
California sublittoral	24	6	0.261	6.26	3	2.75	Clarke et al. (1967)
Lake Nyasa rocky shore	31	3	0.671	20.80	3	2.13	Fryer (1959)
Lake Nyasa sandy shore	37	5	0.298	11.03	3	1.80	Fryer (1959)
Malaysian rain forest	11	3	0.527	5.80	3	1.88	Harrison (1962)
Soil Food Webs							
CPER, CO	17	2	0.294	5.00	7	4.14	Hunt et al. (1987)
Arctic Tundra, AK	20	2	0.305	6.11	9	5.24	Doles et al. (2000)
Lovinkhoeve, NL—Integrated	19	2	0.228	4.33	6	3.66	de Ruiter et al. (1993a)
Lovinkhoeve, NL—Conventional	18	2	0.248	4.47	6	3.74	de Ruiter et al. (1993)
Horseshoe Bend, GA—No-till	14	2	0.209	2.92	3	2.33	Hendrix et al. (1986)
Horseshoe Bend, GA—Conventional	14	2	0.209	2.92	3	2.33	Hendrix et al. (1986)
Kjettslinge, S—B0	18	2	0.261	4.71	4	2.91	Andrén et al. (1990)
Kjettslinge, S—B120	18	2	0.261	4.71	4	2.91	Andrén et al. (1990)

Food Web Summary Statistics
(Mean ± 1 SD)

	S	S_R	C	SC	FCL_{Max}	FCL_{Mean}
Briand food webs	18.08 ± 9.13	3.45 ± 1.97	0.44 ± 0.18	7.05 ± 3.18	4.03 ± 1.35	2.73 ± 0.76
Soil food webs	17.24 ± 2.19	2.00 ± 0.00	0.25 ± 0.04	4.40 ± 1.06	5.00 ± 2.33	3.41 ± 0.99
All food webs	17.94 ± 8.37	3.21 ± 1.88	0.41 ± 0.18	6.61 ± 3.09	4.19 ± 1.57	2.84 ± 0.83

$$C = \frac{2L}{S(S-1)} \quad (3.1)$$

The numerator is 2L, as each trophic interaction is bidirectional, including the effect of the prey on the predator and the effect of the predator on the prey, as represented in a community matrix. The connectance values for our soil food webs are 0.25 ± 0.04 (see Table 3.3).

Monte Carlo analyses involving models of simple food webs demonstrate quite definitively that as connectance increases for food webs of a given richness, the likelihood of the food web being stable, i.e., returning to equilibrium following a minor disturbance, declines (Gardner and Ashby, 1970).

Gardner and Ashby (1970) challenged the tenet that complex systems are more likely to be stable than simpler systems using an eigenvalue-based analysis of randomly constructed interaction matrices that possessed four, seven, or ten interacting nodes (analogous to species in a food web) with varied degrees of complexity. Complexity was measured by connectance, C, defined as the number of interactions among nodes relative to the number of possible interactions (Equation 3.1). Monte Carlo simulations that sampled the interaction strengths, i, of the matrices over a fixed range have revealed that as the connectance of the models increased, the likelihood that the models were stable, i.e., $\lambda_{max} < 1$, have decreased, with a critical value of C being approximately 0.13 (Figure 3.3).

May (1972) adapted this approach to interaction matrices (A) that represented ecological communities and drew a similar conclusion. He set the diagonal elements (α_{ii}) of the matrices to -1, and the off-diagonal elements (α_{ij}, α_{ji}) to be selected randomly from a distribution centered at zero with mean i^2, where i is the average

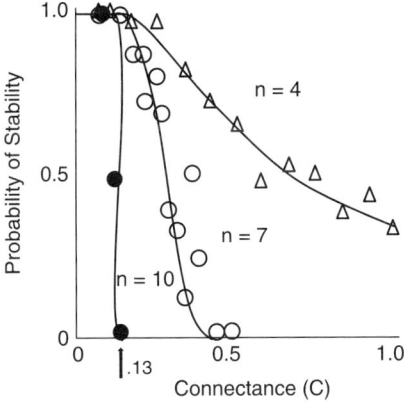

Figure 3.3 The probability that a model system with different numbers of interacting nodes (n = 4 △; n = 7 ○; n = 10 ●) and connectance (C) will be stable (redrawn from Gardner and Ashby, 1970). The connections of these results to ecological food webs was first discussed by May (1972).

interaction strength. May found that the probability P(S, C, i) that a community matrix constructed in such a manner was stable was near certain (P → 1) if

$$i\sqrt{SC} < 1 \qquad (3.2)$$

and unstable (P→0) if

$$i\sqrt{SC} > 1.$$

May further noted, as shown in Figure 3.3, that the transition between stable and unstable states became more abrupt with increased diversity (S), connectance (C), and average interaction strength (i). If diversity and complexity were constrained by dynamic stability in this manner, then we would expect to find a decline in connectance with increased diversity. This relationship seems to be in play when we include our soil food webs in a larger collection of published webs (Figure 3.4).

3.3.2.3 Omnivory

An omnivore is an organism that obtains energy from more than one trophic level. We use omnivory as a device to reveal some broader issues in the study of food webs and the limitations placed on them based on the assumptions we make. Pimm and Lawton (1978) made four predictions about omnivory for systems in general under the condition that the model food webs they represent be dynamically stable: (1) omnivory should be rare based on the instabilities that it induced in simple food

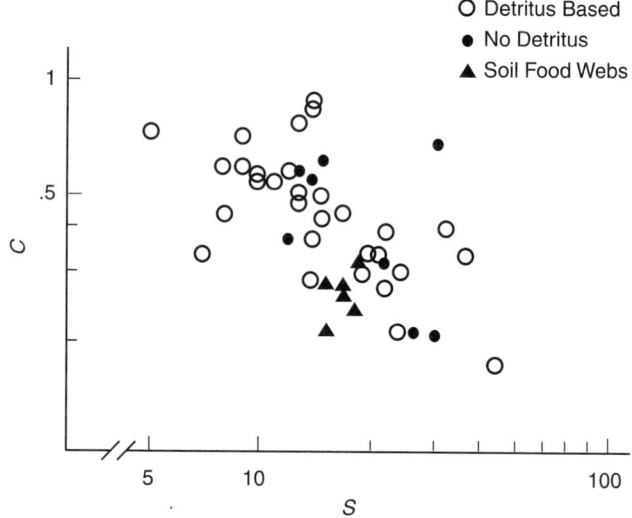

Figure 3.4 The relationship between the connectance (C) and the diversity (S) of the unique soil food webs (▲) and the 40 terrestrial and aquatic food webs (●) compiled by Briand (1983).

chains, (2) omnivores should feed more often at adjacent trophic levels, (3) omnivory should be more common within insect-dominated systems, and (4) omnivory should be more common in systems with donor-controlled dynamics.

Our observations of the CPER food web with respect to omnivory would seem to support predictions 2–4 proposed by Pimm and Lawton (1978), as omnivory is common among soil organisms (Moore et al., 1988), and more common than the CPER description reveals. The top predators in the soil system—like those in others—are omnivores, given that they have adapted to feeding on organisms of different morphologies regardless of where they might have obtained their energy, e.g., predaceous mites and predatory nematodes feed on nematodes in different trophic levels (Figure 3.1). Many of the consumers in the food web are omnivores as well, much more so than depicted in the description. Recall that the CPER description places species into functional groups that have rather restrictive diets. For example, the Collembola and cryptostigmatid mites are classified as being fungal feeders, yet as we noted (see Table 3.1), they are known to feed on bacteria, nematodes, pollen, detritus, roots, and algae. Many genera of protozoans feed on bacteria and fungi.

Our main points center on predictions 3 and 4, which we will then use to address Prediction 1. Insect-dominated systems are discussed in Prediction 3, but this could easily be modified to include Chelicerata and Crustacea, as well as Insecta (to include the Apterygota). We suspect that the reasons for observing a higher than expected incidence of omnivory in insect-dominated systems than in vertebrate-dominated systems contains elements of, but goes beyond, the suggested importance of pseudo-interference (Beddington et al., 1978, May and Hassell, 1981) and the reported lower interaction strengths between parasitoids and hosts (Pimm, 1982). Analyses that follow demonstrate that the placement of the omnivorous interactions within the architecture of the food web, differences in the energetic efficiencies, body sizes, spatial distributions and home range, access to available productivity, and foraging behaviors should be considered as well.

Prediction 4 focused on the inherent stabilizing influence and robustness of donor-controlled systems. We argue that most ecosystems contain a donor-controlled component through the internal and external inputs of detritus. We further argue that the primary production of most ecosystems is not consumed by a herbivore, but rather is returned to the donor-controlled portion of the system as a form of detritus that is consumed by detritivores and subsequently their consumers and predators. In keeping with the observations of Odum (1969), the donor-controlled components, internal and external inputs and cycling of detritus, become more important to the stability of the system at later successional stages, i.e., mature ecosystems, to the point that ecosystem-level productivity (P) is at steady state with ecosystem-level respiration (R).

This brings us back to Prediction 1, which asserts that omnivory should be rare. If we view rarity in terms of the number or proportion of omnivorous interactions relative to all of the realized interactions in the food web, then indeed this might be the case. If we view rarity in terms of the incidence of omnivory within food webs, then we might conclude that it is common. In either case, given the caveats raised above, caution is in order when conducting such analyses, as clear choices were made in the construction of the food web to include the dominant food sources and not the entire diets (e.g., the Collembola and cryptostigmatid mites; see Table 3.1).

3.3.3 Spatial and temporal averaging

The description of the CPER soil food web is by no means the definitive description of the soil food web of the North American shortgrass steppe, but rather is one of many possible descriptions. Our aim in using it was to develop an approach that could be adapted for any system. The issues we raise below could easily be interpreted as being overly critical or dismissing the value of the description. As more information becomes available, we envision changes in the description (Hendrix et al., 1986, Andrén et al., 1990, de Ruiter et al., 1993a, Doles, 2000).

In fact, the CPER food web represents an amalgamation of different studies conducted over different spatial and temporal scales. As such, the description that best presents what might be expected could be encountered at any point along a landscape and at any time. This is true for the vast majority of food web descriptions, as they are static descriptions of spatial averages. We know that food webs take on different configurations over space and time in response to the heterogeneous arrangements of resources and consumers on landscapes and to stochastic and systematic changes in the physical, chemical, and climatic conditions in time.

We could argue that the traditional approach has not framed the study of food webs in the appropriate spatial and temporal contexts. The traditional approach assumes that the underlying spatial processes are unimportant or that averaging over space adequately captures the outcome of disturbance on food web structures and hence ecosystem function. A second line of research suggests that spatial processes such as the movement and dispersal of organisms affect the architecture of the food web over a landscape or season, mediate the relationships between structure and function, and can affect the extent of a disturbance and the response of the food web to disturbance (Bolker, 2003, McCann et al., 2005, Webb and Levin, 2005).

Spatially and temporally localized interactions can create feedbacks (e.g., Levin, 1998) that mediate changes in food web structure and function that affect species diversity and dynamics (sensu Pimm, 1982, Webb, 2007). For example, a plant species consumed at one location is deposited as fecal material at another location, with the nutrients being released via decomposition to affect the growth of another plant species. Likewise, the seasonal succession of growth and senescence of plants and nutrients initiate responses in the food webs, changing the presence and abundance of consumers and the patterns of energy flow and interactions within the food web. The seasonal migration of species among habitats exhibited by the continental scale movements of birds has demonstrated how space and time interact. For example, the foraging success of migratory birds in one habitat can influence breeding success in another, and thereby the structure and stability of coupled geographically disparate ecosystems (Jefferies et al., 2004). We will explore these possibilities in greater detail in Parts II and III.

3.4 Summary and conclusions

We constructed connectedness webs using the food web of the Central Plains Experimental Range (CPER) of the shortgrass steppe in Colorado, USA as representative example. Several underlying assumptions in preparing the description are invoked. The description is based on functional groupings of species that are aggregated using criteria borrowed from niche theory: food source, feeding mode, habitat selection, and life history. The strength of the approach stems from its simplicity and linkage to theory, inclusion of organic and inorganic components, living and dead organic matter, and the adoption of biomass carbon and biomass nitrogen as common currencies. The weakness in the approach stems from its use of aggregates of species, and the use of temporal and spatial averaging.

We compared the diversity, the lengths of food chains, and connectance of the CPER food web and six additional soil food web descriptions to published webs from other systems. The structures of the soil food webs were similar to the published webs. The soil food webs were in the mid-range in terms of diversity compared with other published webs. Both the mean and maximum food chain lengths were comparable with those of the other published webs. The connectance of the soil food webs appears to be consistent with those from other systems with similar diversity. We will fall back on the similarity in the architectures of our soil food webs to others presented in this chapter as one rationale for generalizing our results.

4

Energy flux food webs

4.1 Introduction

The connectedness descriptions presented in Chapter 3 focused on the relationships among functional groups and the trophic interactions among the groups. In this chapter, the connectedness description of the Central Plains Experimental Range (CPER) soil food web will serve as the starting point for constructing an "energy flux" description (energy flow sensu Paine, 1980). If the connectedness food web description depicts a simple cartoon of who-eats-who within the community, the energy flux food web description provides an accounting ledger of the distribution and transfer of matter and energy within the community. In addition to the fluxes among functional groups within the food web, the description and model can be used to estimate the rates of mineralization (e.g., inorganic carbon respiration and nitrogen excretion) for the individual functional groups and the web as a whole.

The energy flux description requires estimates of the biomass for each functional group and basal resource and estimates of the rates of the flows of matter (carbon, C, or nitrogen, N) among the functional groups and resources within the food web. There are several ways to obtain this information. We will advocate an approach that combines field and laboratory data and models. The biomass estimates are based on field measurements of the populations and resources. The estimates of energy flux, expressed as either carbon flow or nitrogen flow, are derived using a variation of the simple model based on the physiologies of the species within the functional groups that accounts for the fate of the consumed matter that we presented in Chapter 2.

Throughout this chapter we use the CPER and other soil food webs to illustrate how the descriptions are constructed and to explore patterns in the distribution of biomass and flow of matter. We will compare the estimates provided by the model in terms of the overall carbon and nitrogen mineralization rates with observed rates. We end with a discussion of both the utility and limitations of the energy flux description.

4.2 Biomass and physiological parameters

An energy flux description requires estimates of the biomass and key physiological traits of the populations of species or functional groups in the food web. The estimates of the physiological parameters and population sizes in biomass (kg C ha^{-1}) of the different functional groups for the CPER food web are presented in Table

4.1, and the biomass estimates for all seven soil webs in Table 4.2. The estimates of biomass are obtained from field estimates of the population abundances of the functional groups, converting these to the desired mass units directly, or by morphometric or other indirect means. The estimates need to reflect the spatial and temporal scales that the description represents. The physiological traits can be estimated directly, through field and laboratories studies, or obtained from the literature for representative taxa (Hunt et al., 1987). For the soil food webs, the estimates of biomass represent the average amount of biomass present within a hectare over several sampling dates during a growing season. Although the actual biomass present on a given date may sometimes vary strongly from these averages, the objective of the construction was to estimate the general importance of the various groups in the soil food webs to C and N cycling, rather than to describe seasonal dynamics. Most data collected at the various sites refer to the top 10 or 25 cm depth layer.

Table 4.I. Physiological parameters and biomasses (kg C ha^{-1}) for the different functional groups for the CPER food web. The biomasses for the other food webs are given in Table 4.2.

Functional Group	C:N	Turnover Rate (yr^{-1})	Assimilation Efficiency %	Production Efficiency %	Biomass (Kg C Ha^{-2})
Predatory mites	8	1.84	60	35	0.160
Nematophagous mites	8	1.84	90	35	0.160
Predatory nematodes	10	1.60	50	37	1.080
Omnivorous nematodes	10	4.36	60	37	0.650
Fungivorous nematodes	10	1.92	38	37	0.410
Bacteriophagous nematodes	10	2.68	60	37	5.800
Collembola	8	1.84	50	35	0.464
Non-cryptostigmatid mites	8	1.84	50	35	1.360
Cryptostigmatid mites	8	1.20	50	35	1.680
Amoebae	7	6.00	95	40	3.780
Flagellates	7	6.00	95	40	0.160
Phytophagous nematodes	10	1.08	25	37	2.900
Arbuscular mycorrhizal fungi	10	2.00	100	30	7.000
Saprobic fungi	10	1.20	100	30	63.000
Bacteria	4	1.20	100	30	304.000
Detritus	10	0.00	100	100	3000.000
Roots	10	1.00	100	100	300.000

Most of these physiological parameter values were used for all webs in Figure 4.2 as well (Hunt et al., 1987, de Ruiter et al., 1993b). Exceptions were the specific death rates of microorganisms and protozoa in the Kjettslinge webs, which are adjusted to values matching the respiration rates given by Andrén et al. (1990). For the Kjettslinge webs, the C:N ratio of detritus has two values, i.e., 10 for the substrate for bacteria and 20 for the substrate for the fungi.

Table 4.2. Biomass estimates (kg C ha^{-1}) for the functional groups in the different food webs. Values refer to the 0–25 cm depth layer, except for the Horseshoe Bend webs (0–15 cm).

	CPER	LH–IF	LH–CF	HSB–NT	HSB–CT	KS–B0	KS–B120
Microorganisms							
Bacteria	304	245	228	440	690	740	900
Fungi	63	3.27	2.12	160	150	1500	2300
VAM	7	–	–	–	–	–	–
Protozoa	–	–	–	–	–	–	–
Amoebae	3.78	18.9	11.5	40^2	50^2	110^2	34^2
Flagellates	0.16	0.63	0.53	–	–	–	–
Nematodes							
Herbivores	2.90	0.35	0.19	0.40	0.50	0.18	0.29
Bacteriovores	5.80	0.36	0.30	0.46	1.40	0.45	0.50
Fungivores	0.41	0.13	0.08	0.12	0.08	0.20	0.12
Predators^1	1.08	0.06	0.06	–	–	0.44	0.44
Arthropods							
Herbivorous herbage arthropods	–	–	–	–	–	0.10	0.14
Predatory herbage arthropods	–	–	–	–	–	0.15	0.19
Herbivorous macroarthropods	–	–	–	–	–	0.19	0.19
Microbivorous macroarthropods	–	–	–	–	–	0.25	0.25
Predatory macroarthropods	–	–	–	–	–	0.49	0.49
Predatory mites	0.16	0.08	0.06	0.20^3	0.04^3	0.18^4	0.28^4
Nematophagous mites	0.16	0.006	0.004	–	–	–	–
Cryptostigmatid mites	1.68	0.003	0.007	0.80	0.22	–	–
Non-cryptostigmatid mites	1.36	0.04	0.02	0.90	0.39	–	–
Bacteriovorus mites	–	0.0003	0.001	–	–	–	–
Fungivorous Collembola	0.46	0.38	0.47	0.30	0.09	0.17^5	0.17^5
Predatory Collembola	–	0.008	0.03	–	–	–	–
Annelids							
Enchytraeids	–	0.21	0.43	0.10	0.30	4.20	3.40
Earthworms	–	63.5	–	100	20	13	13

B0, without fertilizer; B120, with fertilizer; CF, conventional farming; CPER, Central Plains Experimental Range (Colorado, USA); CT, conventional tillage; HSB, Horseshoe Bend (Georgia, USA); IF, integrated farming; KS, Kjettslinge (Sweden); LH, Lovinkhoeve (the Netherlands); NT, no tillage; VAM, vesicular–arbuscular mycorrhizae.
[1] Including predators and omnivores.
[2] Including amoebae and flagellates.
[3] Including all predatory arthropods.
[4] Including all predatory microarthropods.
[5] Including all microbivorous microarthropods.
– Not found.

The mapping from individuals to biomass presents several hidden assumptions. Short of weighing each individual, converting population sizes based on individuals to population sizes based on biomass relies on conversion factors or functions based on detailed studies of the mass and morphometric measures of representative species, or through the integration of functions that relate a process to a population size. For the soil food webs, biomass values were constructed by multiplying the species numbers with estimates of the average individual body sizes. The C and N content of organisms was calculated by assuming that 50% of dry weight is C, and using the C:N ratios in Table 4.1.

4.3 Feeding rates and mineralization rates

Apart from the distribution of biomass within the food web, an energy flux description provides estimates of the feeding rates and mineralization rates. From our discussion in Chapter 2, we could use one of two approaches to estimate these rates: an ecosystem approach that involves simple concepts of the conservation of matter and mass balance (O'Neill, 1969); or a more community ecological approach that involves the formulation of predation rates and ensuing functional response of the predator to changes in prey density (Holling, 1959, 1965). We will use both approaches in this book, but for the energy flux description we will use the approach presented by O'Neill (1969) to estimate feeding rates and mineralization rates. In Chapter 5, we will link the two approaches, enabling us to catalog the distribution biomass and energy flux as part of the energy flux food web description presented here, and to study dynamic properties as part of the functional food web as well.

4.3.1 Feeding rates, mass balance, and energy budgets

The feeding rate describes the rate that living prey or nonliving detritus is consumed. Establishing feeding rates in the field or in the laboratory using experiments of pairwise or multispecies combinations is impractical. O'Neill (1969) devised an approach based on the conservation of matter and energy and the assumption that system is at steady state. The approach requires estimates of the population sizes, death rates, and energetic efficiencies of the species in a connectedness description.

In this scheme, we assume that matter is conserved; so all consumed biomass should be accounted for. In Chapter 2 we presented the fundamentals (see Figure 2.3). The scheme provides a means of quantifying the amount of materials consumed, the fate of the materials, and the rates that they are processed. Matter that is consumed is processed by the consumer, and split into different fractions with their own rates (Figure 4.1). The assimilated fraction is the portion of consumed matter that is used by the consumer for biomass production and maintenance. Biomass production may include growth, repair, and the production of offspring. The maintenance fraction is used for metabolic activities and is the portion of the food that is released to the environment, or mineralized, in an inorganic form—CO_2 and nitrogenous waste (e.g., urea, ammonium). The unassimilated fraction represents biomass left behind by the consumer (e.g., leavings and orts), or that is ingested by the consumer and then egested (e.g., expelled from a vacuole or passed through a gut as feces). In either case, the unassimilated fraction is a form of detritus.

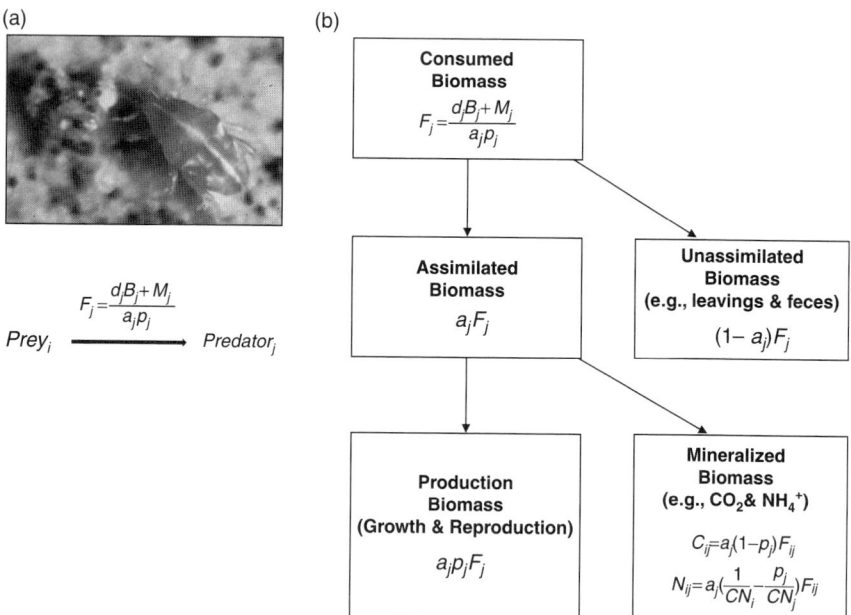

Figure 4.1 A review of the schematic for the fate of consumed matter developed by O'Neill (1969) and presented in Chapter 2. (a) Consumption of prey$_i$ by a predator$_j$ at steady state is estimated as $F_j = \frac{d_j B_j + M_j}{a_j p_j}$. For the predator to maintain their steady-state biomass B_j, they must consume enough prey, to offset losses due to mortality (natural death, $d_j B_j$, death from predators, M_j, or otherwise), assimilation efficiency (a_i), and production efficiency (p_i). Photo © John C. Moore. (b) For every trophic interaction in the food web a quantity of prey, F_j, is killed or consumed by the consumer. A fraction, $a_j F_j$, is ingested and assimilated by the consumer, and a fraction, $(1-a_j)F_j$ is either unconsumed or egested. Of the assimilated fraction ($a_j F_j$) a fraction is immobilized into new biomass, $a_j p_j F_j$, in the form of growth or offspring, and a fraction is mineralized as metabolic waste, $(1-p_j)a_j F_j$. For predators with multiple prey the estimate for F_j is divided into separate fluxes based on weighing factors using equation 4.2.

The second important assumption in this approach is that the system is at steady state. Coupled with the first assumption, the production of each population must balance the rate of loss through natural death and predation. In other words, in order to maintain the observed population abundances, the feeding rates need to be sufficient to keep the abundances at their respective levels. Specifically, the feeding rates for each population should be sufficient to generate levels of production that are large enough to offset all losses due to predation, natural mortality, and energetic inefficiencies.

With the mass-balance and steady-state assumptions discussed above, the feeding rates can be estimated as:

$$F_j = \frac{d_j B_j + M_j}{a_j p_j} \quad (4.1)$$

Using the dimensions for our soil food webs, F_j, refers to the feeding rate of group j (kg C ha^{-1} yr^{-1}), d_j to its specific death rate (yr^{-1}), B_j to the average annual population size (kg C ha^{-1}), M_j to the death rate due to predation (kg C ha^{-1} yr^{-1}), a_j to the assimilation efficiency, and p_j to the production efficiency. We can see from Equation 4.1 that the numerator is the sum of $d_j B_j$ and M_j, representing the total loss that the population experiences due to natural death and death due to predation. If all consumption were transformed into biomass, then consumption would simply have to equate to the numerator. We know that this is not the case. From Figure 4.1 it is clear that to offset the losses and maintain the steady-state population densities the feeding rate F_j needs to be sufficiently high to cover the added losses from excretion and mineralization. To account for these losses we divide the numerator in Equation 4.1 by the product of the assimilation efficiency, a_j, and the production efficiency, p_j.

The feeding rate, F_j, presented in Equation 4.1 is the overall feeding rate of a predator required to maintain its population density at steady state for the time step over which the biomass was averaged or that it depicts. For predators that feed on a single prey or resource, F_j would be sufficient. For predators feeding on more than one prey type, the overall feeding rate F_j needs to be parsed among the different prey types (F_{ij}). This is done by assuming that the predator feeds on a prey type according to the relative abundance of this prey type, and on any possible prey preference, w_{ij}, as follows:

$$F_{ij} = \frac{w_{ij} B_j}{\sum_{k=1}^{n} w_{kj} B_k} F_j \quad (4.2)$$

In Equation 4.2, w_{ij} refers to the preference of predator j for prey i over its other prey types and n is the number of trophic groups. All prey items are assigned weights of one or greater, $w_{ij} \geq 1$, whereas nonprey items are assigned weights of zero, $w_{ij} = 0$. When all w_{ij} are equal to 1, then all prey are eaten according to their relative abundance, but when $w_{ij} > 1$, e.g., $w_{ij} = 2$, then this particular prey type is eaten twice as much as one would expect given its relative abundance.

The values for the parameters used in Equations 4.1 and 4.2 for our soil food webs, with the exception of M_j and w_{ij}, are listed in Tables 4.1 and 4.2. To estimate B_j, we used the average biomass for a growing season obtained from repeated sampling in the field, adjusted to reflect an annual average. We estimated the specific death rates, d_j, as the inverse of the life spans of the different species scaled to years (i.e., used to generate turnover time). The assimilation and production efficiencies were obtained from the literature for representative taxa. Establishing accurate estimates of feeding preferences, w_{ij}, can be as challenging as establishing functional responses. Approaches have included field observations, manipulative studies, food choice or preference experiments, studies involving isotopes, and assessments of gut contents. For some predator–prey combinations weighing factors are available (Hunt et al., 1987); for all other interactions we use $w_{ij} = 1$.

We can obtain estimates of the losses due to predation, M_j, for each species from the model itself. With the exception of the top predator, the values for M_j are in fact the feeding rates corresponding to a particular F_{ij}. To estimate the M_j we start with the top predators (i.e., $j = 1$, the predatory mites in the soil food webs), which by definition are not consumed by other organisms and suffer only from natural death. This means that for the top predator $M_1 = 0$, and all the remaining necessary parameter values are available (Tables 4.1 and 4.2). Hence, to estimate the feeding rates within the food web we start with the overall feeding rate of the top predator, F_1 (Equation 4.1), which then is split up over all its prey types (Equation 4.2). In the next step, the feeding rates are calculated for the prey consumed by the top predator. Here, the feeding rate, F_{i1} from prey i to the top predator equals its death rate due to predation, M_i. In this way all M_j values become available through the calculations in the former steps.

4.3.2 Mineralization rates

Another way to track material flows in food webs is in terms of mineralization rates. Recall that mineralization is the process in which matter is transformed from an organic for to an inorganic form. For our fundamental equation of life (Equation 1.1) this entails the oxidation of glucose to CO_2. Mineralization rates are central in ecosystem function, as they determine the rates with which nutrients become available for plant uptake, as well as the rates with which greenhouse gases such as CO_2 are emitted to the atmosphere. To obtain estimates for mineralization rates, the feeding rates are rewritten following the same energy budget principles discussed above. Given that the biomass and feeding rates are expressed in terms of carbon, carbon respiration can be calculated as follows:

$$C_{ij} = a_j(1 - p_j)F_{ij} \qquad (4.3)$$

Here, C_{ij} (kg C ha^{-1} yr^{-1}) refers to the carbon that is respired as a result of the feeding event corresponding to F_{ij}. The other symbols have the same meaning as in Equations 4.1–4.2.

To derive nitrogen mineralization from the feeding rates, we have to convert from units of carbon to units of nitrogen using C:N ratios of resources and consumers. When the two C:N ratios are the same the conversion is simple, as N-mineralization (N_{ij}) equals carbon C_{ij} divided by the common C:N ratio. But if the two C:N ratios are different, then we assume that the consumer mineralizes N to the point that the C:N ratio of the consumer biomass production equals the required consumer C:N ratio. This leads to the following equation:

$$N_{ij} = a_j \left(\frac{1}{CN_i} - \frac{p_j}{CN_j} \right) F_{ij} \qquad (4.4)$$

In this equation, N_{ij} (kg N ha^{-1} yr^{-1}) refers to the N-mineralization from F_{ij}, the CN_i (kg C kg N^{-1}) and CN_j (kg C kg N^{-1}) to the C:N ratios of the prey and the predator respectively, and the other symbols have a similar meaning as in Equations 4.1–4.3.

4.4 Energy flux descriptions

We now have the tools to develop an energy flux description, or more correctly, a *collection* of energy flux descriptions. The energy flux description that is based on the connectedness description using the approach outlined above for the CPER food web is presented in Figure 4.2 with similar descriptions for the additional soil food webs presented in Figure 4.3. This form of the energy flux description provides steady-state estimates of the biomass for each group and flux rates among groups in the food web. Figures 4.4 and 4.5 depict alternative descriptions, using all seven of our soil food webs. In these descriptions the biomasses of each group and the feeding rates for each predator–prey interaction are ordered by their trophic position.

Several patterns emerge from these descriptions. First, the distribution of biomass and the flow of material are not uniform within the food webs. The systems possess pyramids of biomass, in which the biomass is concentrated at the base of the food webs and decreases with each trophic level. Second, the distribution of the feeding rates within the web exhibit trophic pyramids. The flows at the base of the food web are much larger than those at the top. A high proportion passes from detritus through the bacteria and fungi, and their consumers, relative to plant roots and herbivores. Third, the distribution of the N-mineralization rates mirrors those of the biomass and feeding rates. In Table 4.3 the mineralization rates are given for each functional group in all seven webs. Similarly to the biomass and feeding rates, N-mineralization rates also follow the structure of a trophic pyramid, with low mineralization rates at the top of the food web and high mineralization rates at the lower trophic levels. But this pyramidal pattern is not as pronounced as that for biomass and feeding rates. For example, microorganisms (bacteria and fungi) make up more than 90% of the biomass in the CPER food web, but contribute near 50% of the

80 • Energetic Food Webs

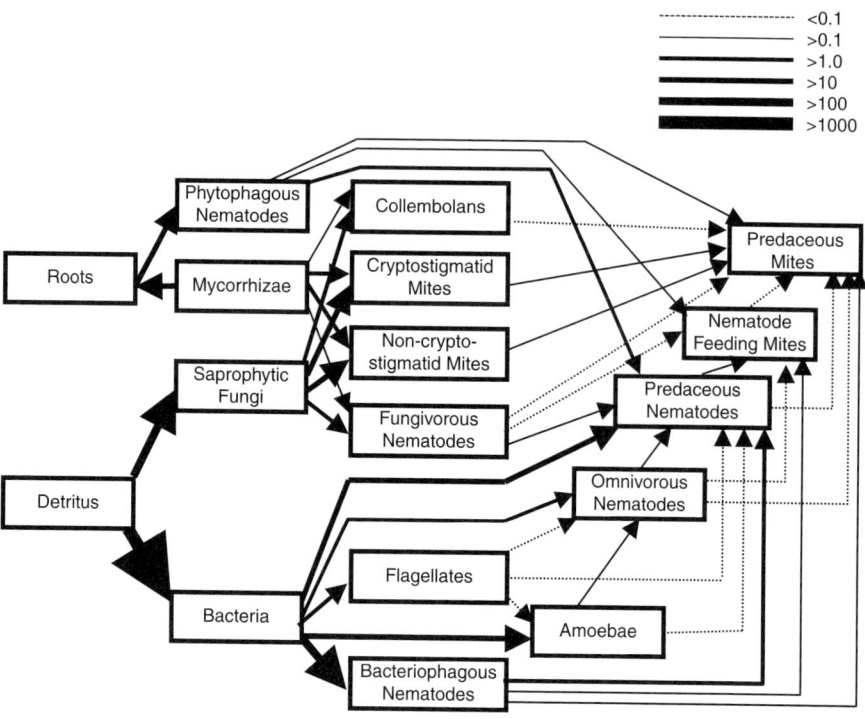

Figure 4.2 The energy flux description of the Central Plains Experimental Range (CPER) soil food web (after Hunt et al., 1987). The thickness of the arrows indicates the amount of material transferred (kg C ha^{-1} yr^{-1}).

N-mineralization. A similar deviation in pyramidal structures can be seen in the food webs from the agricultural sites. This is due to differences in C:N ratios, growth rates, and the influence of trophic interactions. For example, bacteria have a low C:N ratio (4) compared to that of the detritus they feed on (10), leading to a relative low N-mineralization rate per consumption rate, whereas groups that feed on bacteria, e.g., protozoa and nematodes, have higher C:N ratios (7 and 10, respectively), leading to relatively high N-mineralization rates per consumption rates. Also, most faunal groups have higher growth rates than the microbial populations.

It is important to note that there are many instances, particularly in aquatic systems, where the distributions of biomass, feeding rates, and mineralization rates with trophic position take on different configurations, namely inverted trophic pyramids. In the chapters that follow we will discuss how these patterns might arise, the relationship between the distribution of biomass with trophic position, and the general importance of biomass pyramids for the stability of food webs. This will require us to reformulate the information in the energy flux food web description into the functional food web description.

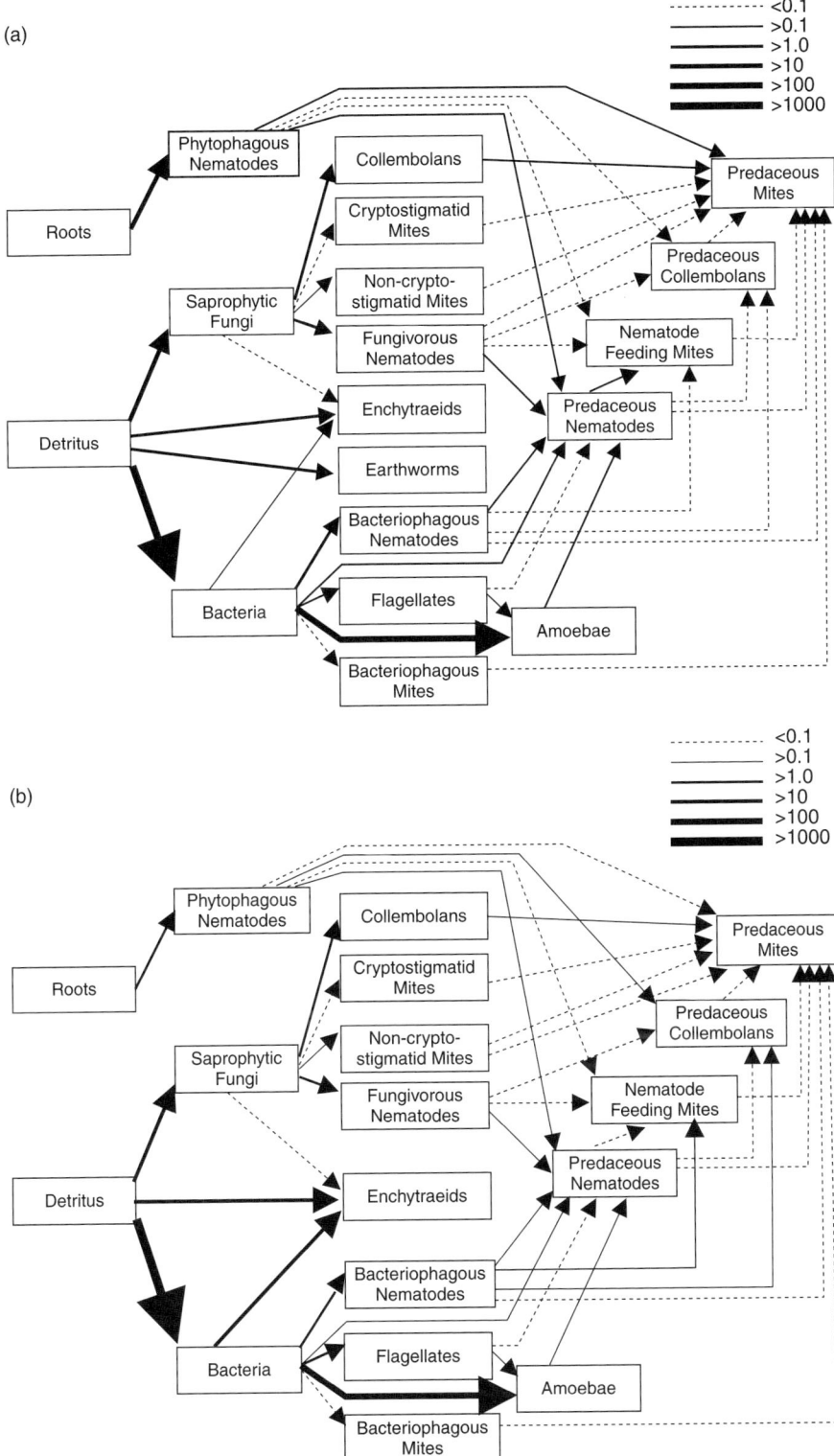

Figure 4.3 *Continues overleaf.*

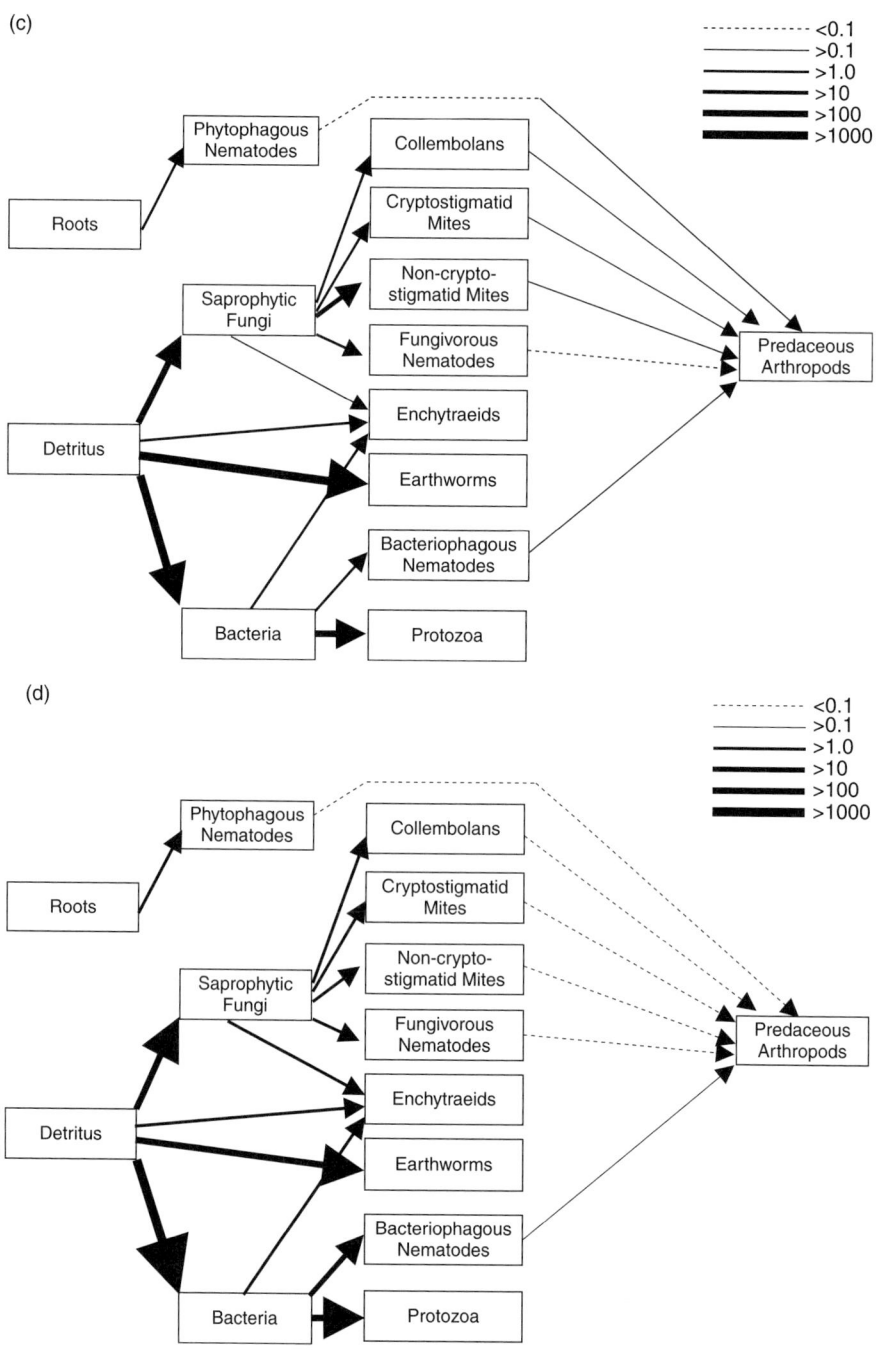

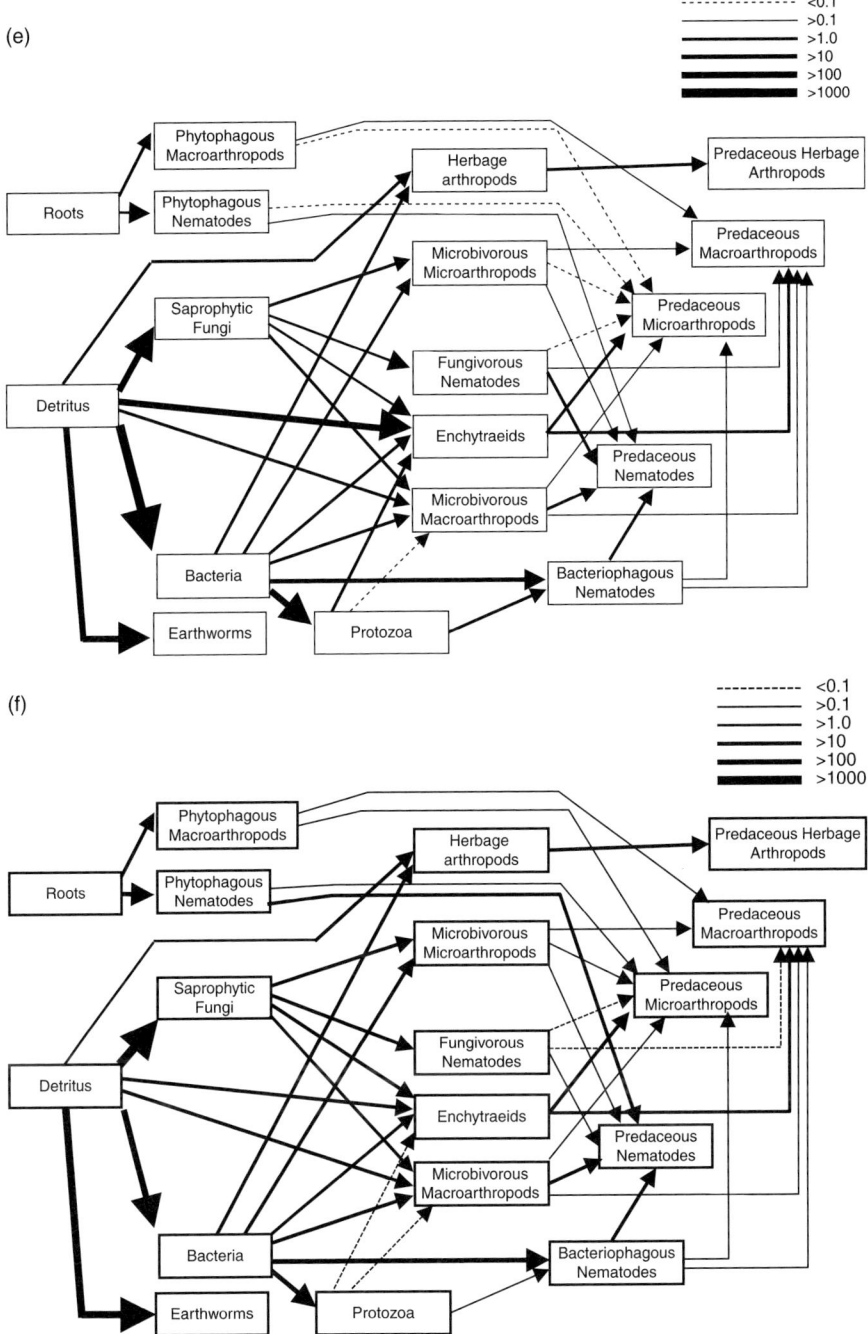

Figure 4.3 The energy flux descriptions of: (a) integrated management winter wheat, Lovinkhoeve Experimental Farm (Marknesse, The Netherlands); (b) conventional management winter wheat, Lovinkhoeve Experimental Farm; (c) no-tillage mixed crop rotation, Horseshoe Bend Research Site (Athens, Georgia, USA); (e) conventional-tillage mixed crop rotation, Horseshoe Bend Research Site; (d) barley without fertilizer (B0), Kjettslinge Experimental Farm (Uppsala, Sweden); and (f) with fertilizer (B120), Kjettslinge Experimental Farm. The thickness of the arrows indicates the amount of material transferred (kg C ha^{-1} yr^{-1}).

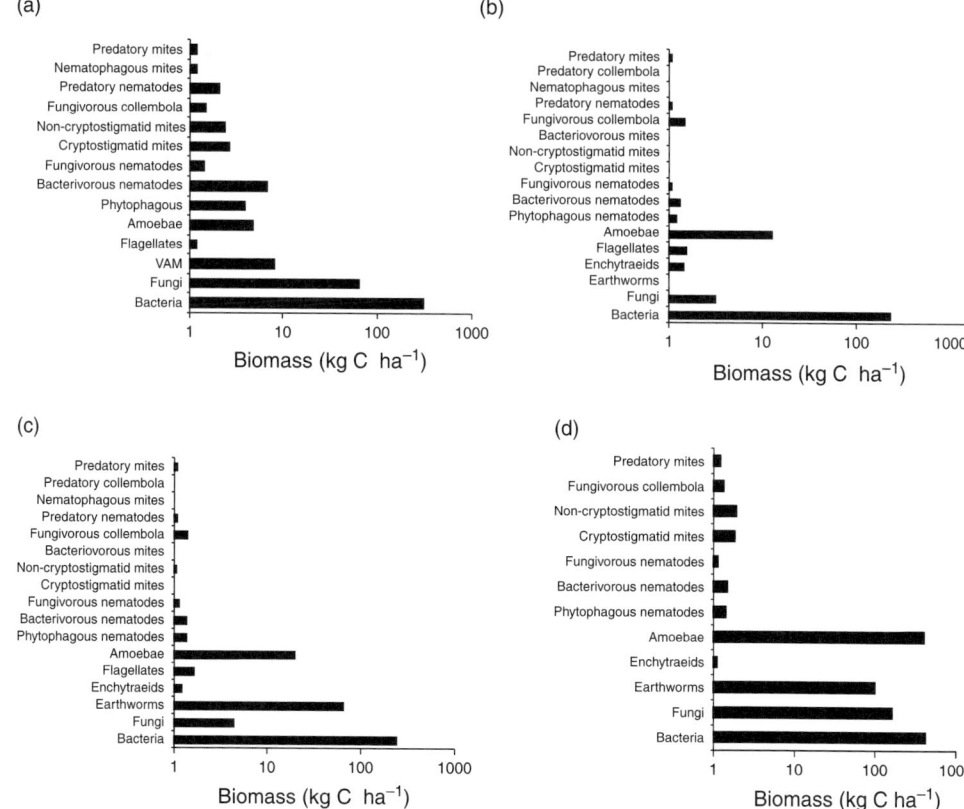

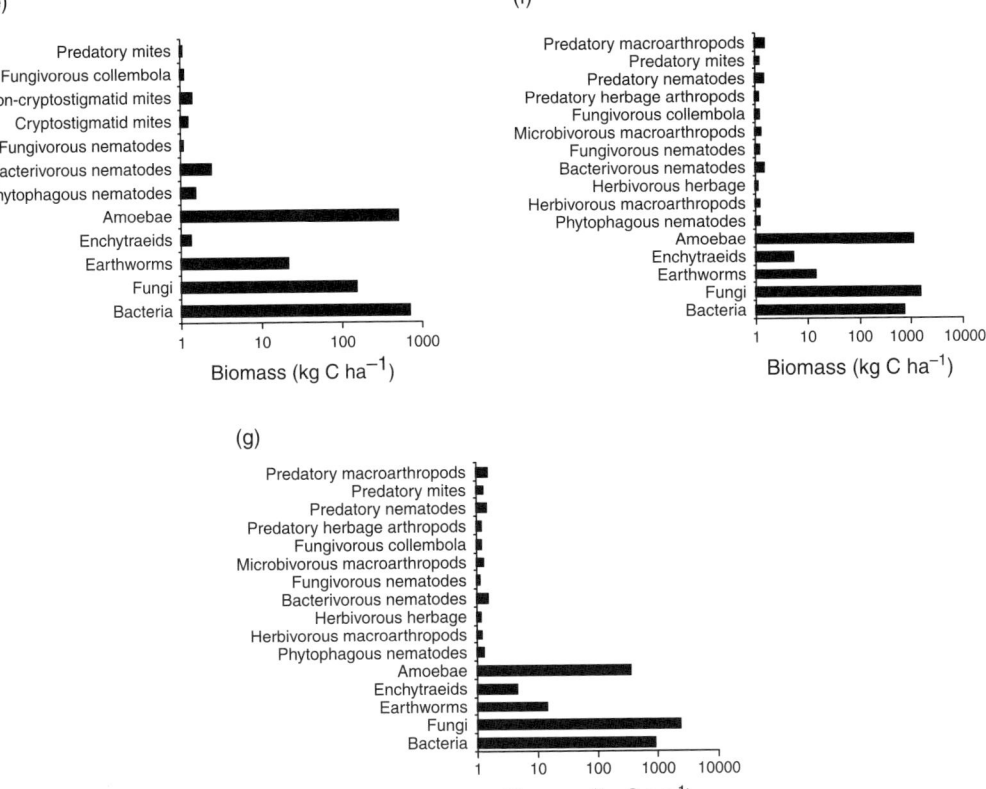

Figure 4.4 The distribution of biomass (kg C ha^{-1}) among functional groups by trophic position within the Central Plains Experimental Range (CPER-Nunn, CO) food web and the six additional soil food webs. The dashed lines separate the low trophic position soil-microbial feeding annelids from other groups. (a) CPER (Nunn, CO); (b) integrated management winter wheat, Lovinkhoeve Experimental Farm (Marknesse, The Netherlands); (c) conventional management winter wheat, Lovinkhoeve Experimental Farm; (d) no-tillage mixed crop rotation, Horseshoe Bend Research Site (Athens, Georgia, USA); (e) conventional-tillage mixed crop rotation Horseshoe Bend Research Site; (f) barley without fertilizer (B0), Kjettslinge Experimental Farm (Uppsala, Sweden); and (g) with fertilizer (B120), Kjettslinge Experimental Farm.

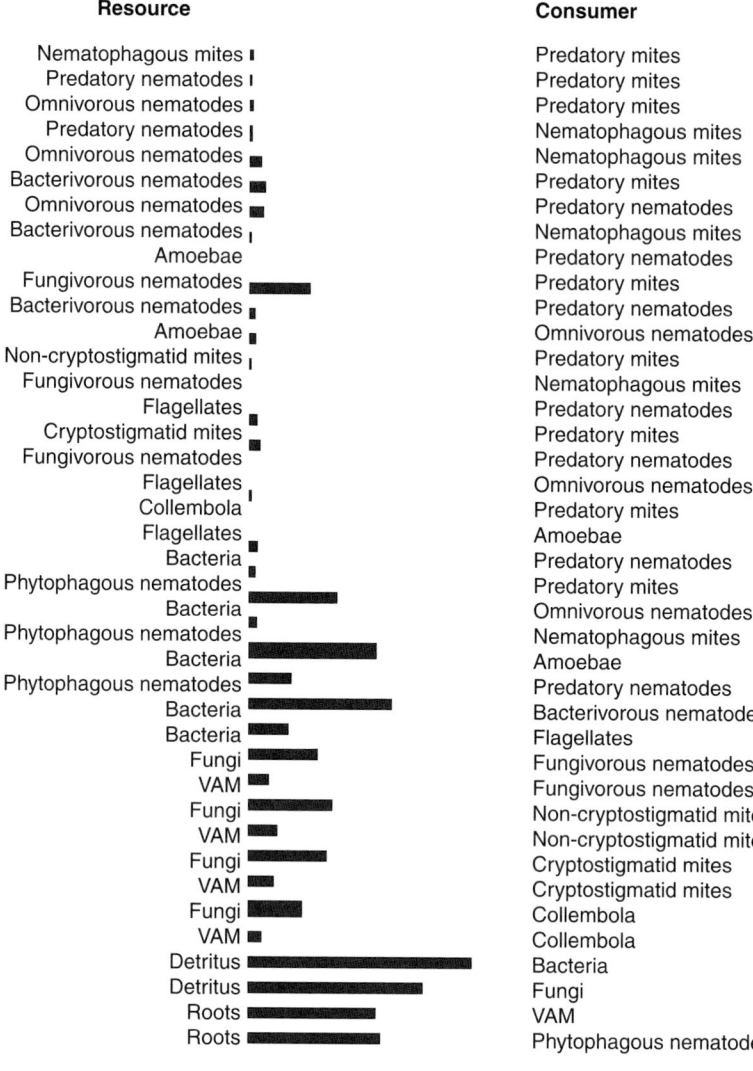

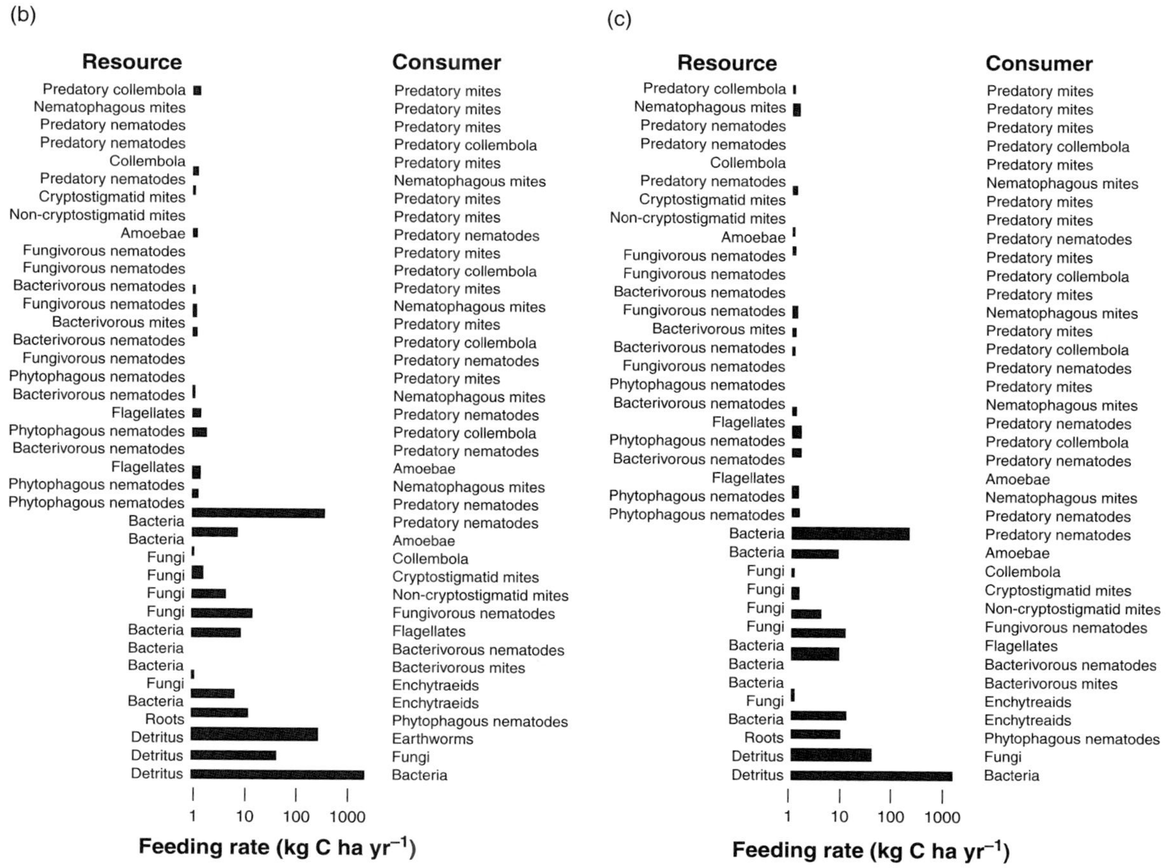

Figure 4.5 *Continues overleaf.*

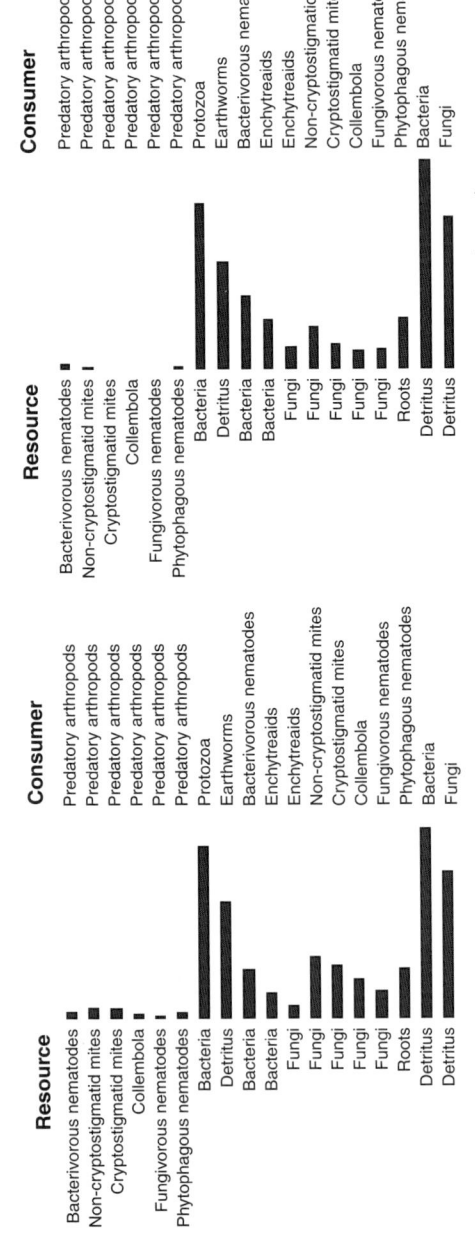

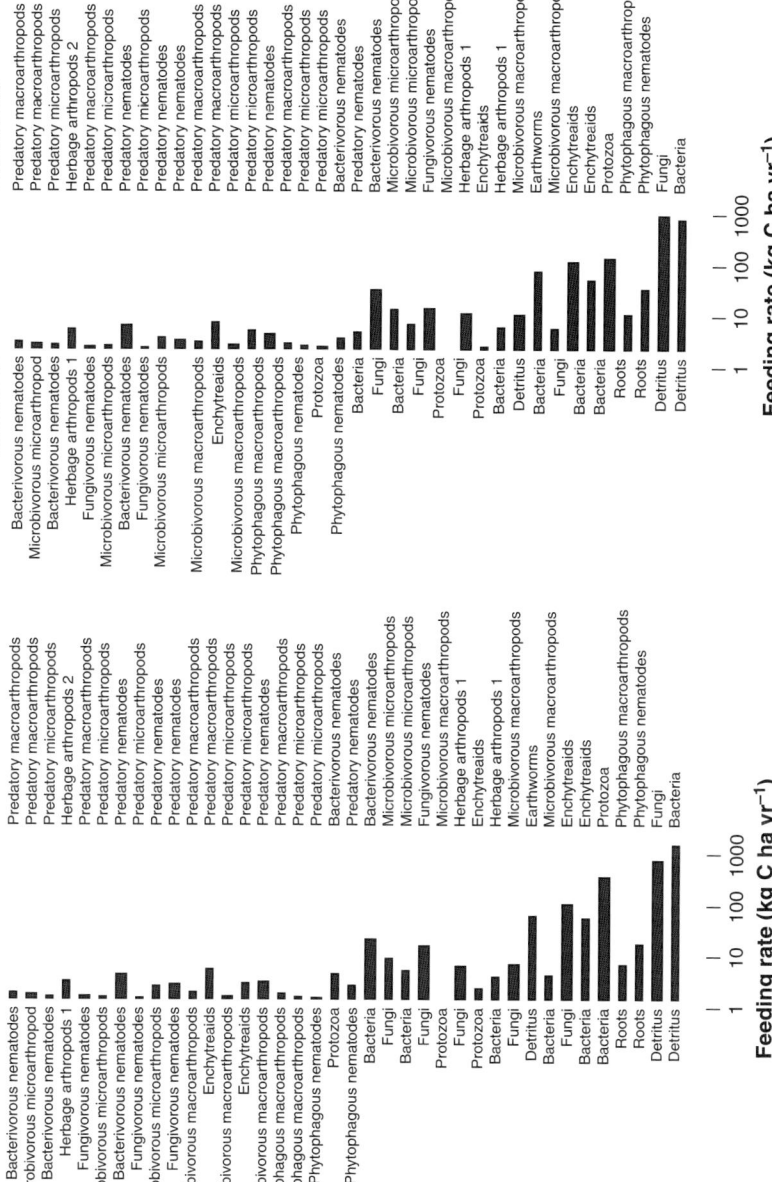

Figure 4.5 Feeding rates (kg C ha^{-1} yr^{-1}) among the functional groups within the soil food webs arranged by relative trophic position of the consumer and resource. (a) CPER (Nunn, CO); (b) integrated management winter wheat, Lovinkhoeve Experimental Farm (Marknesse, The Netherlands); (c) conventional management winter wheat, Lovinkhoeve Experimental Farm; (d) no-tillage mixed crop rotation, Horseshoe Bend Research Site (Athens, Georgia, USA); (e) conventional-tillage mixed crop rotation Horseshoe Bend Research Site; (f) barley without fertilizer (B0), Kjettslinge Experimental Farm (Uppsala, Sweden); and (g) with fertilizer (B120), Kjettslinge Experimental Farm.

Table 4.3. N-mineralization rates (kg ha^{-1} yr^{-1}) of the functional groups in the different food webs. Values refer to the 0–25 cm depth layer, except for the Horseshoe Bend webs (0–15 cm).

	CPER	LH–IF	LH–CF	HSB–NT	HSB–CT	KS–B0	KS–B120
Microorganisms							
Bacteria	45.44	51.59	40.15	97.40	136.95	37.31	20.50
Fungi	7.61	0.91	0.82	14.66	12.67	19.03	23.61
VAM	4.27	–	–	–	–	–	–
Protozoa							
Amoebae	11.05	54.69	33.37	115.71[2]	144.64[2]	54.51[2]	16.80[2]
Flagellates	0.49	2.20	1.73	–	–	–	–
Nematodes							
Herbivores	1.12	0.16	0.12	0.11	0.10	0.004	0.06
Bacteriovores	12.93	0.88	0.88	0.87	2.27	2.03	2.38
Fungivores	0.22	0.08	0.06	0.05	0.03	0.26	0.16
Predators[1]	0.35	0.06	0.07	–	–	0.25	0.25
Arthropods							
Herbivorous herbage arthropods	–	–	–	–	–	0.23	0.28
Predatory herbage arthropods	–	–	–	–	–	0.08	0.10
Herbivorous macroarthropods	–	–	–	–	–	0.02	0.02
Microbivorous macroarthropods	–	–	–	–	–	0.42	0.41
Predatory macroarthropods	–	–	–	–	–	0.34	0.33
Predatory mites	0.06	0.03	0.02	0.08[3]	0.01[3]	0.12[4]	0.18[4]
Nematophagous mites	0.05	0.003	0.002	–	–	–	–
Cryptostigmatid mites	0.37	0.001	0.002	0.23	0.05	–	–
Non-cryptostigmatid mites	0.44	0.015	0.01	0.35	0.12	–	–
Bacteriovorus mites	0.0004	0.001	–	–	–	–	–
Fungivorous Collembola	0.15	0.16	0.19	0.12	0.03	0.44[5]	0.42[5]
Predatory Collembola	0.003	0.01	–	–	–	–	–
Annelids							
Enchytraeids	–	0.31	0.42	0.12	0.38	3.32	2.56
Earthworms	–	7.62	–	12.00	2.40	1.56	1.56

B0, without fertilizer; B120, with fertilizer; CF, conventional farming; CPER, Central Plains Experimental Range (Colorado, USA); CT, conventional tillage; HSB, Horseshoe Bend (Georgia, USA); IF, integrated farming; KS, Kjettslinge (Sweden); LH, Lovinkhoeve (the Netherlands); NT, no tillage; VAM, vesicular–arbuscular mycorrhizae.
[1] Including predators and omnivores.
[2] Including amoebae and flagellates.
[3] Including all predatory arthropods.
[4] Including all predatory microarthropods.
[5] Including all microbivorous microarthropods.
– Not found.

The pattern of interactions and material flow within the webs hints at a compartmentalized architecture. The web exhibits a dependence on both primary production and detritus at the base with separate assemblages of consumer-resource interactions that are united by a group of predators at the top. The detritus assemblage can be further divided into an assemblage based on bacteria and their consumers, and one based on fungi and their consumers. We will explore the importance of the compartmentalized architecture, the reliance on both primary production and detritus and the pyramidal structures to the development and stability of the food webs in Parts II and III.

The models are remarkably accurate at predicting system-level processes. Recall that Equations 4.3 and 4.4 can be used to obtain estimates for all mineralization rates for each trophic interaction. By summing up these rates, we can obtain estimates of mineralization per functional group (Table 4.3) and of the overall rates through the web. These estimates of overall mineralization rates can be compared with independent estimates of observed mineralization rates. For the agricultural webs (Figure 4.2), it is possible to compare the overall C- and N-mineralization rates as calculated by the model with measured C- and N-mineralization rates. For these sites carbon respiration was measured by means of laboratory incubations, while, for the nitrogen mineralization, various techniques were used. N-mineralization was measured through in situ measurements at the Lovinkhoeve sites (Bloem et al., 1994) and through N budget analyses at the Horseshoe Bend and Kjettslinge sites (Stinner et al., 1984, Paustian et al., 1990). The comparison demonstrates that the calculated C-mineralization rates using the model are close to the observed rates for all six food webs (Figure 4.6). In the cases of the Kjettslinge webs, the specific death rates of microorganisms have been adjusted to the respiration rates (see de Ruiter et al., 1993b with reference to Andrén et al., 1990). The comparison of the calculated N-mineralization rates for the food webs from the Lovinkhoeve and Kjettslinge sites shows calculated rates close to the observed rates (Figure 4.6); however, the calculated N-mineralization rates in the Horseshoe Bend webs are high compared with rates obtained from N budget analyses (Stinner et al., 1984). The relatively high nature of the calculated rates may have been due to an underestimation of the C:N ratio of the substrate for microorganisms. Despite these departures, the model tends to calculate C-mineralization and N-mineralization rates in line with the observed rates (Figure 4.6).

The energy flux descriptions have clear limitations. The output is a descriptive, phenomenological account of material flow that arises from the direct interactions among groups and, as described above, at steady state. Although the steady-state assumption may seem like a limitation, the reality is that the description can be made dynamic by altering how we estimate the flux rates. For example, if a site were sampled repeatedly in a time series to capture the dynamics, instead of using the average biomass for each group as estimates of the steady-state biomass in Equation 4.1 to estimate the flux rates, the estimates for biomass for the individual dates could be used along with an additional term to reflect the change in population size from one date to the next (O'Neill, 1969).

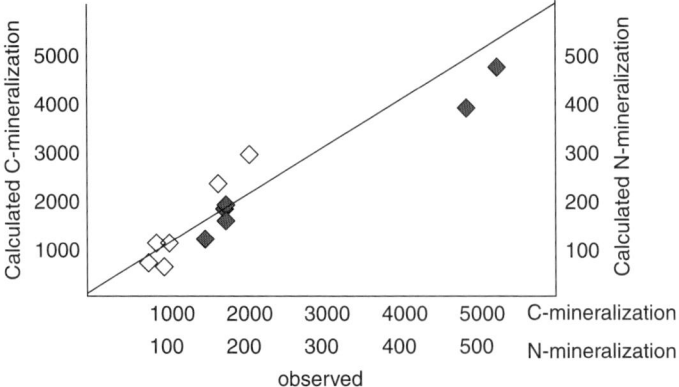

Figure 4.6 Comparison between measured and calculated mineralization rates. Values refer to kg C ha^{-1} yr^{-1} for the 0–25 cm depth layer, except for the Horseshoe Bend webs (0–15 cm); see Table 4.3. Open symbols, N-mineralization; and closed symbols, C-mineralization.

The reliance on direct interactions is more problematic. For example, the contributions of individual functional groups to N-mineralization, as presented in Table 4.3, only refer to N-mineralization resulting directly from the consumption and processing of matter by the group. Most groups may indirectly contribute to nutrient cycling through affecting the functioning of other groups in the web. In the soil food webs small arthropods may consume detritus directly, creating a greater surface area for microorganisms, or may transport microorganisms among microsites. From a food web perspective, the contributions may additionally be influenced by population-dynamical effects exerted by other groups in the web, through indirect as well as direct pathways of trophic interactions. The effects through these indirect pathways may frequently dominate the direct effects, as has been shown in theoretical (Yodzis, 1988) and experimental (Paine, 1980) food web studies.

4.5 Summary and conclusions

By means of relatively simple modeling it is possible to calculate feeding rates and concomitant mineralization rates in food webs to generate an energy flux food web description. The description requires a connectedness description of the food web outlining the trophic interactions among species or groups of species, estimates of the steady-state biomasses, assimilation efficiencies and production efficiencies, specific death rates, and C:N ratios for each species, and estimates of the feeding preferences of predators for prey. In addition to the feeding rates and mineralization rates, the description and underlying model provides estimates of the contributions of the functional groups to specific processes.

The energy flux descriptions of our soil food webs reveal important patterns that are shared by other descriptions. First, the food webs possess structural attributes that we will later demonstrate are important to their development and stability. These features include: (1) a reliance on living primary production and non-living detritus as basal resources, (2) pyramidal arrangements in their biomass, energy flux rates, and mineralization rates with trophic position, and (3) a compartmentalized architecture based on basal resources, trophic interactions, and material flow.

Energy flux descriptions require accurate characterizations of the food web structure, identifying the important species and groups as well as the interactions among them. If we were to stop with the energy flux description, these results would also parallel important conclusions made by others (Paine, 1980, Polis and Strong, 1996). Namely, the magnitudes of biomass and the magnitude of energy flux (amounts and rates) are interesting, but in-and-of themselves tell us little about the magnitude of a species' contribution to the dynamics of the food web. We know from empirical studies that the contributions of soil populations to the cycling of energy, matter, and nutrients depends not only on their own population sizes (biomass), or rates of energy use (energy flux), but also on their ability to influence the functioning of the organisms with which they interact. Microbial populations make up, by far, most of the soil biomass, and will therefore have a dominant contribution to soil processes. The soil fauna, particularly the predators, represent a small fraction of the overall biomass and energy flux, yet, as empirical studies have demonstrated, may make an important contribution to the ecosystem processes, not only directly through the consumption and processing of material, but also indirectly through influencing the dynamics, and the amount of energy processed by their prey (e.g., Coleman et al., 1983, Moore et al., 1988, Verhoef and Brussaard, 1990). These types of indirect effects have been demonstrated to be an important component of numerous ecosystems.

The energy flux descriptions revealed several important patterns in food web architecture and processes related to the distribution of biomass and patterning of flow rates within the food web. The basis for these patterns can be explained in large part by fundamental energetic principles of mass balance, energetic efficiencies, and differences in flow rates of materials. How they relate to dynamic stability is unclear at this time. In the next chapter, we will use the information from the energy flux food web description to construct a functional food web, providing us with the tools we need to study the energetic basis for relationships among food web structure, function, and stability.

5

Functional webs

5.1 Introduction

A functional food web description portrays the impact of each species on the population sizes and dynamics of the other species in the food web. In this chapter we will derive a functional food web description from the energy flux food web descriptions of our soil food web that we developed in Chapter 4. The functional food web description includes several metrics that allow us to study the importance of individual species (or functional groups) and the species in aggregate to the structure and stability of the food web. We will start by developing a simple model of the trophic interactions that melds the differential equations developed in Chapter 2 with the modeling approach developed in Chapter 4 to estimate energy flux rates. From this fusion of community-based and ecosystem-based models we will estimate interaction strengths among the functional groups and additional metrics.

The concept of interaction strength will be vetted, as it is central to the functional food web description, but has many meanings in the literature. We will derive estimates of the interaction strengths by describing the population dynamics of the functional groups in terms of differential equations, and then derive the *per capita* (or per biomass) of species upon one another (May, 1973, Pimm, 1982). Next, we will equate the interaction strengths to the feeding rates we had derived in Equations 4.1 and 4.2. As these interaction strengths are the entries of the Jacobian matrix (May, 1973; and see Chapter 2), they are a means to evaluate the stability of the webs, and the role of the functional groups in food web stability. We will explore other indices of stability that can be directly derived from the biomass estimates and feeding rates we presented in Chapter 4.

5.2 Interaction strengths

The vectors of the connectedness webs represent trophic interactions. The flow of matter and energy along the vectors were the currency of the energy flux description. For the functional food web, the trophic interactions are defined in terms of interaction strength. Empiricists and theoreticians have captured the meaning of interaction strength in different ways (Laska and Wootton, 1998, Berlow et al., 2004). We have summarized some of the different approaches that were discussed in a review by Berlow et al. (2004) with others in Table 5.1. In reviewing these approaches we see that, although similar in intent, the metrics used to estimate interaction strength are indeed different.

Table 5.1. Different theoretical and empirical metrics for interaction strength used in food webs (adapted from Berlow et al., 2004).

Interaction Strength Metric	What it Measures	Level of Measurement	References
Community (Jacobian) matrix	Partial derivative of one species' growth rate with respect to small changes in another species' abundance (units: t^{-1})	Property of individual link	Moore et al. (1993), de Ruiter et al. (1995), Moore et al. (1996), Schmitz (1997), Ives et al. (1999), Krause et al. (2003), Neutel et al. (2007)
Interaction matrix	Interaction coefficients (α_{ij}) in a Lotka–Volterra multispecies competition model; can be generalized to the partial derivative of one species' per capita growth rate with respect to small changes in another species' abundance (units: $n^{-1} t^{-1}$)	Property of individual link	Kokkoris et al. (2002)
Inverse interaction matrix	Change in the equilibrium density of one species in response to a change in the carrying capacity of another species (units: nt); total direct and indirect effects of one species on another	Whole-system response	Bender et al. (1984), Yodzis (1988)
Nonlinear functional response	Number of prey consumed as a function of prey density and predator density or predator–prey ratios; "top-down" measure of consumption intensity; various units	Property of individual link	Beddington (1975), Abrams and Ginzburg (2000), Ruesink (1998)
Relative prey preference	Fraction of a predator's maximum consumption rate that is targeted to a specific prey item; top-down measure of consumption intensity	Property of individual link, or whole-system property	Yodzis and Innes (1992), McCann et al. (1998)
Maximum consumption rate	Measures maximum consumption per unit time on fixed abundance of prey; measures "top-down" potential consumption intensity	Property of individual link	Sala and Graham (2002)
Biomass flux	Absolute or relative magnitude of biomass flowing from prey to predator per unit time	Property of individual link	Benke et al. (2001), Bersier et al. (2002), Cohen et al. (2003)

(*continued*)

Table 5.1. Continued

Interaction Strength Metric	What it Measures	Level of Measurement	References
Change in population variability	Effect of changing the abundance of one species on the pattern of population variability of another species	Whole-system response	Borrvall et al. (2000), Solé and Montoya (2001), Dunne et al. (2002)
Secondary extinctions	Number of species that go extinct as a result of perturbing a given species	Whole-system response	Many field experiments where response variables are untransformed
"Paine's Index"	"Absolute prey response" standardized by some measure of prey abundance; measured either as a per capita effect or a species-level effect	Whole-system response	Paine (1992)
Log response ratio	Log of the ratio of prey abundance "with" versus "without" predators; measured either as a per capita effect or a species-level effect	Whole-system response	Berlow (1999), Laska and Wootton (1998)
Statistical correlation	Measures magnitude of correlation between change in one species and change in another	Whole-system response	Wootton (1994), Pfister (1995), Ives et al. (1999)
Frequency of consumption	Frequency of hosts that are parasitized (e.g., parasite prevalence)	Whole-system response	Hawkins and Cornell (1994), Müller et al. (1999), Montoya et al. (2003)

To the empiricist, functional food web descriptions are products of extensive field manipulations and removals of selected species with follow-up measurements to assess the responses of the remaining species (Paine, 1980, 1992). This approach can be traced to the seminal work of Paine (1966) on the role of individual species in shaping and maintaining communities. Here, interaction strength is a whole-system property represented by magnitudes of the changes in the densities of all other species within the food web that a trophic interaction elicited when it was altered. Although not explicitly stated, an equilibrium or steady state is implied in the approach for the studies for the results to have meaning.

Theoreticians have been more precise in their definitions of interaction strength and generally limit it to the property of an individual link between species, with an explicit link to a steady state. Examples include the interaction coefficients within the interaction matrix (Levins, 1968) or the elements of the related Jacobian matrix from the system of equations that describe the dynamics of the food web (May, 1973). As such, each interaction between two species possesses two estimates of interaction strength—one for each species. Stability and steady state serve as the bases when interpreting the functional food web description.

The extent to which energy flow can be used to develop a functional food web has been a point of contention. The controversy centers on whether or not there is a relationship between the flow of matter and energy and interaction strength that tells us anything about how communities are structured and respond to disturbance. Some have made the connection directly by defining interaction strength as the flow of matter from one species to another (Benke et al., 2001, Bersier et al., 2002, Cohen et al., 2003), or indirectly by defining interaction strength as a prey preference (Yodzis and Innes, 1992, McCann et al., 1998), a consumption rate (Sala and Graham, 2002), or a functional response (Beddington, 1975, Abrams and Ginzburg, 2000, Ruesink, 1998). Others question whether there is a link between energetics and interaction strength in complex food webs. Field studies in which species are removed or their densities are significantly altered have demonstrated unequivocally that trophic interactions that involve large quantities of material flow do not necessarily directly relate to their importance to the stability or recovery of the web (Paine, 1980, 1992, Polis and Strong, 1996). Still others have argued that in complex food webs, energetics, population dynamics, and stability are deeply interrelated, and that interaction strength embodies each (e.g., Ulanowicz, 1986, Hairston and Hairston, 1993, Moore et al., 1993). From this point of view, it is not the direct correlation between energy flow and interaction strength, but the patterning of interactions in a systems context that matters to structure and stability. Our leanings are clearly with the latter.

5.2.1 Population models

We will use the coefficients of the Jacobian matrix A as presented in Chapter 2 as our metric for interaction strength. To calculate the interaction strengths among the functional groups in the food webs we first describe the population dynamics of the functional groups in terms of differential equations. The equations we used are of the Lotka–Volterra form, repeated here for the convenience of the reader. The equation for primary producers is:

$$\frac{dX_i}{dt} = r_i X_i - \sum_{j=1}^{n} c_{ij} X_i X_j \qquad (5.1)$$

where X_i represents the population density (kg C ha^{-1}) of the primary producer, X_j represents the population densities (kg C ha^{-1}) of the herbivores, r_i is the specific growth rate (yr^{-1}) of the primary producer, and c_{ij} is the consumption coefficient for a herbivore on the primary producer, when $i \neq j$ and is the intraspecific competition coefficient when $i = j$. For now we will work with the simple linear functional responses with constant Holling type I linear coefficients. In later chapters we will discuss more complex, if not realistic, functional responses.

The equation describing the dynamics of the consumers (detritivores, herbivores, predators) is:

$$\frac{dX_i}{dt} = -d_i X_i + \sum_{j=1}^{n} a_i p_i c_{ji} X_j X_i - \sum_{j=1}^{n} c_{ij} X_i X_j \qquad (5.2)$$

Consumers are assumed to die at a specific death rate of d_i (yr^{-1}), which can be approximated as the inverse of their life spans. Consumer growth is a function of the amount of prey or detritus consumed ($c_{ij} X_i X_j$) adjusted by assimilation efficiency, a_i (fraction), and production efficiency, p_i (conversion fraction of resource biomass to consumer biomass), as in Equation 5.1. Once again we have opted to use the Holling type I functional response here but will expand our discussion to more complex responses in later chapters.

5.2.2 The Jacobian matrix

Now that we have described our food webs based on a set of differential equations we can define the "per capita" interaction strengths (May, 1972, Pimm, 1982). For our examples, the term "per capita" should be read as "per biomass," given that we have used biomass and not individuals as our currency. The interaction strengths, α_{ij}, are the elements of the Jacobian community matrix A representation of our model, defined as the partial derivatives of the differential equations near equilibrium (see Chapter 2 for its derivation):

$$\alpha_{ij} = \left(\frac{\partial \frac{dX_i}{dt}}{\partial X_j} \right)^* \qquad (5.3)$$

Applying this approach to the set of equations for the Central Plains Experimental Range (CPER) food web yields a 17 x 17 dimension Jacobian matrix (Figure 5.1).

The paired off-diagonal elements of this matrix correspond to the paired values for the interaction strengths that correspond to each vector in the description, and the diagonal elements correspond to the effects of the dynamics of each group on its own dynamics. For now we will focus on the off-diagonal elements. We will revisit the diagonal elements in our discussions below (sections 5.2.3 and 5.3.3). The elements in the upper off-diagonal region section of the matrix correspond to the α_{ij} elements depicting the effects of the dynamics of consumers j on the dynamics of prey i. The elements in the lower off-diagonal section of the matrix correspond to the α_{ji} elements depicting the effects of the dynamics of prey i on the dynamics of consumer j. Using Equation 5.3 we can see that the α_{ij} elements and the α_{ji} elements have each taken the following common form:

$$\alpha_{ij} = -c_{ij}X_i^* \quad (5.4)$$

$$\alpha_{ji} = a_j p_j c_{ij} X_j^* \quad (5.5)$$

5.2.3 Estimating interaction strength from energy flux

We will use information from the energy flux food web (Chapter 4) to obtain values for the interaction strength α_{ij} elements and the α_{ji}, using a few assumptions and simple substitutions. Recall that our estimates of energy flux rates, F_j, were based on simple mass-balance equations with parameters that we could obtain from the field and laboratory measurements or from the literature, where $F_j = \frac{d_j B_j + M_j}{a_j p_j}$. A quick glance at Equations 5.4 and 5.5 reveals that in part we can use the same parameter values as were used in our equation for the flux rates (F_j), as d_j, a_j and p_j have the same meanings in both representations. A closer look indicates some important differences that will require some assumptions. For the model based on the differential equations presented in Equations 5.1 and 5.2, the population size or biomass estimates at equilibrium, X_i^*, are obtained by setting our set of differential equations to zero and simultaneously solving for all X_i. This is possible, if not unruly, for large systems of equations. The simpler alternative would be to equate the field estimates for the biomasses of our groups at steady state, B_i, that we used in the energy flux models to the theoretical equilibrium values, X_i^*, that we would obtain from our differential equations. That is, let $X_i^* = B_i$.

Now the only unknown parameter for the off-diagonal elements of the Jacobian matrix from the differential equations is c_{ij}, which is part of the predation rate. The term $c_{ij}X_iX_j$ in our differential equations is therefore the feeding rate of group j on group i, hence $c_{ij}X_iX_j$ is equal to F_{ij} (Equation 4.1). Given that the dimension of c_{ij} is $X_j^{-1}t^{-1}$, $c_{ij}X_j$ corresponds to the linear type I functional response (Holling, 1965). So if we substitute the values of F_{ij} for $c_{ij}X_iX_j$ we have a means of parameterizing the off-diagonal elements of the Jacobian matrix and nearly complete parameterization for our model based on the differential equations. We will address the diagonal terms to complete the parameterization of our models in section 5.2.4.

(a)

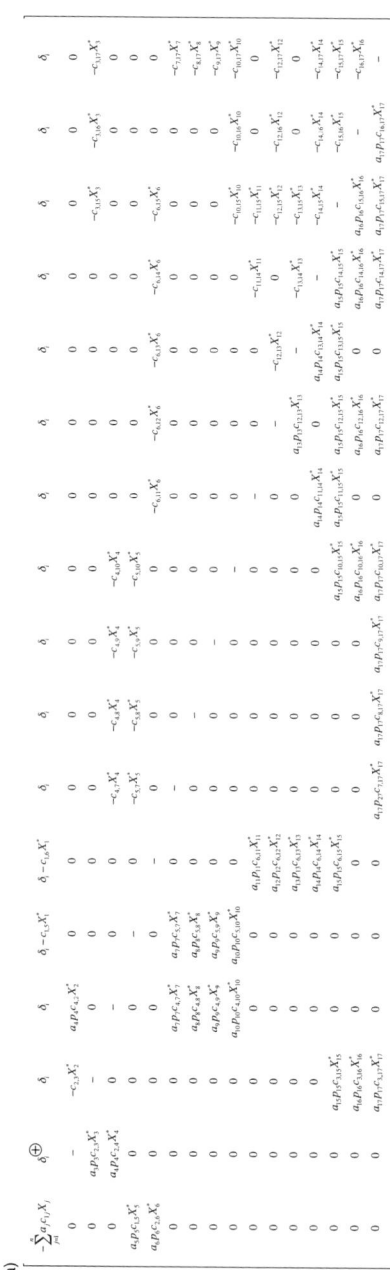

⊕ $\delta_i = \sum_{j=1}^{n}(1-a_{ij})c_{ij}X_j^* + \sum_{j=1}^{n}(1-a_{ji})c_{ji}X_j^* + d_i$

Key to the functional groups of the CPER soil food web

1	Detritus	10	Fungivorous Nematodes
2	Plant Roots	11	Flagellates
3	Phytophagous Nematodes	12	Bacteriophagous Nematodes
4	Mycorrhizal Fungi	13	Amoebae
5	Saprophytic Fungi	14	Omnivorous Nematodes
6	Bacteria	15	Predaceous Nematodes
7	Collembolans	16	Nematode Feeding Mites
8	Cryptostigmatid Mites	17	Predaceous Mites
9	Non-cryptostigamtid Mites		

(b)

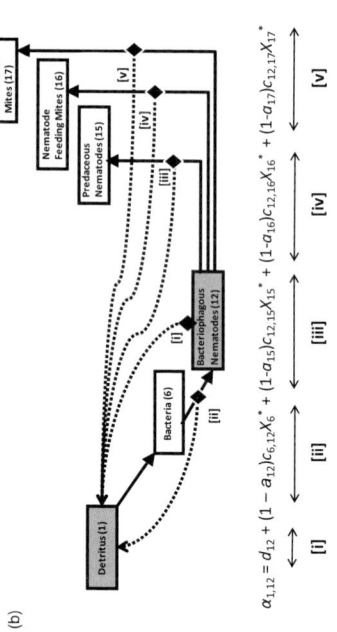

$a_{1,12} = d_{12} + (1 - a_{12})c_{6,12}X_6^* + (1-a_{15})c_{12,15}X_{15}^* + (1-a_{16})c_{12,16}X_{16}^* + (1-a_{17})c_{12,17}X_{17}^*$

Figure 5.1 (a) The Jacobian matrix for the soil food web for Central Plains Experimental Range, Colorado, USA. (b) An example of the components that go into a matrix element depicting the impacts of trophic interactions on the dynamics of detritus. In this example, the detritus (group 1) does not have a direct impact on the dynamics of bacteriophagous nematodes (group 2), as bacteriophagous nematodes do not feed directly on detritus, hence $a_{12,1} = 0$. The bacteriophagous nematodes do impact on the detritus directly ($a_{12,1}$) through their contributions to detritus through death (i), the unassimilated fraction of bacteria they consume (ii), and through the unassimilated fractions of their biomass released by their predators (iii, iv, and v).

With the assumptions outlined above, we can derive values for the off-diagonal interaction strengths from the energy flux rates by equating the death rate of group i due to predation by group j at equilibrium, $c_{ij}X_i^*X_j^*$, to the average annual feeding rate, F_{ij}, and by equating the production rate of group j due to feeding on group i, $a_jp_jc_{ij}X_iX_j$, to $a_jp_jF_{ij}$. We further assume that the theoretical population sizes (X_i^*, X_j^*) at equilibrium are equal to the observed time averaged population sizes, B_i, B_j. The *per capita* effects of predator j on prey i are then represented as follows:

$$\alpha_{ij} = -c_{ij}X_i^* = -\frac{F_{ij}}{B_j} \tag{5.6}$$

and the *per capita* effect of prey i on predator j as

$$\alpha_{ji} = a_jp_jc_{ij}X_j^* = \frac{a_jp_jF_{ij}}{B_i} \tag{5.7}$$

Had we structured our differential equations to include more complex predation rates that lead to more complex functional responses we could have applied the same approach, but would need to have invoked additional assumptions to derive any additional parameters. We will leave this exercise to the readers.

Our Jacobian matrix A is still incomplete in two regards. We have not dealt with the off-diagonal elements associated with detritus and have not completely satisfied how we would estimate the diagonal elements. To address the first of these, we need to recognize that not all elements connected to the detritus pool are purely trophic. Apart from losses due to consumption, detritus increases in abundance as it receives material via the autochthonous return of dead organisms and unassimilated fractions of consumption:

$$\frac{dX_D}{dt} = R_D + \sum_{i=1}^{n}\sum_{j=1}^{n}(1-a_j)c_{ij}X_iX_j + \sum_{i=1}^{n}d_iX_i - \sum_{j=1}^{n}c_{Dj}X_DX_j \tag{5.8}$$

where X_D, X_i, and X_j are the densities of detritus and the living resources and consumers, respectively, R_D is the allochthonous input of detritus, d_i is the specific death rate of the functional groups, a_j refers to the assimilation efficiency, p_j to the production efficiency, and c_{Dj} and c_{ij} to the consumption coefficients between detritus and detritivores and between living resources and consumers, respectively. This implies that there is a series of mostly positive terms in the matrix (effects of all groups on detritus) without negative counterparts. Some of the effects on detritus are negative—i.e., those of the organisms feeding on detritus. Values for these elements are obtained by following exactly the same procedure as that for estimating the trophic interaction strengths by taking the partial derivatives (Equation 5.3) of the differential equation that describes the dynamics for detritus (Equation 5.8) and subsequently use the substitutions as given in Equations 5.6 and 5.7. This leads to the following two equations. For the nondetritivore organisms, the effect on detritus is:

$$\alpha_{Dj} = d_k + \sum_{j=1}^{n}(1-a_j)c_{kj}X_j^* + \sum_{j=1}^{n}(1-a_k)c_{jk}X_j^* \qquad (5.9)$$

and for the organisms that feed on detritus it is:

$$\alpha_{Dj} = d_k + \sum_{j=1}^{n}(1-a_j)c_{kj}X_j^* + \sum_{j=1}^{n}(1-a_k)c_{jk}X_j^* - c_{Dk}X_D^* \qquad (5.10)$$

The Jacobian matrix is still incomplete, as we do not have values for the diagonal terms referring to the self-limitation terms, α_{ii}. The diagonal terms take the form, $-c_{ii}X_i$, except for the diagonal term for detritus. The partial derivative of the detritus equation to detritus itself generates a negative term:

$$\alpha_{DD} = -\sum_{j=1}^{n} a_j c_{Dj} X_j^* \qquad (5.11)$$

This seems counterintuitive but follows from the fact that the input to detritus is independent of the amount of detritus, in contrast to all other groups, in which growth and death rates are density dependent (see also DeAngelis, 1992). An important feature is that we can estimate the diagonal term for detritus in the matrix using the approach discussed above for the off-diagonal terms.

In all other cases, the diagonal terms present a challenge, because unlike the other parameters in the model we have no direct observations or simple means of approximating the values for the self-limitation terms, c_{ii}. In the present approach we assume that the death rates of organisms are in part density dependent and in part density independent. The empirical values for the death rates we use refer to all death except mortality due to predation (Hunt et al., 1987). This implies that the diagonal terms can be modeled as $-s_i d_i$, where s_i is the fraction of the death rate that is density dependent. One approach to estimating the diagonal elements sets the s_i at a fixed value or set of values in some proportion to the magnitude of specific death rates (d_i). An alternative approach estimates the diagonal elements by finding the smallest value of s_i needed to achieve stability, by meeting the condition that the Jacobian matrix is stable, as discussed in Chapter 2. In this case, the value of s_i can serve as an index of critical values of stability. We will use both approaches, and discuss them in greater detail below.

5.2.4 Stability

Stability is assessed by evaluating the eigenvalues of the Jacobian matrix A for the system near equilibirum. If the real parts of the eigenvalues of the matrix are negative, then trophic interactions are stabilizing and the deviations from equilibrium decay. To account for variation in the data used to parameterize the models and the sensitivity of the models to changes in parameter values we evaluated stability using a Monte Carlo approach. Several matrices are constructed based on the connectedness description. The off-diagonal elements of the matrices are parameterized using the approaches discussed above and by selecting values from set intervals, and the eigenvalues are evaluated to assess stability. In the cases in which the diagonal elements, $-s_i d_i$, are selected by fixing s_i, stability is expressed in terms of the probability of the Jacobian matrix being stable based on the Monte Carlo trials. In the example presented below we set s_i at three different levels: 1, 0.1, and 0.01. This produced diagonal elements with values of $s_i d_i = -d_i, -0.1 d_i$, or $-0.01 d_i$. The specific natural death rates (d_i), which can be obtained experimentally or from the literature, include all nonpredatory losses that can be expected in populations in their natural environment. In this approach we assume that s_i is the same for all groups, except detritus, and evaluate the stability of the system by increasing s_i in an incremental manner (de Ruiter et al., 1995). This probability can be used in a comparative way.

Alternatively, the values of s_i, or the average value of s_i from a given Monte Carlo trial, can be used to index stability or sensitivity to change. This approach estimates the diagonal elements by finding the smallest value of s_i needed to achieve stability, by meeting the condition that the Jacobian matrix is stable, as discussed in Chapter 2. All other parameter values are estimated as discussed above. In these cases, s_i can be used in a comparative way.

5.3 A functional food web for the CPER

Figure 5.2 is a skeletal representation of the functional food web for CPER based on its connectedness and energy flux descriptions. When the description is fully parameterized, each functional group would possess estimates of their biomass and the estimates of the diagonal element in the Jacobian matrix that corresponds to self-limitation. The vectors for each trophic interaction in the connectedness description would possess estimates of energy flux from resource to consumers and prey to predator, and of interaction strength associated with the off-diagonal elements of the Jacobian matrix. For the remainder of this chapter we will use the functional food web description and the underlying model as tools to investigate patterns within food webs that are important to stability, the sensitivity of the food web to changes in the densities functional groups and the nature of trophic interactions, and the relationship between energetics and food web structure and stability.

5.3.1 Energy flux and interaction strengths

The representation of the functional food web presented in Figure 5.2 when fully parameterized reveals that the energetic organization of the food web is woven into the Jacobian matrix. The connectedness structure defined the dimension and positioning of nonzero elements of the Jacobian matrix. The distribution of biomass among species, the physiologies of the species and the flux rates among species

104 • Energetic Food Webs

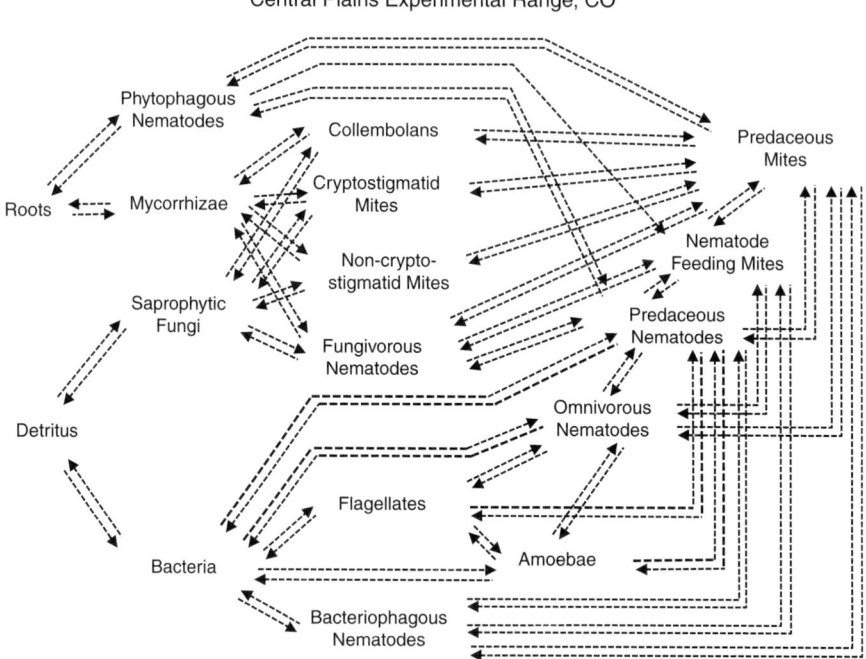

Figure 5.2 A skeletal functional food web description for the functional groups of the soils at the Central Plains Experimental Range (CPER), Colorado, USA, based on its connectedness food web description. The arrows that depicted carbon fluxes in the energy flux description are split to reflect the negative (predator to prey) and positive (prey to predator) interaction strengths (yr^{-1}) associated with each trophic interaction. When fully parameterized, the biomass (kg C ha^{-1}) for each functional group and the magnitudes of the interaction strengths would be included.

within the food web define the magnitudes of the elements. In Figures 5.3 and 5.4 we pull the information from the functional food web, and juxtapose the energy flux associated with each trophic interaction portrayed in the connectedness description of the CPER soil food web and our six other food webs with their corresponding estimates of interaction strength. The trophic interactions are arranged with the resources listed on the left and consumers listed on the right, and with those at the lower trophic positions in the food web at the base and higher trophic positions further up. Three patterns emerge from our estimates of interaction strength, and the relationship between energy fluxes and interaction strengths.

First, the absolute values of the negative α_{ij} elements of the Jacobian Matrix are from one to two orders of magnitude larger than their positive α_{ji} counterparts for a given trophic interaction. We can understand this from Equations 5.6 and 5.7 and from a little information about the energetic structure of the food webs. Assume that the energetic efficiency of the trophic interaction is 10% (i.e., $e_j = a_j p_j = 0.1$) and assume that the food web possesses a steep pyramid of biomass such that B_i at the

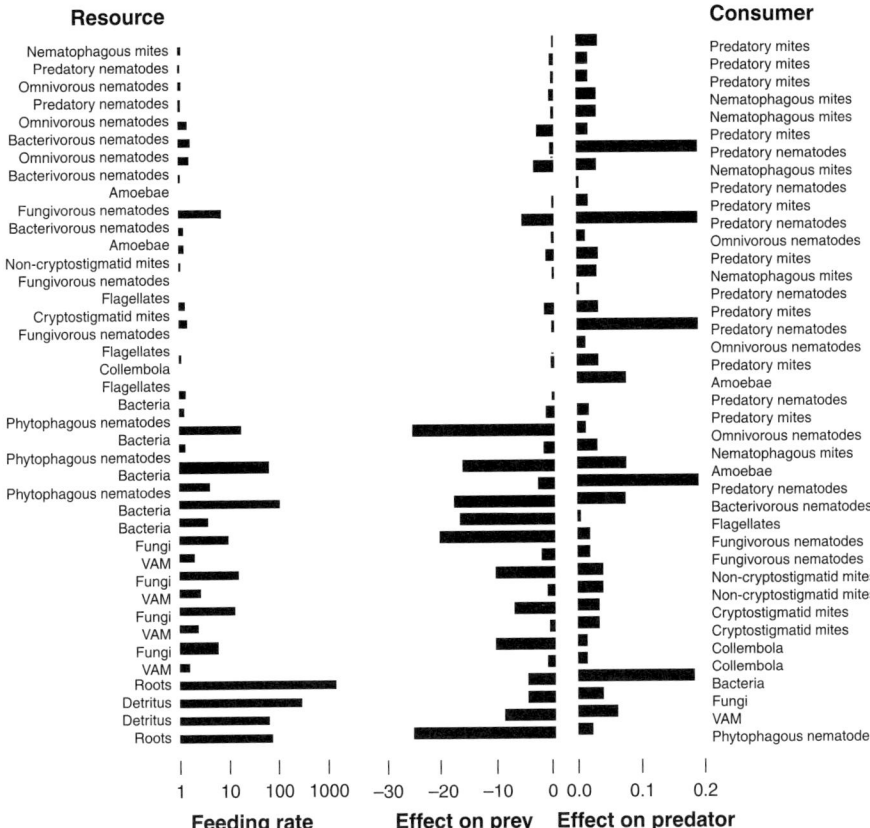

Figure 5.3 The distributions of the carbon flux rates (kg C ha^{-1} yr^{-1}) and interaction strengths (yr^{-1}) with trophic position for functional groups of the Central Plains Experimental Range (Colorado, USA).

resource trophic level is 10 times the biomass of B_j at the consumer trophic level (i.e., $B_i = 10B_j$). At this extreme, $\alpha_{ij} = 100\alpha_{ij}$.

Second, the relative magnitude of the negative elements to those of their positive element counterparts tends to be large at lower trophic positions and diminishes with increased trophic position. Like the relationship between the paired positive and negative elements discussed above, this relationship is also a reflection of the energetic organization of the food web. From Equation 5.6 we can see that the observed decline in the magnitude of the negative elements is a direct result of the distribution of prey biomass with trophic position and the flux rate from prey to predator. In fact, the distribution of the negative elements mirrors that of the energy flux estimates (F_{ij}) scaled by the consumer biomass (B_j). Likewise, the observed pattern in the magnitudes of the positive elements is

(a) Lovinkhoeve Experimental Farm, NL – Integrated Management Winter Wheat

a result of the physiologies of the consumers through their energetic efficiencies ($a_j p_j$) and the relationship of prey and resource biomass (B_i) to consumers and predators (B_j).

A final observation that is related to the first two is that the magnitude of the negative elements (i.e., the effects of predators on prey) is related to the number of prey types on which the predator feeds (Figure 5.5). This relationship could be expected, as the elements are defined in part by the flux rates, which if they were parsed between many prey would on average be lower than if parsed by one or fewer prey, all else being equal. The interesting feature is that the number of prey types per predator increases with the trophic height of the predator, i.e., consumers at lower trophic positions have fewer prey types than those at higher trophic positions. If this pattern is real and not an artifact of the description or idiosyncratic to the CPER and soil food webs, it indicates that the interactions within the food web are more siloed at the base and reticulated at the top.

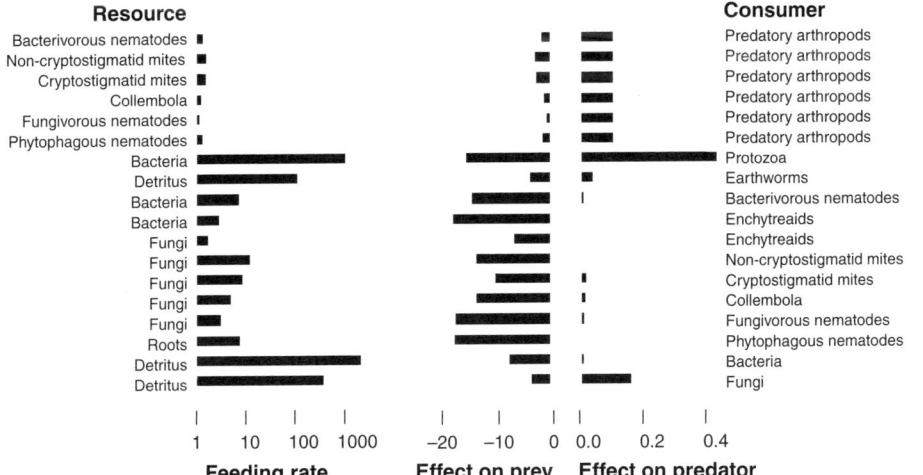

Figure 5.4 *Continues overleaf.*

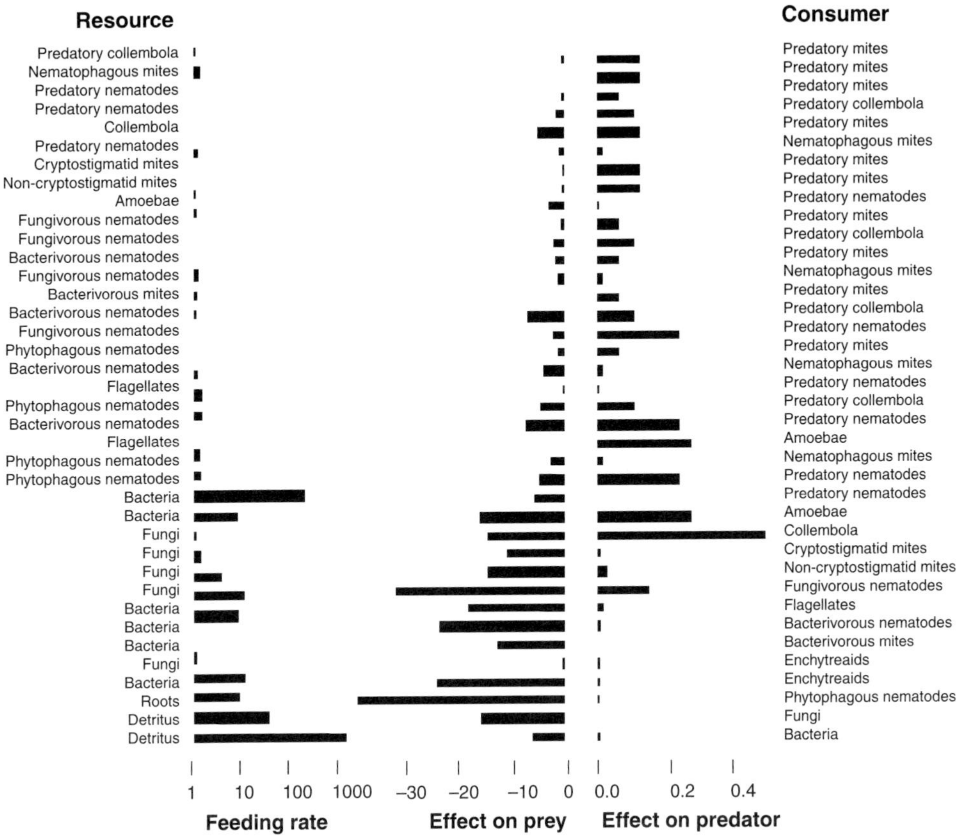

108 • *Energetic Food Webs*

(d) **Horseshoe Bend Experimental Farm, GA – Conventional-Tillage Mixed Crop Rotation**

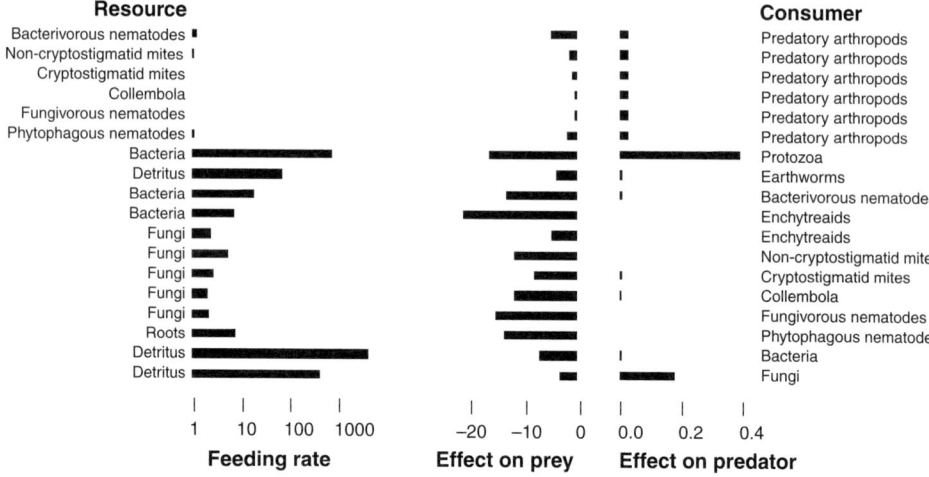

(e) **Kjettslinge Experimental Farm, S – Barley without Fertilizer (B0)**

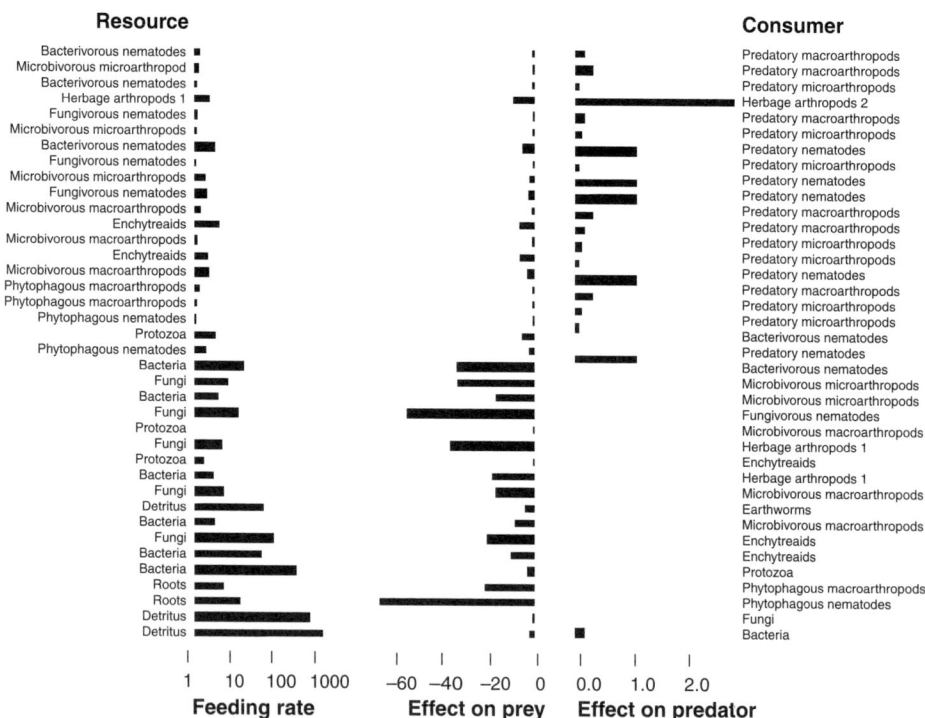

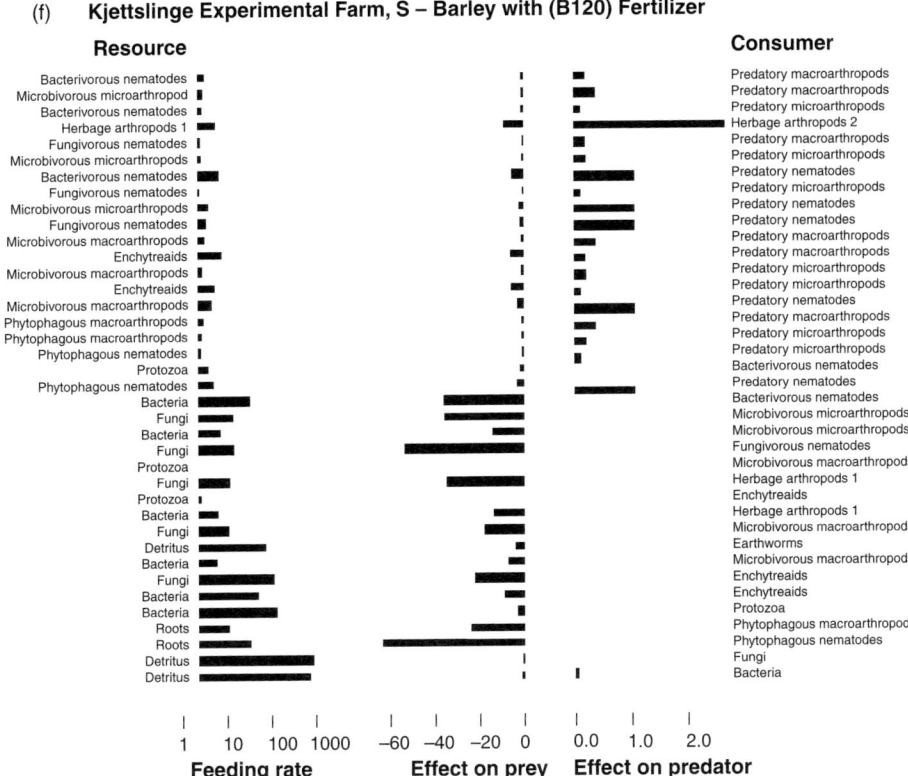

Figure 5.4 The distributions of the carbon flux rates (kg C ha^{-1} yr^{-1}) and interaction strengths (yr^{-1}) with trophic position for the six additional soil food webs. (a) Integrated management winter wheat, Lovinkhoeve Experimental Farm (Marknesse, The Netherlands); (b) conventional management winter wheat, Lovinkhoeve Experimental Farm; (c) no-tillage mixed crop rotation, Horseshoe Bend Research Site (Athens, Georgia, USA); (d) conventional-tillage mixed crop rotation Horseshoe Bend Research Site; (e) barley without fertilizer (B0), Kjettslinge Experimental Farm (Uppsala, Sweden); and (f) with fertilizer (B120), Kjettslinge Experimental Farm.

If we look at the information within the connectedness, energy flux, and functional food web descriptions in their totality, additional information and patterns emerge. Like the distribution of biomass within systems, the relationships of organisms with different physiological types, body sizes, home ranges, and feeding behaviors do not appear to be distributed at random within food webs, but rather patterned in some way. Predators tend to be larger than their prey, and possess larger home ranges than their prey (Figure 5.6). Although we relied on type I functional

110 • Energetic Food Webs

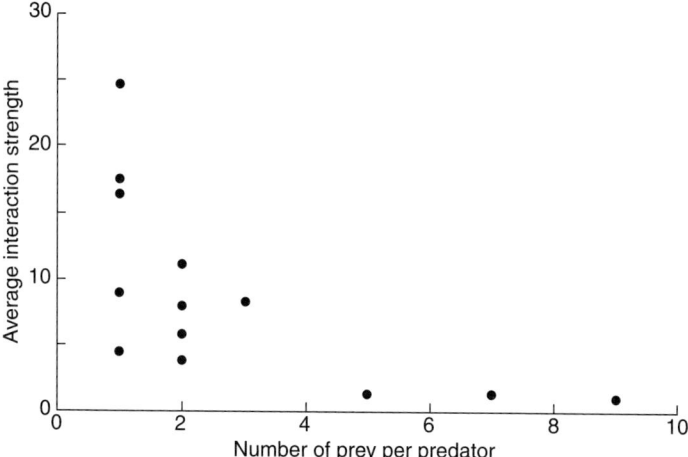

Figure 5.5 The relationship between the effect of a predator on prey, y, presented as the average of the absolute values of the estimates obtained from Equations 5.6 and 5.7, and the number of prey per predator, x, for the functional groups within the CPER soil food web ($y = 14.1e^{-0.347x}$, $r^2 = 0.679$).

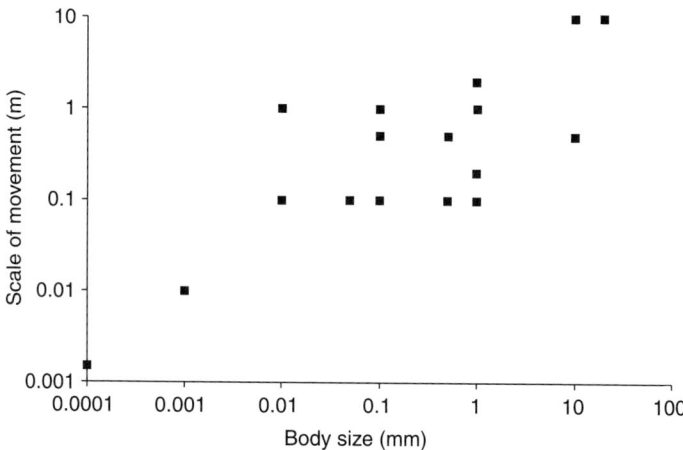

Figure 5.6 The relationship between the home ranges and body sizes of the functional groups within the Central Plains Experimental Range, Colorado, USA, soil food web.

responses in our presentation, a review of the literature suggests that taxa at the trophic interactions involving higher trophic levels might better be characterized using type III functional responses (Table 5.2). These are important points, as all affect to some degree the structure of the Jacobian matrix. As discussed in Chapter 2, both the positioning of nonzero elements and their magnitudes within the Jacobian

Table 5.2. Functional responses of various taxa.

Taxon	Functional Response (Holling type)	References
Protozoa		
Ciliates	II	Taylor (1978), Fenchel (1980)
Protostomes		
Rotifers	II	Nandini and Sarma (1999), Mohr and Adrain (2000)
Nematodes	I, II	Nelmes (1974)
Mollusks		
Gastropods	I, II	Tahil and Juinio-menez (1999)
Cephalopods	I, II, III	Borer (1971), Luck et al. (1979), Mather (1980), Marquez et al. (2007)
Tardigrades	II	Hohberg and Traunspurger (2005), Jeschke and Hohberg (2008)
Arthropods		
Crustaceans	1, II, III	Pichlova and Vijverberg (2001), Taylor and Collie (2003), Muschiol et al. (2008)
Collembola	II	Melville (2003)
Insects	II, III	McCoull et al. (1998), Fonseca et al. (2000), Greenberg et al. (2001), Jalahi et al. (2010)
Deuterostomes		
Echinoderms	I, II	Leonard (1989)
Chordates		
Amphibians	II	Viertel (1992)
Fish	I, II	Gulbrandsen (1991), Hart and Klumpp (1996)
Birds	II, III	Neilsen (1999), Redpath and Thirgood (1999)
Mammals	II, III	Marshal and Boutin (1999), Sundell et al. (2000)

matrix are important determinants of the dynamics and stability of the system. In Part II of the book we will discuss the implications of these patterns in more detail.

5.3.2 Mortalities from predation

The functional food web description and its underlying model can be used to analyze the extent to which the species affect one another through trophic interactions. Recall that the model takes the effects of trophic interactions into account by including death due to predation in the estimates of the death rates of the organisms. This provides a straightforward way to estimate and compare the fractions of mortality that are due to predation from the natural death rates. This gives an indication of the impact of predation on the dynamics of the populations, but it also indicates the degree that the feeding rates and mineralization rates of the functional groups are influenced by predation (Moore and Hunt, 1988).

112 • Energetic Food Webs

Table 5.3. Fraction of mortality due to predation, and the direct and indirect contributions of the functional groups to N-mineralization in the food webs from the CPER and the Lovinkhoeve conventional farming system (LH–CF).

	CPER				Lovinkhoeve			
	A	B	C	D	A	B	C	D
Microorganisms								
Bacteria	33	52.47	–	–	43	51.57	–	–
Fungi	34	8.79	–	–	79	1.05	–	–
Protozoa								
Amoebae	1	12.76	18.52	(1.45)	0	42.87	62.63	(1.46)
Flagellates	4	0.56	0.81	(1.45)	12	2.23	3.13	(1.41)
Nematodes								
Herbivores	53	1.30	–	–	70	0.15	–	–
Bacteriovores	31	14.94	21.41	(1.43)	48	1.13	1.43	(1.26)
Fungivores	38	0.25	0.71	(2.78)	57	0.08	0.12	(1.50)
Predators	10	0.40	8.54	(21.11)	26	0.09	0.90	(14.29)
Microarthropods								
Fungivorous Collembola	8	0.17	0.60	(3.47)	26	0.24	0.34	(1.37)
Cryptostigmatid mites	11	0.42	1.38	(3.22)	35	0.003	0.004	(1.37)
Non-cryptostigmatid mites	8	0.50	1.76	(3.48)	26	0.012	0.02	(2.00)
Nematophagous mites	8	0.06	1.24	(21.60)	26	0.002	0.03	(20.00)
Predatory mites	0	0.06	1.55	(22.50)	0	0.03	0.63	(20.42)

A, fraction of mortality due to predation (%).
B, relative contribution to N-mineralization (%).
C, relative reduction in N-mineralization after group-deletion (%).
D, ratio between the decrease in N-mineralization (kg ha^{-1} yr^{-1}) and the direct contribution to N-mineralization (kg ha^{-1} yr^{-1}).

The fractions of mortality of functional groups due to predation and the direct and indirect contributions of the functional groups to N-mineralization are summarized in Table 5.3 for two food webs: those of the CPER and those of the Lovinkhoeve Experimental Farm (conventional practice). For the CPER food web the fraction of mortality due to predation is approximately 30% for the microbial populations, whereas for protozoa it amounts to less than 5%. Among nematodes, the fraction ranges from 10% for predators to 22–53% for the other groups, and among microarthropods the fraction ranges from zero for predaceous mites to 11% for oribatid mites. In the web from the Lovinkhoeve, as an example of the agricultural sites, the death rates due to predation are generally higher than those in the CPER web: bacteria 43%, fungi 79%, nematodes 25–70%, and microarthropods 0–35%. These estimates of predation pressure indicate that trophic interactions may have a considerable impact on the functioning of many functional groups and on their contributions to mineralization and soil mineralization as a whole, which are disproportionate to their relative biomass and predation rates. This latter point echoes those made by others (Paine, 1980, Polis and Strong, 1996).

5.3.3 Functional groups and food web stability

Here we will try to identify which species or interactions play a key role in stability. For this analysis we will carry out a press-perturbation analysis (sensu Yodzis, 1988) in order to see which perturbations are likely to influence food web stability. In many respects this corresponds in a way to the single press perturbations that have been studied experimentally in the field (e.g., Paine, 1980, Paine, 1992, Wootton, 1994) as well as to the theoretical exercises using models (Yodzis, 1988, McCann et al., 1998). The press perturbations are applied by changing the values of one particular pair of the matrix elements. These changes are modeled by sampling values from the interval $[0, 2\alpha_{ij}]$. In this analysis the values of s are set at a value of 1% above their critical values; this means that the matrix is stable, but relative near the point that they become unstable. In doing so we make the matrix relatively sensitive to parameter change. We will henceforth refer to the effect of changing the values of the matrix elements on stability as the "impact on stability." The impact on stability is calculated as the probability, based on the Monte Carlo trials, that the matrix becomes unstable. We conducted these press perturbations on all trophic interactions in all seven of our soil food webs. Figure 5.7 depicts the impact on stability for each predator–prey interaction when ordered by their trophic positions.

The results reveal two important observations about the relationship between the impact of an individual species on the stability of a food web and its contribution to energy flow, as gauged by its feeding rates. The first is that the impact on stability can vary strongly between interactions; some have no effect or minimal effect, whereas others have relatively large effects—nearly 50% in some instances. Comparing the different webs, there seems to be no clear pattern in which interaction will be important to stability (Figure 5.7). In most cases, particular interactions have a similar impact in different webs, but in a few cases it also appeared that interactions have a strong impact on stability in one web and a small impact in another web. For example, predatory mites feeding on Collembola have a small impact in the CPER web and a large impact in both food webs for the Lovinkhoeve, but these impacts on stability are not correlated with feeding rate. Both these observations are in agreement with the experimental observations by Paine (1980) and reviews by Polis and Strong (1996), in that some interactions are seen as key components in food web functioning, whereas others are unrelated to energy flux.

5.4 Summary and conclusions

We generated functional webs using the structure and information from connectedness and the energy flux food web descriptions. The functional food webs included: (1) the interaction strengths, or the per biomass effects of the functional groups on the dynamics of the other functional groups in the food web, (2), the fractions of mortality due to predation, and (3) the impact of each interaction on food web stability. The task required that we develop a set of differential equations that

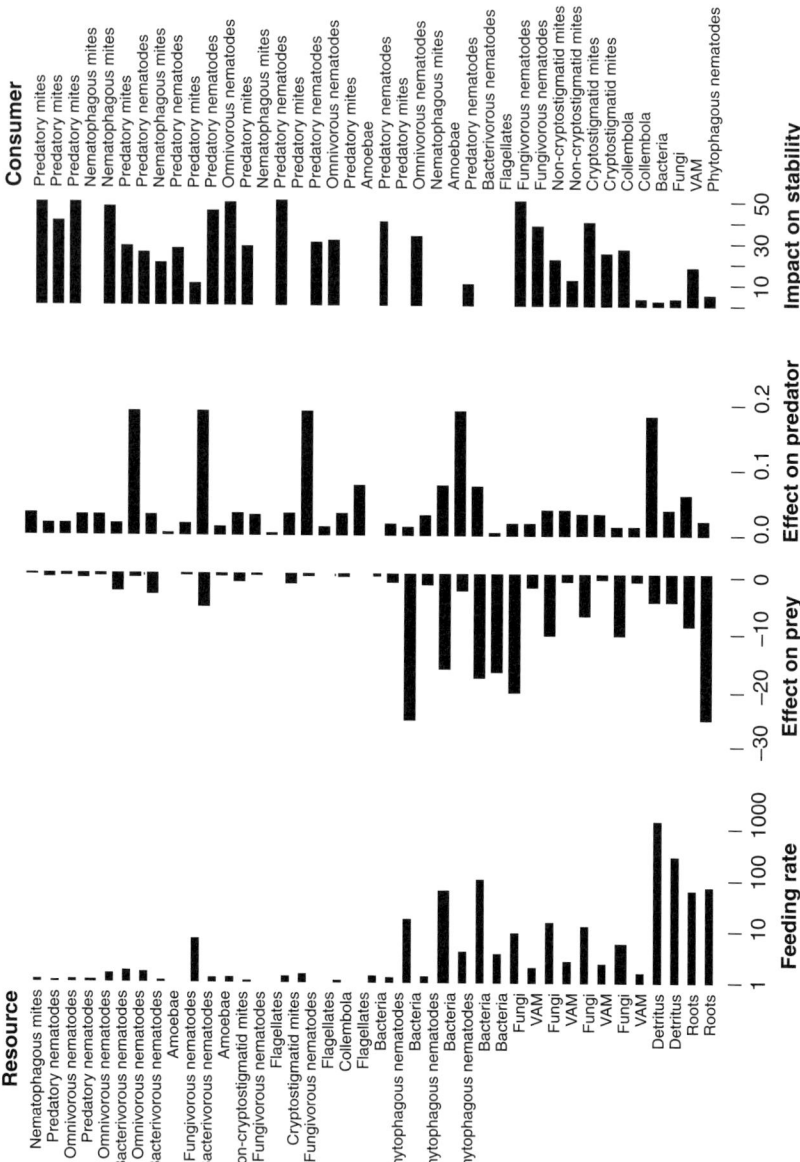

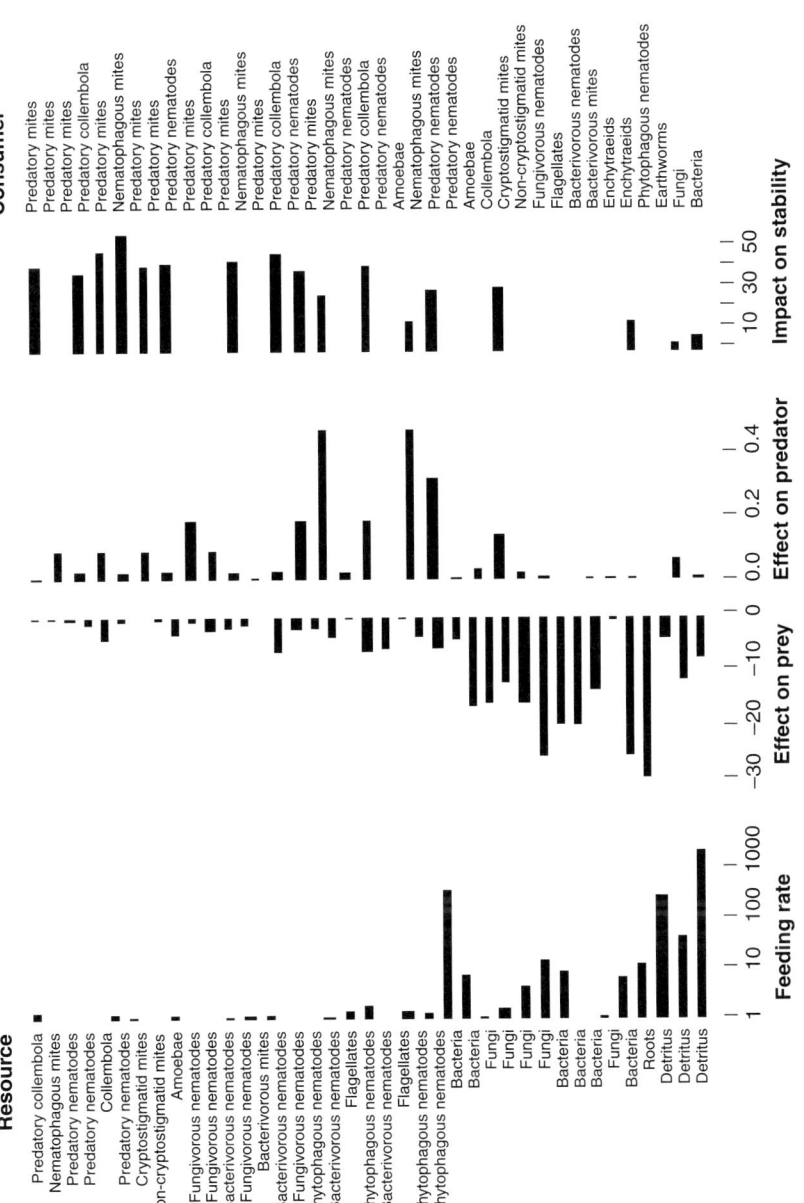

Figure 5.7 *Continues overleaf.*

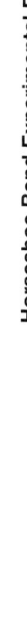

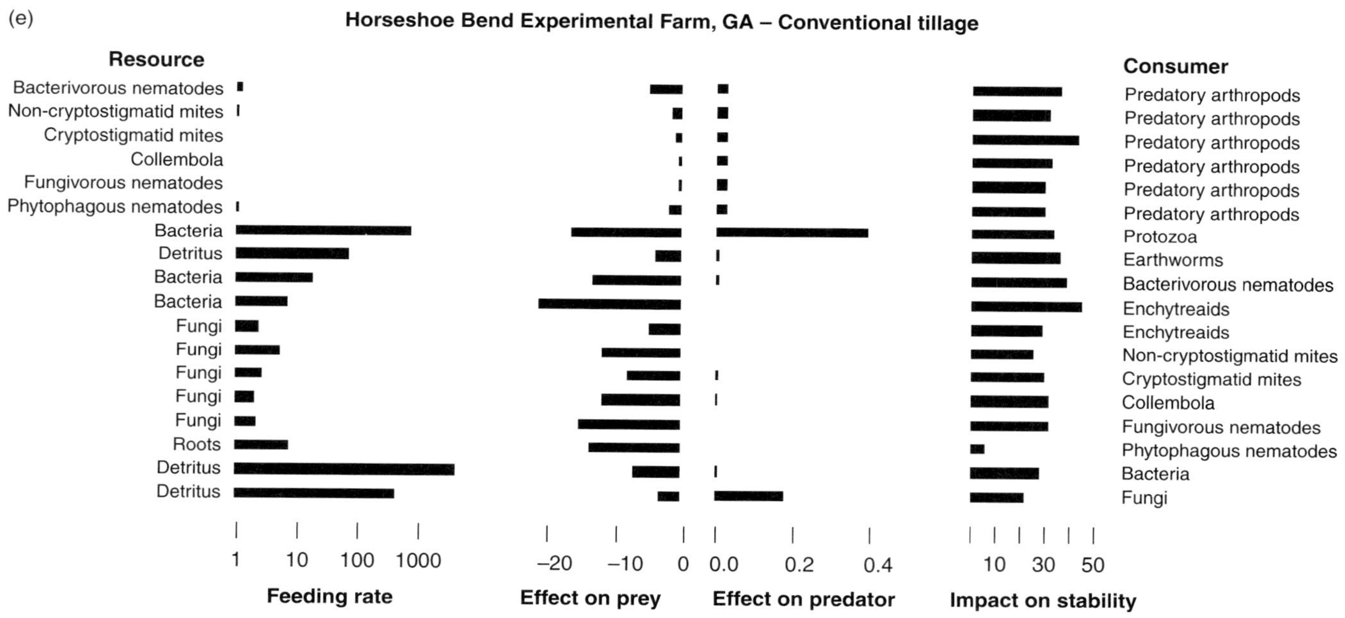
(e)
Horseshoe Bend Experimental Farm, GA – Conventional tillage
Resource
Bacterivorous nematodes
Non-cryptostigmatid mites
Cryptostigmatid mites
Collembola
Fungivorous nematodes
Phytophagous nematodes
Bacteria
Detritus
Bacteria
Bacteria
Fungi
Fungi
Fungi
Fungi
Fungi
Roots
Detritus
Detritus
1
10
100
1000
Feeding rate
–20
–10
0
Effect on prey
0.0
0.2
0.4
Effect on predator
10
30
50
Impact on stability
Consumer
Predatory arthropods
Predatory arthropods
Predatory arthropods
Predatory arthropods
Predatory arthropods
Predatory arthropods
Protozoa
Earthworms
Bacterivorous nematodes
Enchytreaids
Enchytreaids
Non-cryptostigmatid mites
Cryptostigmatid mites
Collembola
Fungivorous nematodes
Phytophagous nematodes
Bacteria
Fungi

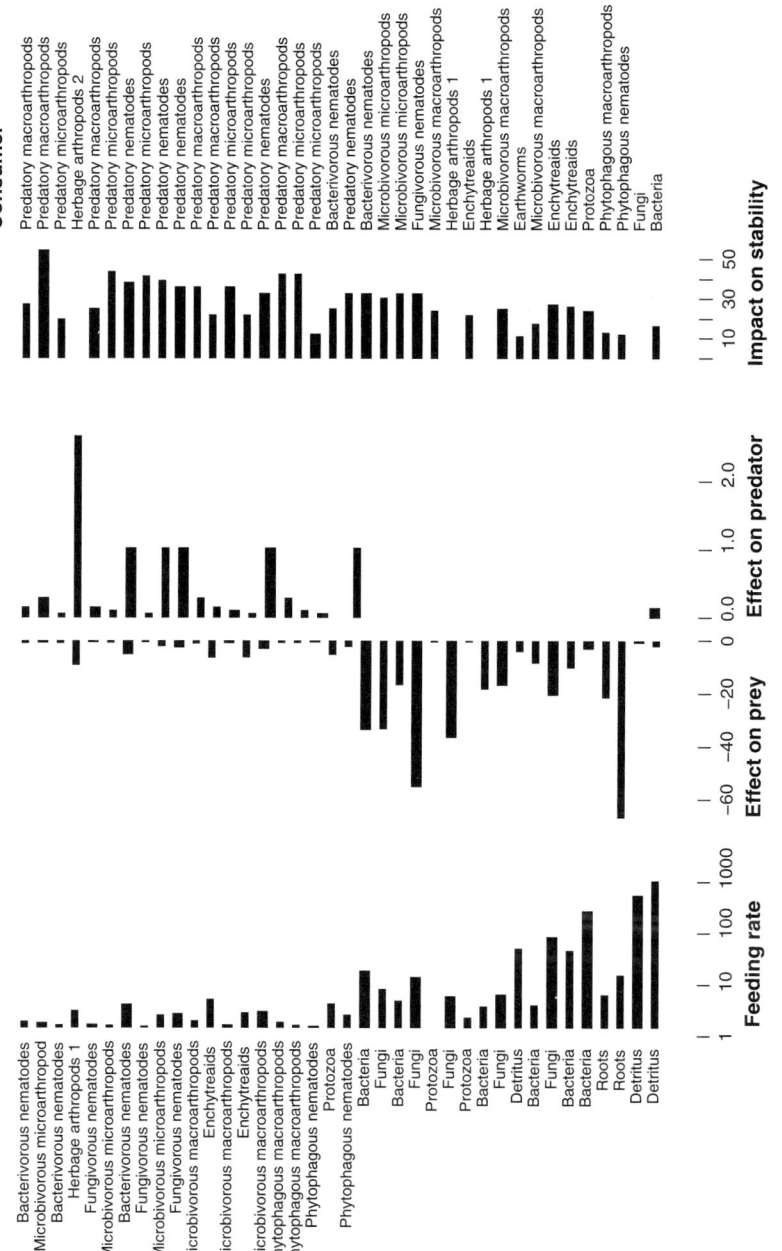

Figure 5.7 *Continues overleaf.*

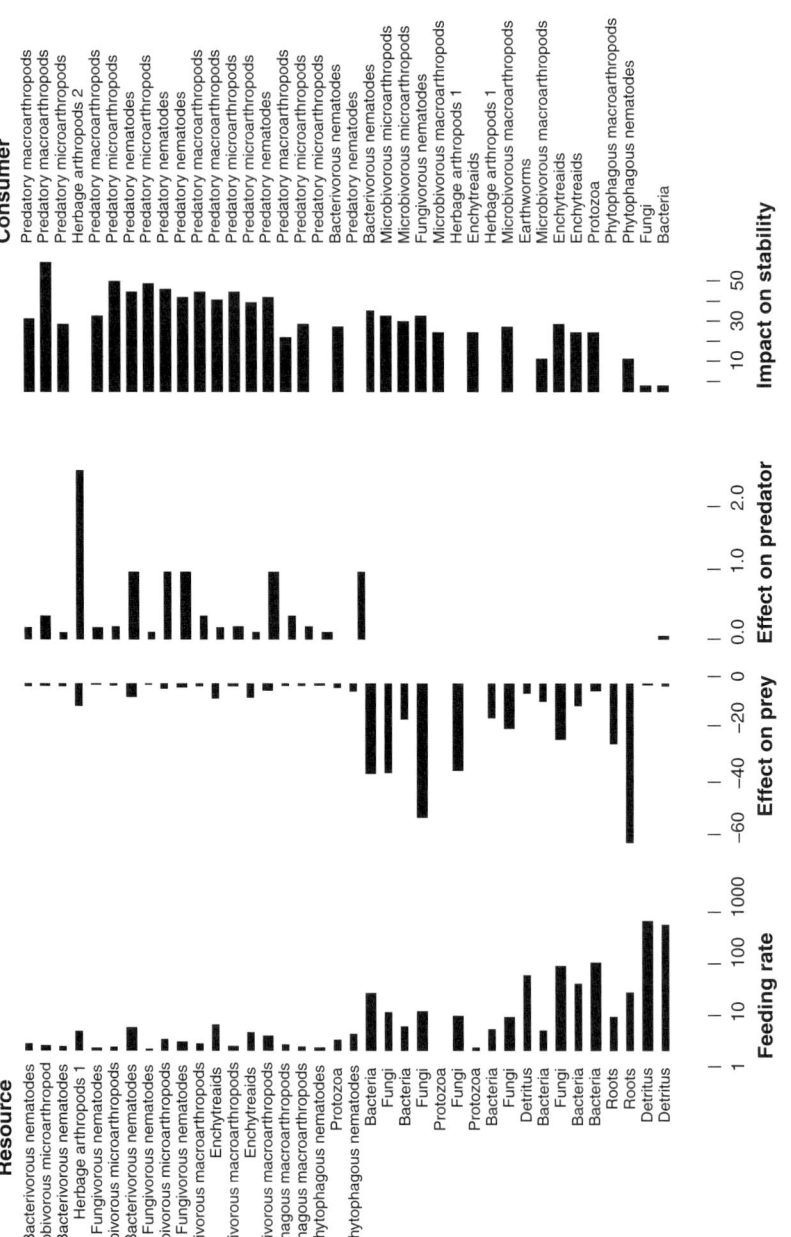

Figure 5.7 Impacts on stability for each trophic interaction ordered along trophic position. (a) CPER (Nunn, CO); (b) Integrated management winter wheat, Lovinkhoeve Experimental Farm (Marknesse, The Netherlands); (c) conventional management winter wheat, Lovinkhoeve Experimental Farm; (d) no-tillage mixed crop rotation, Horseshoe Bend Research Site (Athens, Georgia, USA); (e) conventional-tillage mixed crop rotation Horseshoe Bend Research Site; (f) barley without fertilizer (B0), Kjettslinge Experimental Farm (Uppsala, Sweden); and (g) with fertilizer (B120), Kjettslinge Experimental Farm.

described the trophic interactions among functional groups. Equations were developed to accommodate the dynamics of primary producers, the dynamics of detritus, and the dynamics of consumers. We opted to use familiar equations of the Lotka–Volterra form with predation terms that generated type I functional responses. The approach we developed could have involved more complex predation. From the set of differential equations we estimated the interaction strengths using the criterion developed by May (1973). By equating the predation rates from the equations to the energy flux rates we were able to derive the interaction strengths.

Several energy-related patterns emerged from our analysis of seven soil food webs that have direct impacts on the magnitudes of the elements (i.e., interaction strengths) within the Jacobian matrix. The distribution of biomass, flux rates, and paired interaction strengths were not randomly distributed within the food webs. The pattern in the distribution of interaction strength when arranged by increasing trophic position displayed a distinct asymmetry. The negative effects of predators on their prey were large at the base the web and diminished at higher trophic positions. This pattern mirrored the pyramid of biomass and flux rates with increased trophic position. The positive effects of prey on their predators increased from the base of the food web to the upper trophic levels.

We ended by assessing the impact of individual trophic interactions on the stability of the food webs. We found no relationship between the biomass of the species or the flux rates between the species involved in the trophic interactions and the impact that the interactions have on the stability of the food web. This would seem to bolster the argument made that material fluxes (read energy flux) within the web do not have a significant impact on stability. We will argue that this absence of a correlation between feeding rates and impact on stability does not mean that material and energy flux, interaction strength, and stability are unrelated. To the contrary, they are inextricably interrelated, as the stability of the web depends directly on the patterning of interaction strengths, which are derived from energy flux. This is a topic we will study in more detail in Part II.

Part II

The dynamics and stability of simple and complex communities

In Part II we explore different aspects of the energetic organization of communities and their responses to enrichment and disturbances. Two energetic properties of community organization include the distribution of biomass within the community and the flow of matter and energy within and through the community. We ask whether there are patterns in these energetic properties, and if so, are they related to community dynamics and stability? We will start with a study of the soil food webs presented in Chapters 3–5. In some respects these represent communities with complex architectures: arguably not the most tractable of systems to study. The complexity in these systems provides enough degrees of freedom—or, put another way, is rich enough to provide as study of patterns. Our method will be to use the complex models to generate patterns, and then deconstruct the patterns in more tractable pieces with simpler models.

In Chapter 6 we investigate several interesting patterns in the soil food webs that involve the biomasses of functional groups, the fluxes of matter between functional groups, and the paired interaction strengths associated with each trophic interaction when arranged by trophic position. The energy flux food web descriptions of our soil food webs possessed the familiar pyramid of biomass distribution when the biomasses of functional groups were arranged by increased trophic position. Not surprisingly, this pattern was mirrored by the estimates of flux rates of matter, given their reliance on the biomasses of the interacting groups, with high flow rates occurring at lower trophic positions and lower flow rates at higher positions. The trophic interactions are arranged in groups or assemblages of species, termed energy channels, based on dominant flows of energy and matter. The Jacobian matrices of the functional descriptions for these same food webs possessed an interesting asymmetric arrangement in the weighting of the paired matrix elements with trophic position for each interaction.

We altered the paired interaction strengths of individual trophic interactions within the Jacobian matrix and then all the paired interaction strengths within the matrix and followed their impacts on the dynamics and stability of the food webs. Manipulating the magnitudes of the interaction strengths of individual interactions changed the steady-state biomasses of the populations and the energy flux rates from the prey species to the consumer species. Manipulating all interactions within the food web effectively moves biomass from lower to upper trophic positions and blurs the distinction between energy channels, making energy flux more homogeneous within the web.

The impacts of these manipulations on the dynamics and stability of the food web reveal a tension between the importance of individual species and the importance of system-level energetic organizational structure. On the one hand, certain species are shown to greatly affect the stability of the system, whereas others are benign in their impact. Regardless of their magnitudes, these impacts do not correlate with the magnitudes of the biomasses of the species involved or the energy fluxes among the species. On the other hand, manipulating all interactions within a food web invariably affected its dynamics and stability, as models parameterized with field data are more stable than their manipulated counterparts.

Using the connections between interaction strengths and energetics that we established in Part I, we show that the manipulations used in the modeling exercise not only repositioned the interaction strengths, but also altered the energetic organization of the community. Biomass and energy flux in the models following the manipulations was positioned at the upper trophic positions, and energy flux through energy channels was homogenized in ways that were found to be destabilizing. Under the steady-state assumptions, the new energetic configuration implies an increase in productivity and flux rates. These changes could be induced solely or by a combination of enrichment (detritus inputs or primary productivity), and increased reproductive capacity, energetic efficiency, or consumption at the base of the food web coupled with the imposition of a bottleneck at the top of the food web. In other words, individual species possess the capacity to both accelerate productivity and the flux of biomass through the food web and also ameliorate these destabilizing factors through self-regulation, which impedes their growth rates and dampens the flow of material through the web. The manipulations exposed these capacities and their consequences on the energetic organization, dynamics, and stability of communities.

In Chapter 7 we explore how enrichment (aka productivity) at the base of the food web in the form of inputs from primary production or from detritus affects the vertical distribution of biomass within a community and its dynamics. This topic has been discussed in the literature in terms of the length of food chains and trophic levels in the lexicons of community ecology and ecosystem ecology, respectively. Here we develop sets of simple and complex model food chains based on primary producers or detritus and study their structure and dynamics along a gradient of productivity. Four important concepts emerge. First, productivity needs to be sufficiently large to support consumers at densities greater than zero over time, beyond which the biomass in the system increases and accumulates at the upper trophic levels. In other words, the feasibility of a system ($F_{[SYS]}$) is governed in part by the level of productivity ($F_{[NPP]}$). Second, as productivity increases, the dynamic properties of the system change as well. In some cases the models remain stable but develop oscillations on their return to a steady state. In other cases, the asymptotic behavior of the models transitions from stable steady state to limit cycles and even chaos. The capacity of the system to cope with these transitions leads to the argument that the feasibility of the system ($F_{[SYS]}$) is also governed in part by dynamics ($F_{[\lambda]}$). Third, the primary-producer-based models and detritus-based models are similar in their responses to increased productivity, although the points along the productivity gradient where the transitions in structure and dynamics occur may differ. Finally, the physiologies and life histories of the individual species, and degree of intraspecific competition within the populations of individual species, and the positions of individual species within the community can alter the transition points in structure and dynamics along the productivity gradient. This is an important point to scaling that can explain the ambiguities that arise when seeking patterns in community architecture and dynamics, when the life histories, turnover rates, energetic efficiencies, and sizes (i.e., foraging patterns, movement patterns, and home ranges) of constituent species differ.

In Chapter 8 we discuss compartments or assemblages of species within communities and their impacts on the distribution of biomass, material and energy flow, dynamics, and stability. Theoreticians have demonstrated that models of a given dimension and complexity that possess a compartmentalized structure are more likely to be stable than their randomly constructed counterparts. We identify energy channels as energy-based compartments within food webs that describe the flow of matter and energy through food webs. Energy channels originate with basal resources, are comprised of consumers that feed on the resources and their consumers, and are linked by top predators. We study how the dynamics and stability of the food web are affected by the flux rates through energy channels and the preferences and physiologies of the predators that link them.

In Chapter 9 we knit together our findings in the previous chapters to explore how the energetic organization of communities relates to species diversity. We focus on the observed patterns of species richness along latitudinal gradients at local, regional, and continental scales as a way to broach the topic. The often-cited pattern is the hump-shaped curve, where diversity for a given taxon is highest at mid-latitudes on a continental scale. Putative factors used to explain this pattern include: (1) productivity and available production for consumption by higher trophic positions, (2) the spatial extent (area) of mid-latitudes and the home ranges and movement patterns of consumers, and (3) time, evolution, and age of mid-latitude landmasses. Subsequent work demonstrated the scale-dependent nature of the relationship, but the factors cited above come into play nonetheless.

We expand and formalize our definition of feasibility to include both a structural and a dynamic component. The feasibility of a system ($F_{[sys]}$) includes the initial condition that there be sufficient energy inputs so that all population densities remain greater than zero ($F_{[NPP]}$) and the condition that the resulting dynamics are likely to persist in time ($F_{[\lambda]}$). The feasibility of a system is defined as the intersection of the structural and dynamic components, where $F_{[sys]} = F_{[NPP]} \cap F_{[\lambda]}$. This relationship, coupled with our results on energy channels, provides the basis for a model of community growth that optimizes species diversity at intermediate levels of productivity.

6

Energetic organization and food web stability

6.1 Introduction

We ended Part I with an analysis of our soil food webs that revealed apparently conflicting ideas regarding the role of energetics in food web stability. On the one hand we find that there is no correlation between the magnitudes of the individual feeding rates between groups within the food web and the role groups play in the stability of the food web. This seems to confirm the results from empirical studies such as those of Paine (1980, 1992) and reviewed by Polis and Strong (1996). On the other hand, the Jacobian matrix and the arrangement of its elements embody the stability character of a system (May, 1973). In our calculus, the interaction strengths are the elements of the Jacobian matrix, and they derived from energy fluxes within the food web. Energy-based patterns emerged that map to the magnitudes of the elements and arrangement of the elements within the Jacobian matrix. This would indicate that at the system level, energetic properties of the food webs might govern food web stability. Expecting that individual feeding rates, important components of the system, would correlate directly to stability takes a decidedly additive view of the potential relationship and ignores the concept of emergence. In other words, from a whole-system perspective, should we expect individual feeding rates within a food web to correlate with food stability? A better question might be: To what extent does the patterning of the distribution of biomass and the feeding rates, which form the energetic basis of community architecture, affect dynamics and stability?

In this chapter we will disentangle the observed relationships between the architecture of the food webs, which includes the distribution of biomass and flux rates, and stability, using the tools developed in Part I. We will study the stability of our models by evaluating the eigenvalues of the Jacobian matrices derived from the functional food web descriptions that are parameterized with the estimates of the biomasses, feeding rates, and energy conversion efficiencies of the species or functional groups. We will then link the observed distributions in biomass and fluxes, using theory on matrix stability, in order to see what properties of the observed patterning in interactions strengths may be important to stability. We then take the analysis a step further to try to resolve the conflict discussed above, to show a relationship between the observed patterns in interaction strengths and numerical loops in the Jacobian matrices. This latter finding is used to interpret the

6.2 Energetic organization and stability

The functional descriptions of our soil food webs revealed an asymmetric distribution of the paired interaction strengths (α_{ij}, α_{ji}) associated with each trophic interaction with increased trophic position (see Figures 5.4 and 5.5 in Chapter 5). We present a conceptualization of this pattern and related patterns in the distributions of biomass of functional groups and the flux rates between the functional groups in Figure 6.1. We altered the magnitudes of individual pairs of interaction strengths for each trophic interaction and gauged the impact that the disturbance had on the stability of the food webs. We found no relationship between the energy flux associated with each trophic interaction and the impact of disturbing it on stability. Here we will focus on the overall configuration. What would happen if we preserved the connectedness structure as above, but instead of altering the interaction strengths associated with a single trophic interaction, we altered the energetic organization of the food web?

For this analysis we constructed two classes of matrices. The first class of matrix consists of the Jacobian matrices in which the values are derived from the empirical data; we will refer to this class as the "real" food webs. The second class is constructed by randomly permuting pairs of interaction strengths (α_{ij}, α_{ji}) in the "real" Jacobian matrices repeatedly so that all pairs of elements are permuted at least one time. This results in the matrices that have the same connectedness and sign-structure (0, +, −) as the "real" food webs, and also the same collection of interaction strength values, although in different positions (Yodzis, 1988). We will refer to this class as the "randomized" food webs.

The stability of the food webs is assessed in the usual manner by evaluating the eigenvalues of the Jacobian matrix A. If all the eigenvalues of the Jacobian matrix A are negative, the matrix is stable, otherwise it is unstable. The analyses involved Monte Carlo trials, which randomly sampled the matrix elements from a uniform distribution with intervals [0, $2\alpha_{ij}$], in which α_{ij} is the value as derived from the observations and flux rates, where $\alpha_{ij} = -c_{ij} X_i^* = F_{ij}/B_j$ (see Equations 5.4 and 5.6) and $\alpha_{ji} = a_j p_j c_{ij} X_j^* = (a_j p_j F_{ij})/B_i$ (see Equations 5.5 and 5.7), where X_i^* and X_j^* are the theoretical biomasses of the prey and predators at equilibrium, B_i and B_j are the field estimates of the biomasses of the prey and predators at equilibrium, F_{ij} is the carbon or nitrogen flux rate from prey to predator, and a_j is the assimilation efficiency, p_j is the production efficiency, and c_{ij} is the consumption coefficient for the predator. The diagonal terms were modeled as $-s_i d_i$, where s_i is the fraction of the death rate, which is density dependent (see Chapter 5). For this analysis the s_i were set at various levels of magnitudes proportional to the specific death rates (d_i), with $s_i = 1, 0.1$, and 0.01, and hence $\alpha_{ii} = -d_i, -0.1 d_i$, and $-0.01 d_i$, for all groups equally. Stability is expressed in terms of the probability that the matrix is stable.

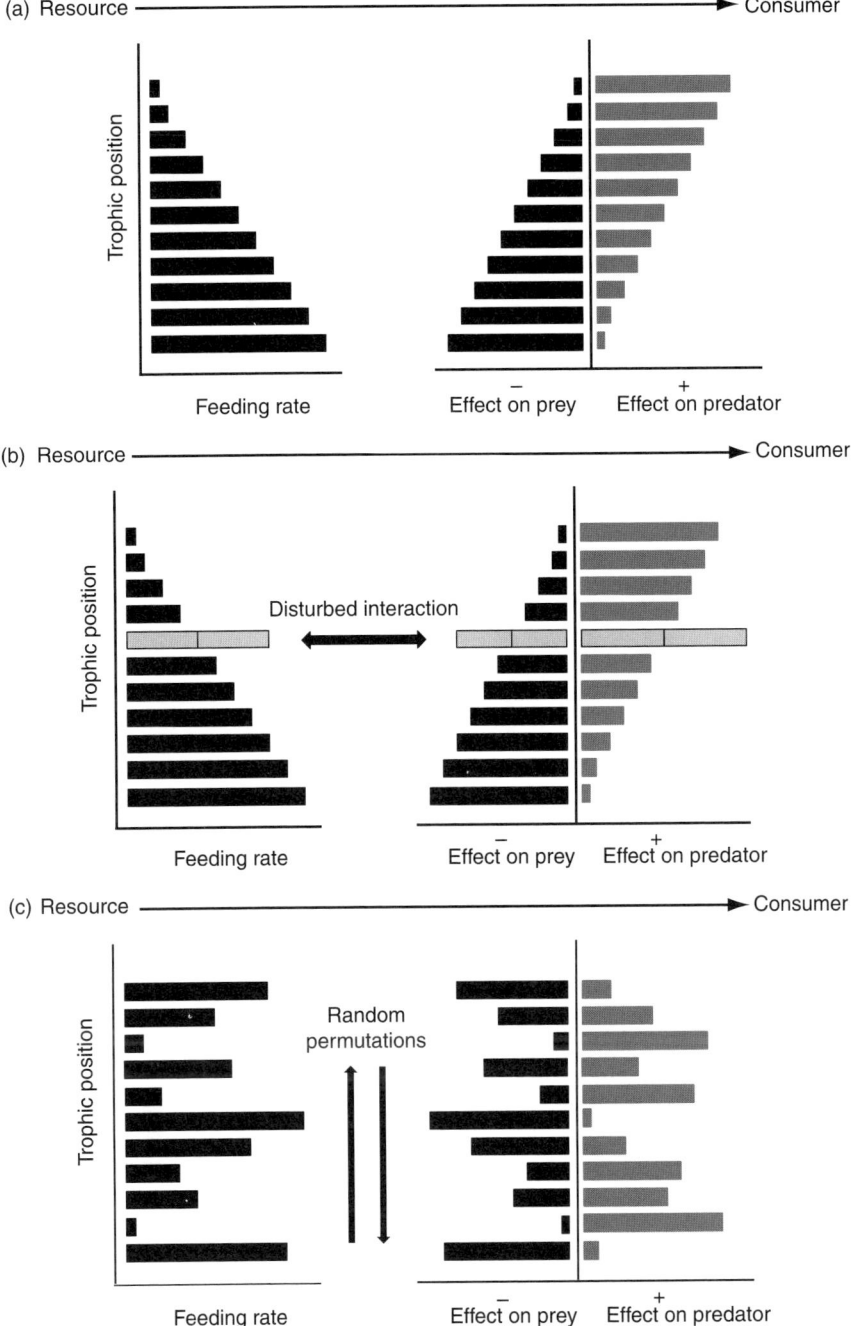

Figure 6.1 Conceptualizations of the observed patterns and alterations of the patterns in the different analyses of stability. (a) Stylized representations of the distributions of energy flux and paired interactions strengths for the CPER and other soil food webs. (b) Example of how an individual trophic interaction within the Jacobian matrix was manipulated to assess the impact of the interaction on stability. (c) Example of a single iteration of a manipulation that repositions the interaction strengths to assess the impact of the overall configuration on stability in a way that preserved the connectedness and sign-structure of the Jacobian matrix.

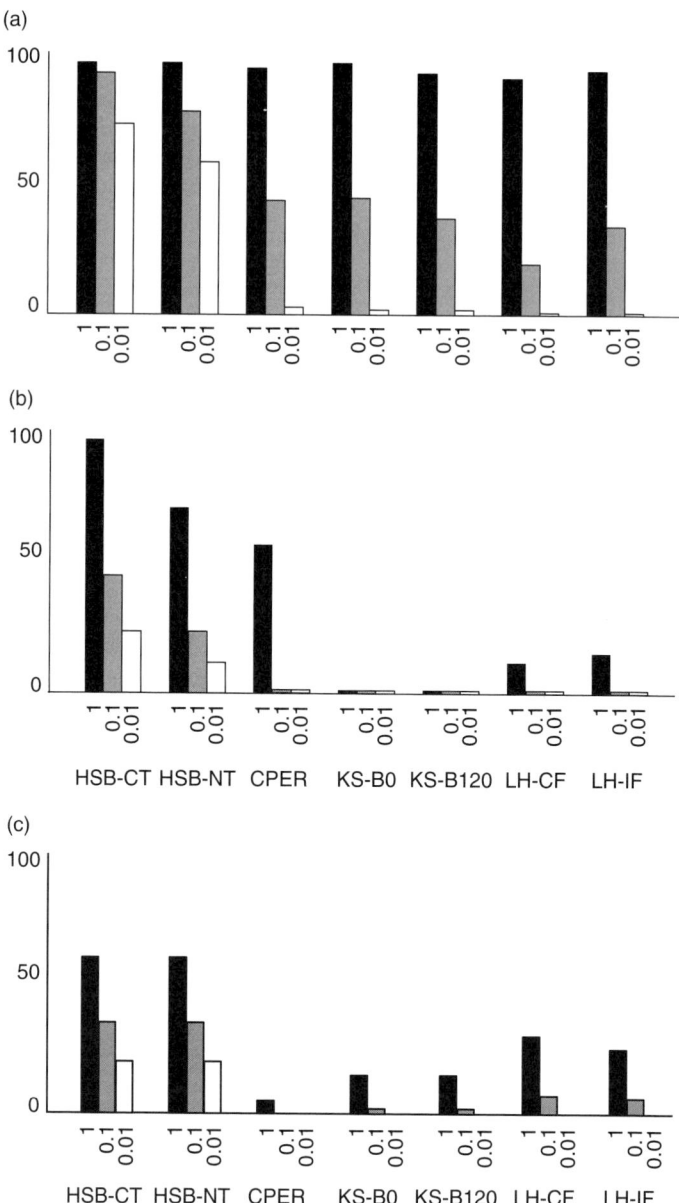

Figure 6.2 Comparison of the stability of Jacobian matrices for the seven soil food webs that included the (a) empirically derived values of interaction strengths ('real') with that of (b) 100 randomizations ('randomized) of these matrices and with that of (c) 100 community matrices constructed using randomized values for biomass, death rates, and efficiencies. The off-diagonal elements of the real and randomized matrices were sampled from a uniform distribution with the interval $[0, 2\alpha_{ij}]$, where the α_{ij} are derived from the observed flux rates (see text). The diagonal elements of the matrices were modeled as $-s_i d_i$, where $s_i = 1, 0.1$, or, 0.01.

In Figure 6.2 we show the likelihood of stability for the "real" and "randomized" Jacobian matrices for all seven webs generated for a series of 100 replications. The comparison shows for each value of s selected, the matrices were more likely to be stable than their randomized counterparts. There is another way of interpreting this. The real webs were more likely to be stable at lower values of s for than the randomized webs, implying that for the real systems stability was less a consequence of the diagonal terms. Stability, therefore, is a direct consequence of the trophic-level-dependent patterning of the interaction strengths, as this patterning is the only difference between the real and the randomized matrices. Moreover, as the patterning of the interaction strengths is directly derived from the energy flux web, these results show that energetic structure governs food web stability by generating stabilizing patterns in the interaction strength.

As a final note, one might argue that the relatively high likelihood of stability of the real matrices compared with the randomized matrices is the consequence of the steady-state assumption used to calculate the feeding rates (which were then used to calculate interaction strength). This aspect will be discussed below, when we sort out the mechanisms behind these results.

6.3 Distribution of interaction strengths: trophic-level-dependent interaction strengths

The analysis presented above demonstrated that the empirically derived interaction strengths generate Jacobian matrices that are much more likely to be stable than their randomized counterparts (Figure 6.2). Given the way in which the randomizations were carried out to create the randomized matrices, we concluded that the stability of the empirical matrices is the result of a particular organization of the interaction strengths in the Jacobian matrices. The permutations that generated the random matrices preserved the sign structure of the matrix and the actual values of the matrix elements from the empirical matrix. From an energetic standpoint, under the steady-state assumption the total amount of standing biomass was also preserved. Given this, why were the real webs more stable than their random counterparts? To answer this question, we will focus on the energy-based patterns that we observed in the energy flux descriptions of the real webs and their connections to the patterns of interaction strength that we observed in the functional descriptions, and dissect how these patterns and connections were affected by the randomizations.

Figure 6.1 presents stylized representations of the distribution of feeding rates and interaction strengths for each trophic interaction when arranged with increased trophic position for our soil food webs. Several patterns are revealed:

1. The food webs possessed pyramidal arrangements of biomass and flux rates with increased trophic position.
2. The food web possessed dominant pathways of fluxes that originated from different basal resources and combined at upper trophic positions by predators.

3. The negative interaction strengths have much larger (in an absolute way) values than the positive interaction strengths.
4. The interaction strengths, both negative and positive interaction effects, depend on trophic level. The negative interaction strengths associated with the effects of the consumer dynamics on the dynamics of the prey possess a pyramidal arrangement, being largest at the base of the food web and diminishing at the upper trophic positions. The positive interaction strengths associated with the effects of the dynamics of the prey on the dynamics of the consumers possess a pattern that was opposite to that for the negative interaction strengths.
5. Taken together, the interaction strengths exhibit an asymmetric arrangement in their distribution by trophic position that is linked to the pyramidal arrangement of biomass and fluxes. This is to be expected, given the equations we used to calculate the interaction strength. The negative interaction strengths are equal to the feeding rates divided by the consumer biomass, whereas their positive counterparts are equal to the same feeding rate, but divided by the consumer biomass and multiplied with the energetic efficiencies. Given that the food webs have trophic biomass pyramids and that the efficiencies have values between 0 and 1, the negative interaction strengths are much stronger (as it appears two orders of magnitude) than the positive interaction strengths.

In the following sections we show that it is this trophic-level-dependent patterning that plays an important role in food web stability. We will study the nature of the stabilizing properties of the observed patterns by asking two questions: (1) What biological mechanisms are responsible for the stabilizing patterning? and (2) can we understand the stabilizing effect of the observed patterning mathematically?

6.3.1 Distribution of biomass and flux rates: the role of energetics

The randomizations effectively redistributed biomass within the food web by moving biomass and fluxes up to higher trophic positions, or by shifting biomass and fluxes from one pathway within the web to another (Figure 6.3). To illustrate the first point, we start with a food chain of length four that possesses a pyramid of biomass with the dynamics represented by a system of differential equations similar to those we have used to this point. The trophic heights of biomass and flux rates within the chain can be indexed by weighting the biomass and flux rates at each trophic level by their steady-state estimates. If we permute the elements of the matrix in the manner used in our analysis, the biomasses and the flux rates move with the elements. In this case, the trophic heights of biomass and the flux rates increase with each permutation.

To explain the shift in biomass and fluxes from one pathway to another requires that we expand our simple representation to two conjoined food chains, each of which possess a pyramid of biomass, and originate from a single or separate resource (Figure 6.3). We will consider two scenarios. In the first scenario, the production at

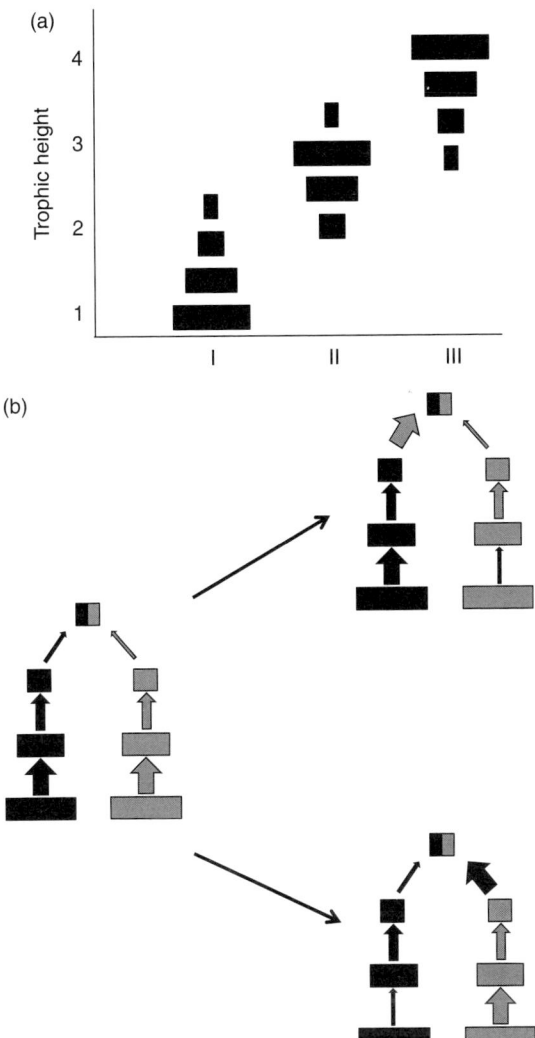

Figure 6.3 The effects of the randomizations on the energetic organization of the food webs. (a) Effect of the randomization on the distribution of biomass. If starting with a pyramid of biomass and flux rates, the randomizations effectively moved biomass to higher positions as indexed by trophic height (mean trophic position weighted by biomass) of the system. (b) Effect of the randomization on the distribution of flux rates. If starting with two parallel energy channels linked by a common predator with equal flux rates in each, the randomizations can shift the flux rates from one channel to another. The shifts are further complicated by the movement of biomass across channels and to higher trophic positions that accompany each randomization.

the bases of the chains are identical, the distribution of biomass and flux rates within the chains are identical, and the predator that joins the chains shows no preference for one chain over the other. In the second scenario, we maintain the pyramids of biomass and flux rates within each chain, but allow for differences between the chains. Furthermore, production at the base of the chains can differ, and predators may prefer one chain to the other. Under these scenarios we can realize simple movement of biomass up one chain or the other, a shift in the relationship in the fluxes between the chains, or the preference the predator has for one chain over the other.

Next we explore the role of the trophic pyramidal structures at steady state in the stability of the seven soil food webs by means of a sensitivity analysis. Here we study the effect of disturbing the pyramid of biomass on food web stability. We start with the observed biomasses for each functional group within the food webs. We then permute the biomasses randomly among the groups. The permuted biomasses are then used as the input of the model that calculates the feeding rates, and subsequently the interaction strengths to parameterize "disturbed" Jacobian matrices. We can make a similar comparison to the one presented in Figure 6.2, but now between the "real" matrices and their "disturbed" counterparts, as opposed to the earlier comparison between the "real" and "randomized" counterparts. We find a pattern that is similar to the one that emerged from the earlier analysis; the real matrices are more likely to be stable than their disturbed counterparts.

Recall that the procedure to construct these disturbed matrices is exactly the same as the one used to construct the real matrices, with the same underlying assumptions. This brings us back to the note we made earlier in the chapter that the high stability of the real matrix might be the consequence of the steady-state assumption used to calculate the feeding rates (which were then used to calculate interaction strength) as opposed to the energy-based patterns. Apparently, this is not the case. The sensitivity analysis confirms our conclusions that energetic properties of groups and their positioning in the food web determine the patterns of interaction strengths, which forms the basis of food web stability. Having said this, we end this section with a final comment on how the randomizations affected the systems that is important and that is not readily apparent. Shifting biomass and fluxes within the chains and webs in the manner described above effectively elevated the productivity or efficiencies of transfer within the systems. The only way to sustain high biomass at high trophic levels given the energetic constraints imposed in the analysis is through increasing energy input at the base, increasing the efficiency of consumers, redistributing productivity within the web, or through some combination of these three. As a result, elevated productivity or a shift in existing productivity within the food web can explain the results of our stability analysis (Figure 6.2). The parallels of these changes to the ways in which natural and anthropogenic disturbance affect systems, through enrichment or other means, are clear. We will address these topics at length in the chapters that follow.

6.3.2 Distribution of interaction strengths: the role of trophic interaction loops

A key attribute of food web structure that helps clarify the relationship between interaction strengths and stability is the weight of the "trophic interaction loops" (sensu Neutel et al., 2002) that we introduced in Chapter 2. A trophic interaction loop describes a pathway of interactions (i.e., not feeding rates) from a species through the web back to the same species without visiting the species more than once; hence a loop is a closed chain of trophic links (Figure 6.4). Trophic interaction loops may vary in length, with the "loop length" being the number of trophic groups visited. Loops may also vary in weight, with the "loop weight" being the geometric mean of the interaction strengths in the loop, defined as the per capita effects of the Jacobian matrices (May, 1973). As discussed in Chapter 2, loop weights can be used as an indicator of food web stability (Neutel et al., 2002).

Figure 6.4 shows an example of a loop of length 3 and a loop of length 4. A loop generated from the trophic interactions generates a pair of loops based on interaction strength. To illustrate this point, we will use the loop of length 3 in Figure 6.4. The first loop works in the clockwise direction and includes one negative effect exerted by the predatory mites on the bacterivorous nematodes and two positive effects exerted by the bacterivorous nematodes on the predatory nematodes and by the predatory nematodes on the predatory mites. The second loop moves in the counter-clockwise direction and includes two negative effects exerted by predatory mites on the predatory nematodes and by the predatory nematodes on the bacterivorous nematodes, and one positive effect exerted by bacterivorous nematodes on predatory mites. Likewise, the example of the trophic interaction loop of length 4 includes a clockwise loop with one negative and three positive effects, and a counterclockwise loop with three negative effects and only one positive effect.

If we take the patterning of the interaction strengths and the buildup of loops that are in consideration, we find an interesting relationship between loop length and loop weight. First, we have seen that the negative interaction strengths are on average much (two orders of magnitude) stronger than positive interaction strengths (Figure 6.1). For this reason, loops that contain relatively many negative effects are potentially the heaviest ones. Second, the longer the loop, the more negative than positive effects it may contain. In the example of Figure 6.4, the loop of length 3 contains a loop with two negative links and one positive link, whereas the loop of length 4 contains a loop with three negative links and only one positive link. Together, this should suggest that the heaviest loop is likely to be a long loop with a majority of negative effects.

In Figure 6.5 we plot the loop weights against the loop lengths in a CPER food web for both the real and the randomized Jacobian matrices. It shows that in the real matrix as a whole the variation in loop weight declines with loop length and that there is a slight decrease in loop weight with increased loop length (Figure 6.5a). The opposite is the case for the randomized matrix. We expect that the weights for

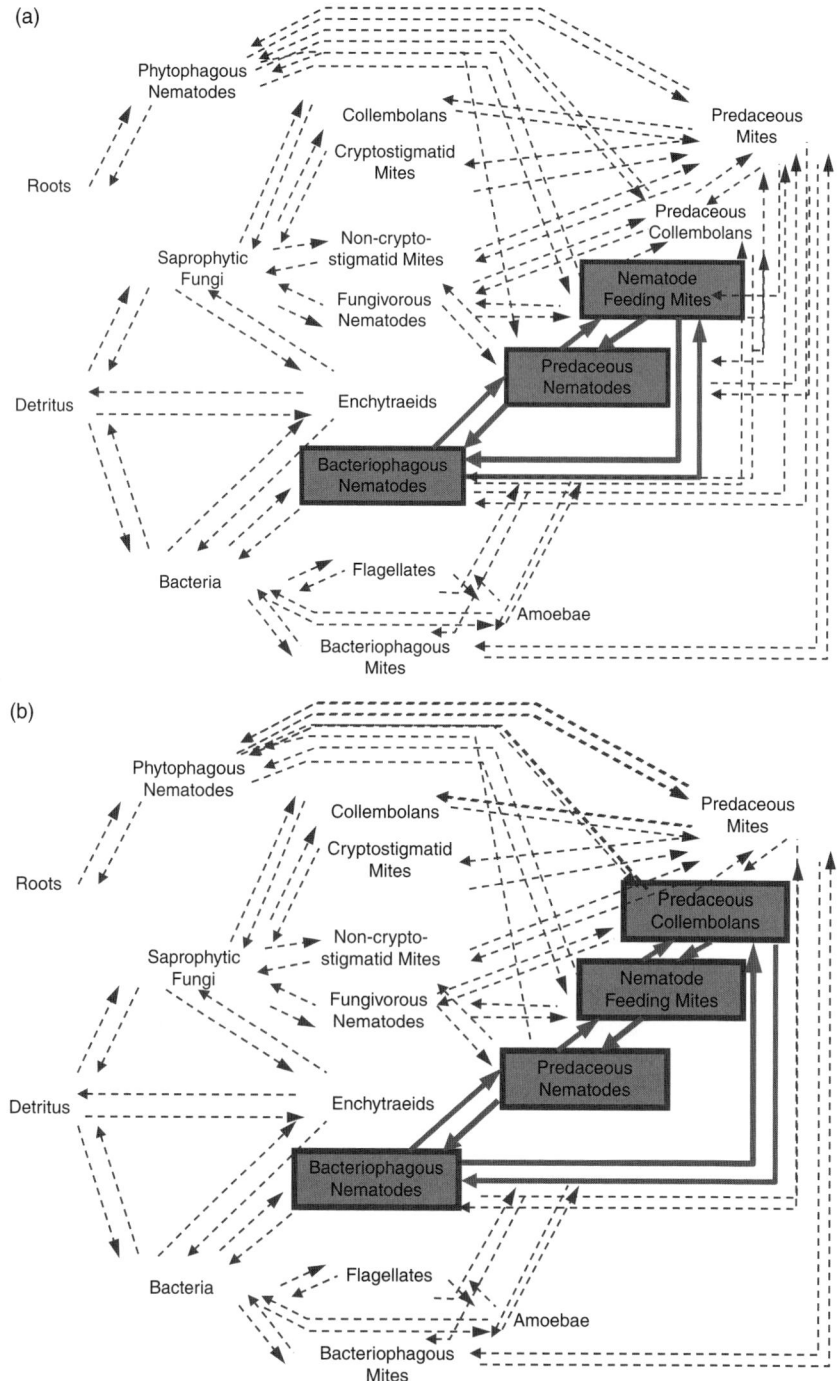

Figure 6.4 Examples of a trophic interaction loop of length 3 and 4 within a soil food web. Each trophic interaction represents a trophic interaction loop of length 2. (a) A trophic interaction loop of length 3 that includes bacteriophagous nematodes, predaceous nematodes, and nematode-feeding mites (predatory mites in text). (b) A trophic interaction loop of length 4 that includes bacteriophagous nematodes, predaceous nematodes, nematode-feeding mites (predatory mites in text), and predaceous Collembola.

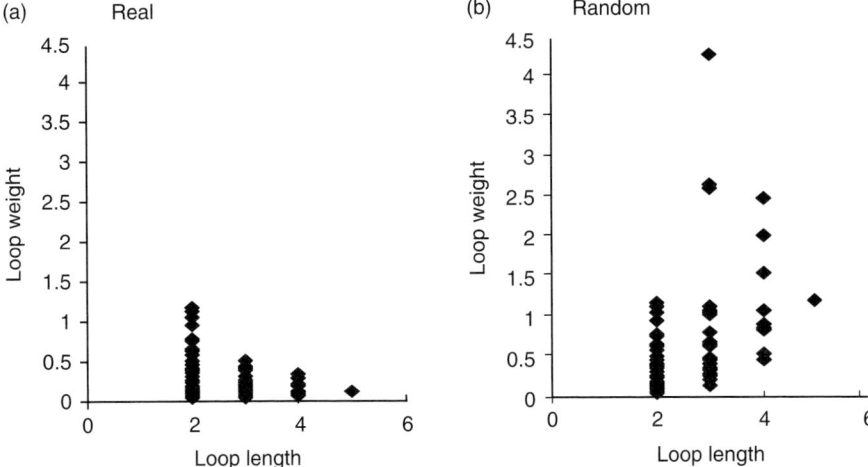

Figure 6.5 Loop length and loop weight and in the CPER food web and randomizations of this matrix. (a) Loop weight versus loop length in the "real" matrix. (b) Loop weight versus loop length in a randomized matrix. Long loops with a relatively small weight, those with many bottom-up effects, are not given in the figure, not being relevant for maximum loop weight (figure from Neutel et al., 2002).

loops of length 2 would be identical in the real and randomized matrices, as the randomization affects the placement in the web but not parameter values used in the metric for weight. For loops of length 3 or greater, the randomized matrix possessed heavier loops than their real matrix counterparts (Figure 6.5b). Moreover, for the real matrix all loop weights for loops of length 3 or greater are relatively low compared with the loop weight in the randomized matrix. Hence, the patterning of the interaction strengths in real matrices (Figure 6.1) is such that for the long loops, which have the potential to include the heaviest loop, the negative effects are in fact relatively weak. In other words, the negative effects are distributed in such a way that the weak links are in the long loops, making the maximum of all loop weights relatively low in the real webs and relatively high in the random webs.

6.3.3 Food web stability: loops and energetics

In this section we link the energetic properties of food web structure we observed to the low maximum loop weight and stability of the real Jacobian matrices. In Figure 6.6 we show the relationship between maximum loop weight and matrix stability. In this figure we use a different measure of stability as the one used above (see Figure 6.2). Here we follow Neutel et al. (2002), who offer an elegant way to overcome the problem of the incompleteness in the Jacobian matrices regarding the diagonal terms. A minimum or "critical" value of diagonal strength (s) is calculated that is required for matrix stability, using the criteria for quasi-diagonal dominance (McKenzie, 1960). The procedure starts with $s = 0$ and then increases s to the value at which the matrix becomes stable. This "critical" value of the parameter s reflects the threshold of the minimum degree for density-dependent death required for food web stability. The critical value also serves as a relative measure of the sensitivity of the food web to make comparisons of stability between different food webs. The smaller the value of s, the more stable the food web. As was the case in Chapter 5, this procedure is not followed for the diagonal term for detritus.

In Figure 6.6, the critical s value is plotted against maximum loop weight for the real Jacobian matrix representing the CPER food web together with ten randomizations of the same matrix in which pairs of interaction strengths were exchanged. The stability of the real Jacobian matrix had a lower loop weight and lower critical value for s than all randomizations. Taken together, the real and randomized matrices show a strong correlation between critical s value and maximum loop weight. In other words, maximum loop weight is an index of matrix stability. To make an energetic connection explicit in this finding, we turn to one of our soil food webs and study the distribution of biomasses, feeding rates, and interaction strength in a caricature of a common type of omnivorous trophic interaction loop of length 3 consisting of an omnivore feeding on two groups on different trophic levels, i.e., on a microbivorous population and a microbial population (Figure 6.7).

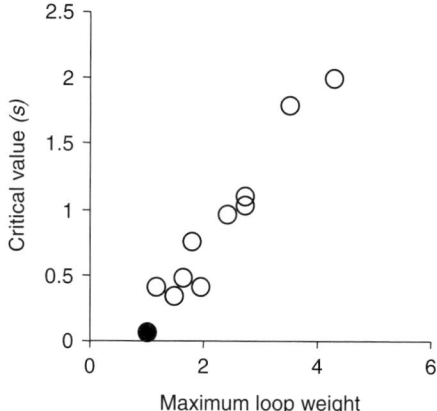

Figure 6.6 The relationship between maximum loop weight and the minimum value of diagonal strength (s) for matrix stability for the CPER food web (●) and 10 randomizations of the CPER food web (○). The maximum loop weight is the heaviest of all loops estimated from the off-diagonal elements within the Jacobian matrix. The diagonal elements are scaled to $-sd_i$ where d_i is the specific death rate of species i, and s is the minimum or 'critical' value of diagonal strength (s) that is required for matrix stability, using the criteria for quasi-diagonal dominance (McKenzie, 1960). From Neutel, et al. (2002). Reprinted with permission from AAAS.

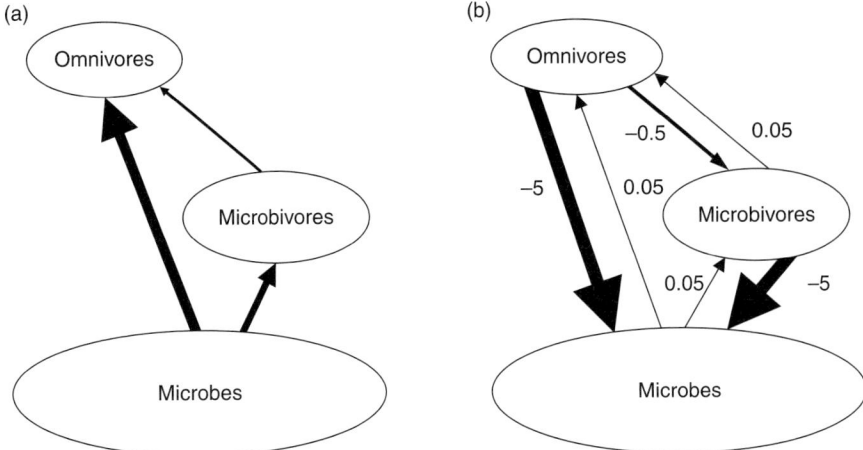

Figure 6.7 Interaction strengths and loop weights in an omnivorous food web. (a) Equilibrium feeding rates and population sizes are estimated from field data and the models. Feeding rates were assumed to be proportional to the population sizes of the prey. (b) Interaction strengths. In the example, efficiencies were set at 0.1 for all species. The loop weights of the two loops of length 3 are $(|-0.5|\cdot|-5|\cdot 0.05)^{1/3} = 0.5$ (anticlockwise loop, starting with species 1) and $(|-5|\cdot 0.05\cdot 0.05)^{1/3} = 0.23$ (clockwise loop).

The omnivorous feeding loop depicted in Figure 6.7 forms two interaction loops. The population sizes are organized in the form of a trophic biomass pyramid, and the omnivore is assumed to feed on the two prey types in accordance with their relative availability. For this set of trophic interactions, the omnivore feeds mainly on the microorganisms, and the negative effect of the omnivore on the microbivores will be weaker than the effect on the microorganisms, as both negative interaction strengths are defined as the feeding rate divided by biomass of the omnivore. The clockwise interaction loop is potentially the heaviest loop, as it is the one that possesses two negative effects. However, given the energetic arrangement of biomass and flux rates, the clockwise interaction loop possesses at least one relative weak negative link, the one from the omnivores on microbivores. The negative effect between the microbivores and the microorganisms is equal to the—"very large"—feeding rate of microbivores on microorganisms divided by the—"relatively large"—population size of the microbivores; therefore it will roughly be about the same strength as the effect of the omnivore on the microorganisms. This follows from the trophic pyramid structure with a relatively low biomass of the microbivores compared with the biomass of the microorganisms. The consequence is that maximum loop weight is kept at a relative low level compared with the random matrix in which, by chance, a loop may occur that contains both strong negative effects. Had the trophic structure included an inverted pyramid of biomass, the clockwise interaction loop, i.e., potentially the heaviest loop, would have a much larger loop weight. The same principle applies to loops longer than 3. Hence, the aggregation of weak links in long loops that enhances stability results from the pyramidal biomass structure in the loop.

6.4 Summary and conclusions

In this chapter we discussed the importance of the energetic organization of food webs and their stability. Using our collection of soil food webs, we demonstrated that the observed pyramidal distributions of biomass, energy flux rates, and interaction strengths, and the organization of the food web into dominant pathways of interactions and fluxes, reflect the energetic organization of these communities and that these patterns enhance their stability. Changes to these patterns had a destabilizing impact on the systems. Specifically, we found that enrichment influences the distribution of biomass and material flux rates within the food web in ways that influence the system stability. Movement of biomass and flux rates from lower trophic positions to higher trophic positions, or shifts in the flux rate within and among the dominant pathways, had a destabilizing influence.

We further explored the interrelationship between energetic properties of food webs, the maximum loop weight interaction loops within food webs and stability. The loop weight analysis confirmed the findings of our study that linked energetic patterns with interaction strengths. The trophic pyramids of biomass and fluxes, which are common to many ecosystems, generate stabilizing patterns in interaction strength that possess low maximum loop weight.

The use of empirically derived models, traditional stability analyses, and the analyses based on trophic interaction loops, proved to be useful tools in assessing structure and organization of the complex networks of ecological relations in whole food webs. In this way we can also see a direct connection between energetic structure and stability, which was not apparent when we tried to assess the stabilizing impact of individual trophic interactions on the web as a whole. The results from these approaches will provide the foundation for the topics in the remaining chapters.

7

Enrichment, trophic structure, and stability

7.1 Introduction

In the previous chapter we studied the relationships between the distributions of biomass and interaction strengths within food webs and their stability using models that were parameterized with field data. The webs possessed two structural features that were related to trophic levels and their stability. From the energy flux descriptions, we found that the food webs possessed a pyramidal distribution of biomass, with the majority of the biomass of the functional groups positioned at lower trophic levels and diminished with successive levels. The functional description revealed that the food webs also possessed an asymmetric arrangement of the paired interaction strengths for each trophic interaction. Here, the per capita effects of predators on prey (negative elements, $-\alpha_{ij}$, of the Jacobian matrix) being relatively large at lower trophic levels and diminishing with successive levels, and their counterpart effects of prey on predators (positive elements, α_{ji}, of the Jacobian matrix) being relative small at lower trophic positions and increasing with successive levels. Our analyses of these pattern demonstrated that there is a strong relationship between the energetic organization of a community and the dynamics and stability of the community. The rate of enrichment at the base of a community and the rate that material flows through a community serve as the link between structure and dynamics.

These results are familiar, as they also align well with the observations by Elton (1927) on the pyramid of numbers and biomass, are germane to the propositions by Hutchinson (1959) and Hairston et al. (1960), and fit the models formulated by Oksanen et al. (1981) and Carpenter et al. (1985), in which the roles of productivity and predation in regulating the structure of ecological communities were formulated. They also align with the dynamic considerations that have been proposed (Rosenzweig, 1971, May, 1976, Pimm and Lawton, 1978, Moore et al., 1993, McCann et al., 2005). Models of simple food chains based on the Lotka–Volterra form and more complicated forms have demonstrated that increased rates of productivity can lead to changes in dynamic states and instability, their sensitivity to which depends largely on the degree of self-limitation of species and food chain length (Rosenzweig, 1971, DeAngelis, 1975, May, 1976, Pimm and Lawton, 1977, Moore et al., 1993, Sterner et al., 1997).

Oksanen et al. (1981) presented the exploitation ecosystem hypothesis (EEH), based on their observations in low-production Arctic systems, which integrated the roles of productivity and consumers in structuring communities at the scale of the trophic level. An important feature of EEH is that the rate of production shapes communities in the bottom-up trophic level manner envisioned by Hutchinson (1959), but the top-down regulation that consumers have on lower trophic positions (Hairston et al., 1960) depends in part on their trophic position in the community relative to top predators. Carpenter et al. (1985) expanded on this theme with the concept of trophic cascades, based on their work in lake systems. The trophic cascade stresses the importance of the top predators in directly regulating not only the densities and dynamics of their prey but also, in doing so, indirectly regulating the density and dynamics of primary producers.

These observations are appealing, as they reinforce our central tenet on the linkage between energetics and dynamics. Productivity affects communities in a bottom-up manner, providing energy at the base to support the community, whereas the organisms at upper trophic levels affect the community in a top-down manner. However, these ideas raise a few questions and issues, not the least of which is that the empirical evidence is lacking for two critical points. Studies have yet to confirm the relation between food chain length and the productivity (Pimm and Lawton, 1977, Briand and Cohen, 1987, Post, 2002), and trophic cascades are rare in terrestrial systems (Strong, 1992). Furthermore, modeling studies and reviews on the topic have offered alternative or contradictory evidence on the role of dynamics in governing food web architecture (Sterner et al., 1997, Post, 2002). These concerns may stem from observations and experiments not covering or operating over steep enough gradients of productivity, the role of detritus all but being ignored (Polis and Strong, 1996, Moore et al., 2004), and the importance of the diversity of production on structuring food webs being underemphasized (Moore and de Ruiter, 1997, Rooney et al., 2006, Rooney et al., 2008). The reason for this lack of agreement between models and observation may also stem from how the models are depicted and the underlying assumptions used in the analyses performed.

In this chapter we focus on the influence of enrichment at the base of a food web or food chain on trophic structure, energy flow, dynamics, and stability. Enrichment will include new living biomass entering the system through primary production, or nonliving biomass entering or recycling as detritus. We will collectively refer to enrichment via primary production and detritus as "productivity." There are several aspects of trophic structure and energy flow to consider. Here we will focus on how enrichment influences the number of trophic levels and the length of food chains, and the vertical distribution of biomass within food webs. We will consider the horizontal distribution and the internal and lateral flow of biomass within and among food chains, respectively, and their relationships to dynamics and stability, separately in the chapters that follow.

7.2 Simple primary-producer-based and detritus-based models

All production has its origins with living autotrophic organisms. The new production that is produced by autotrophs is either consumed by another organism or is released into the environment as nonliving organic matter, which we define as detritus. Here we will assess the influence of productivity on the structure and dynamics of our primary-producer-based food chains and detritus-based food chains using simple differential equations following the Lotka–Volterra form (Figure 7.1). We have defined productivity as the rate of accrual of new biomass through growth and reproduction that occurs at the base of a food chain. We will adopt the units of g C m^{-2} yr^{-1} or its equivalents for real-world comparisons.

We begin by assessing the feasibility and resilience of our simple producer-based and detritus-based models to changes in productivity. In Chapter 2 we compared the feasibility and resilience of simple primary-producer-based and detritus-based food chains over a narrow range of productivity, as these parameters were sampled over the interval (0,1). Here we will explore how feasibility and resilience vary over a wider range of productivity by adjusting the specific rate of increase, r_1, for the

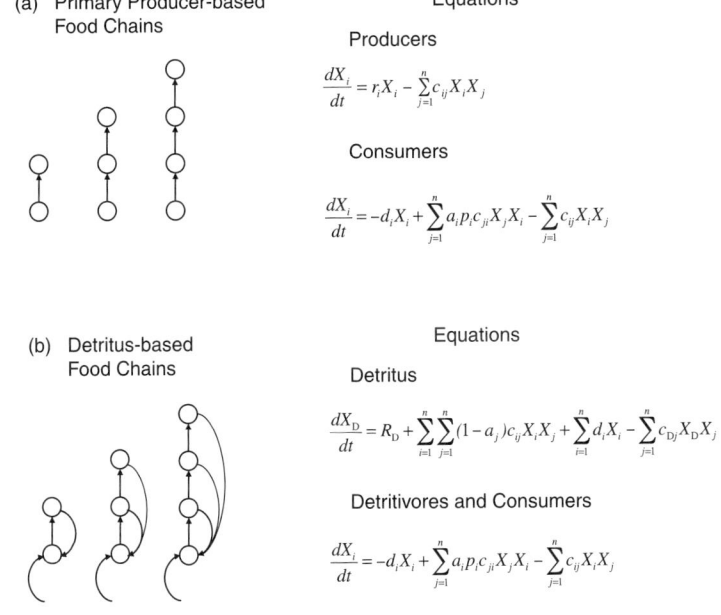

Figure 7.1 Simple food chains and forms of the equations to model them using type I functional responses. (a) Primary-producer-based food chains. (b) Detritus-based food chains. Detritus enters the system from an allochthonous source (R_D) and from autochthonous sources in the form or unassimilated consumption and unconsumed death.

primary-producer-based models and the allochthonous input of detritus, R_D, for the detritus-based food chains at increments of an order of magnitude from 10^{-2} to 10^5 units. All other parameters (death rates, intraspecific competition coefficients, consumption coefficients, and the energetic efficiencies) were selected at random from a uniform distribution, I (0,1).

Our analyses will follow a pattern similar to the one in Chapter 2. For each level of productivity we will construct 1000 food chains for each length and type (primary producer based and detritus based) and determine if the individual food chains are feasible ($X_i^* > 0$ for all i species). For feasible food chains we will calculate the elements of the Jacobian matrix A. We will then assess the stability of the food chains by confirming that all the eigenvalues are negative, and if stable, estimate the return times to establish the resilience of the food chains. We will repeat the analysis to evaluate the sensitivity of the food chains to changes in the energetic efficiencies and growth rates of the living species by constructing and contrasting models that emulate ectotherms and endotherms. Varying the productivity in the manner described above affected the structure and dynamics of both the primary-producer-based and detritus-based food chains (Figure 7.2).

7.2.1 Feasibility

The structure of our food chains along the extended gradient of productivity followed the same pattern that we observed in Chapter 2 for the lower end of the gradient between 0 and 1 units. The feasibility ($X_i^* > 0$ for all i species) of the three-species and four-species detritus-based food chains and all of the primary-producer-based food chains increased along with productivity, reaching their sills (100% feasibility) in order of trophic diversity (Figure 7.2a). By 10^5 units of productivity all of the food chains were feasible. The two-species detritus-based food chains were an exception to this trend, as all were feasible across the entire range of the productivity gradient that we evaluated.

These results arise when satisfying the requisite thermodynamic and mass-balance conditions that must be met for all species to have positive equilibrium densities. If we evaluate the levels of productivity required to maintain positive densities for all species (i.e., feasibility) directly from our equations we find that for both types of models greater levels of productivity are required to maintain the systems with increased numbers of trophic levels. We can do this by deconstructing the equations for the primary-producer-based models into separate sets of equations for two, three, and four species and evaluate the level of r or R, defined as r_{Fi} and R_{DFi} respectively, that are required to maintain each at levels greater than zero.

Oksanen et al. (1981) postulated that along a gradient of productivity the amount of biomass at a given trophic position would increase in steps, with the biomass accumulating at the lower trophic levels until a threshold in biomass is reached so that there is sufficient production to maintain the next trophic level (Figure 7.2). This pattern of development can be explained by evaluating the steady-state density of

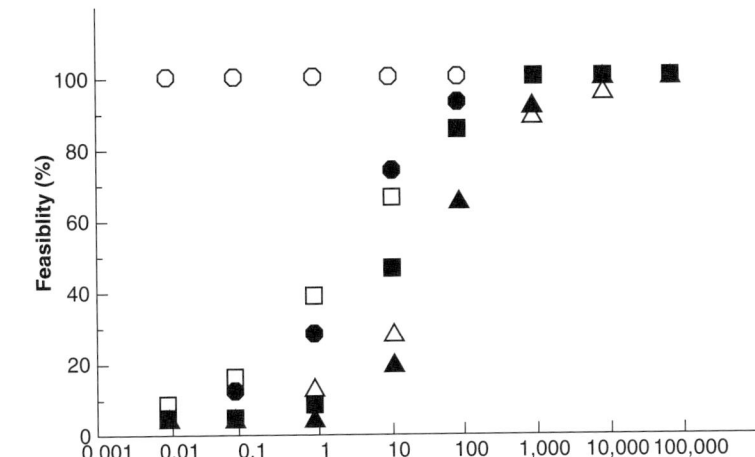
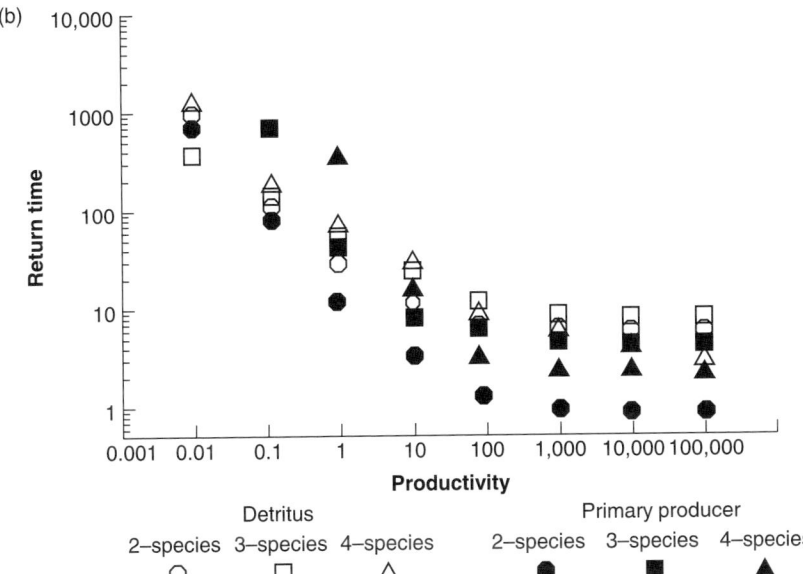

Figure 7.2 The feasibility (a) and return times (b) for primary-producer-based (closed symbols) and detritus-based (open symbols) food chains of two trophic levels (circles), three trophic levels (squares), and four trophic levels (triangles) along a gradient of productivity. Each symbol represents 1000 model runs. Feasibility is the percentage of food chains at a given level of productivity where all species maintained positive densities at equilibrium (Roberts, 1974). The resilience is expressed as the return time (RT), estimated as RT = $-1/\text{real}(\lambda_{max})$ where real(λ_{max}) is the real part of the dominant eigenvalues. The food chains were evaluated at fixed levels of productivity r_i for the primary-producer-based food chains and R_D for the detritus-based food chains beginning at 10^{-2} units and increasing by orders of magnitude to 10^5 units. All other parameters were sampled from a uniform distribution over the interval (0,1). Depending on the assumptions used to establish the death rates and the consumption coefficients, productivity scales between 4 and 36 g C m^{-2} yr^{-1} (redrawn from Moore et al., 1993).

the functional groups at a trophic level in terms of the minimum level of productivity, r_{Fi}, required to maintain the populations at positive levels, i.e., $X_i^* > 0$. For our simple primary-producer-based food chains, this condition is met if $r_1 = r_{F2}$ as follows:

$$r_{F2} > \frac{d_2}{a_2 p_2} \frac{c_{11}}{c_{12}} \qquad (7.1)$$

Interestingly, the threshold of feasibility defined by r_{F2} is equal to $|\alpha_{11}|$, indicating that feasibility is defined by the degree of intraspecific competition within the producer population.

Next we determine the level of productivity that is required to support the third trophic level ($r_1 = r_{F3}$). If we set $X_3^* = 0$ and solve for r_1, then r_{F3} is as follows:

$$r_{F3} = r_{F2} + \frac{c_{11}^2 d_3}{c_{12} c_{23} a_3 p_3} \qquad (7.2)$$

The value for r_{F3} that establishes the level of production required to support a third trophic level is greater than the level required for two levels ($r_{F3} > r_{F2}$). Continuing along this path we find, as illustrated in Figure 7.2a, that $r_{F4} > r_{F3} > r_{F2}$ for the primary-producer-based food chains, and $R_{DF4} > R_{DF3} > R_{DF2}$ for the detritus-based food chains.

7.2.2 Stability and resilience

The mean of the return times of the feasible models from the 1000 trials is presented in Figure 7.2b. The return time is a measure of resilience, defined as the time required for the system to return to equilibrium. All our feasible food chains were locally stable, as all eigenvalues were negative. This was a direct result of the structure and parameterization of the equations meeting the criteria for qualitative stability. The resilience of both types of food chains was a function of both the level of productivity and food chain length. In general, the return times decrease with increased productivity. In all cases they decayed to what appears to be a lower limit. The relationship between return time and food chain length was less clear. For our runs after ten units of productivity the four-species food chains had shorter return times than their three-species counterparts.

To get a better sense of what is happening with the return times along our productivity gradient we need to revisit how they are estimated. Recall that return time (RT) is defined as $-1/\text{real}(\lambda_{\max})$ for $\lambda_{\max}$, where $\text{real}(\lambda_{\max})$ is the real part of the dominant (least negative) eigenvalue (Pimm and Lawton, 1977). For illustrative purposes we will focus on a simple two-species food chain. The dominant eigenvalue for this food chain is:

$$\lambda_{max} = \frac{\alpha_{11} + \sqrt{\alpha_{11}^2 + 4\alpha_{12}\alpha_{21}}}{2} \qquad (7.3)$$

In this case return time should decline and eventually decay to $2/\alpha_{11}$ with increased net primary productivity (NPP) (Figure 7.2b) where

$$\alpha_{11} = -c_{11}X_1^* \qquad (7.4)$$

and

$$X_1^* = \frac{d_2}{a_2 p_2 c_{12}} \qquad (7.5)$$

This occurs because of the decrease of the value of the discriminant (μ) of λ_{max}, where $\mu = \alpha_{11}^2 + 4\alpha_{12}\alpha_{21}$. Recall that the α_{21} within μ depends on productivity as

$$\alpha_{21} = a_2 p_2 c_{12} X_2^* \qquad (7.6)$$

and X_2^* is a follows:

$$X_2^* = [(r_1 a_2 p_2 c_{12}) - (c_{11} d_2)]/(a_2 p_2 c_{12}^2) \qquad (7.7)$$

Given that the NPP was regulated by r_i in our models, as NPP $\rightarrow \infty$, μ approaches 0 and ultimately becomes negative, giving rise to the imaginary part of the critical eigenvalue and the onset of oscillations in densities during the decay of the disturbance as it returns to equilibrium. That said, the return times approach $2/\alpha_{11}$ as NPP $\rightarrow \infty$.

7.3 Trophic structure and dynamics along a productivity gradient

The results presented above using our simple models illustrate the connection between trophic structure and stability. Building on this and the ideas presented by Rosenzweig (1971) with the paradox of enrichment, Oksanen et al. (1981) with EEH, Carpenter et al. (1985) with trophic cascades, and the analysis we presented above, a model of ecosystem development along a gradient of productivity unfolds (Figure 7.3). Consistent with EEH, as productivity increases the biomass accumulates at the highest trophic level, until the time at which the rate of production, or input of energy as it were in the case of detritus, is sufficiently large to support an additional trophic level. Pyramids of biomass result in cases in which top predators are in sufficient abundance to maintain positive densities given their energetic efficiencies relative to the rate of production. These structural features when combined with trophic loops and loop weight provide the connection to dynamics and stability.

To illustrate these points, we will apply the approach discussed above using the simple two- and three-species primary-producer-based food chains presented in Figure 7.1 as examples to identify three important thresholds along a gradient of

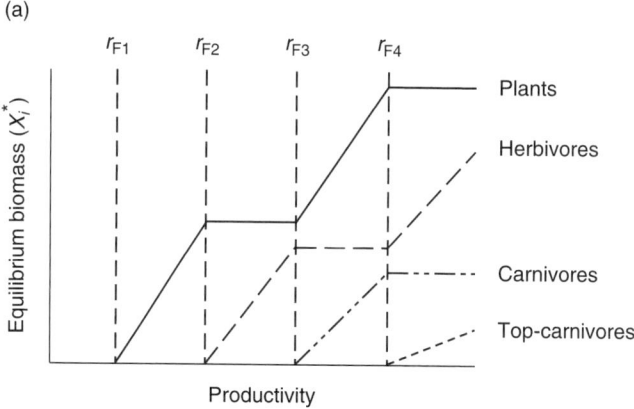

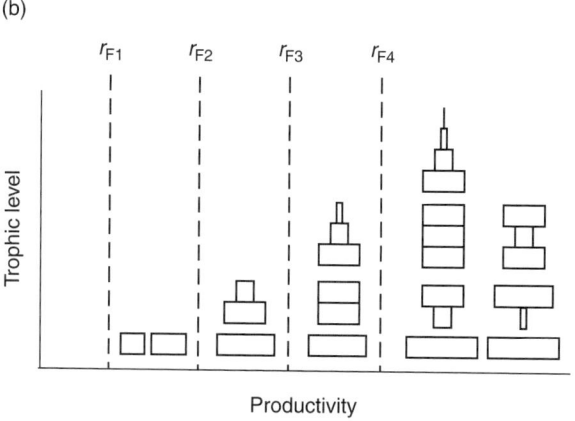

Figure 7.3 Conceptual diagrams of the equilibrium biomass increments (a) and distribution of biomass (b) along a resource availability gradient for four trophic levels. (a) For the equilibrium biomass, the figure shows that there are thresholds in resource availability whereby for a given species there is sufficient energy to support the species and trophic levels (i.e., r_{F1} is the level of productivity at which it is feasible to support one trophic level, r_{F2} is the level at which it is feasible to support two trophic levels, etc.). (b) For the distribution of biomass the thresholds mark the same transitions as above. In the absence of the addition of a trophic level, biomass accumulates at the lower levels to the point that the pyramids of biomass give way to inverted pyramids of biomass or other configurations. Coincidence with the decays of the pyramids of biomass are transitions in dynamic states, r_{Ci}. For the Lotka–Volterra representations, $r_{F2} < r_{F3} < r_{C2} < r_{C3}$.

increasing productivity-one representing a transition in trophic structure of the food chain, one representing a change in the distribution of biomass within the food chain, and one representing a change in the dynamic state of the food chain. The simple two species (Eq. 7.8a,b) and three species (Eq. 7.9a,c) primary-producer-based food chains can be modelled as:

$$\frac{dX_1}{dt} = r_1 X_1 - c_{11} X_1^2 - c_{12} X_1 X_2 \qquad 7.8a$$

$$\frac{dX_2}{dt} = -d_2 X_2 + a_2 p_2 c_{12} X_1 X_2 \qquad 7.8b$$

and

$$\frac{dX_1}{dt} = r_1 X_1 - c_{11} X_1^2 - c_{12} X_1 X_2 \qquad 7.9a$$

$$\frac{dX_2}{dt} = -d_2 X_2 + a_2 p_2 c_{12} X_1 X_2 - c_{23} X_2 X_3 \qquad 7.9b$$

$$\frac{dX_3}{dt} = -d_3 X_3 + a_3 p_3 c_{23} X_2 X_3 \qquad 7.9c$$

where X_1, X_2, and X_3 represent the population densities of primary producer, herbivore, and predator respectively expressed as biomass (e.g., units of g C m^{-2}), r_1 is the specific growth rate (time^{-1}) of the primary producer, c_{11} is the self-limitation coefficient for the primary producers, while d_2 and d_3 (time^{-1}), a_2 and a_3 (%), p_2 and p_3 (%), c_{12} and c_{23} ([g C m^{-2}]$^{-1}$ time^{-1}) are the specific death rates, assimilation efficiencies, production efficiencies, and consumption coefficients for the herbivores and predators. The three thresholds are defined as follows:

1. r_F, the point at which an additional trophic level is energetically feasible;
2. r_c, the point at which the pyramid of biomass disappears;
3. r_λ, the point at which oscillations ensue.

7.3.1 Transitions in trophic structure

The first two thresholds, r_{Fi} and r_{Ci}, indicate that trophic structure is a function of productivity, and that food chains, and by extension food webs, will undergo transitions in their trophic structure along a gradient of productivity. First, the food chain has to be feasible in the sense that there is sufficient productivity within the ecosystem so that all species maintain positive steady-state densities, i.e. all $X_i^* > 0$. This condition is met if $r_1 = r_{F2}$:

$$r_{F2} > \frac{d_2}{a_2 p_2} \frac{c_{11}}{c_{12}} \qquad (7.10)$$

The threshold at r_{F2} is equal to $|\alpha_{11}|$, indicating a direct relationship to the degree of intraspecific competition on the part of the producer. Second, given that the two-species food chains are feasible, the equilibrium density of the species at the upper trophic level must exceed that of the lower trophic position, i.e., $X_2^* > X_1^*$. For our food chain this occurs when $r_1 = r_{C2}$:

$$r_{C2} > \frac{d_2}{a_2 p_2}\left(1 + \frac{c_{11}}{c_{12}}\right) \tag{7.11}$$

The level of productivity at which the three-species food chain is feasible ($r_1 = r_{F3}$) is:

$$r_{F3} = r_{C2} - \frac{d_2}{a_2 p_2} - \frac{c_{11}^2 d_3}{c_{12} c_{23} a_3 p_3} \tag{7.12}$$

The level of productivity at which three-species food chain exhibits a cascade of biomass (r_{C3} when $X_3^* > X_2^*$) is:

$$r_{C3} = r_{F3} + \frac{d_3}{a_3 p_3 a_2 p_2} \tag{7.13}$$

The level of productivity at which the three-species food chain first exhibits a cascade is greater than the level of productivity at which the chain becomes feasible. Furthermore, the point at which the three-species chain becomes feasible is near the point at which the two-species food chain cascades, i.e., $r_{F2} < r_{F3} < r_{C2} < r_{C3}$. If we extend this reasoning to four levels and beyond, the difference between the productivity that insures feasibility of four levels and the rate that initiates a cascade for each trophic position precludes the widespread existence of trophic cascades of biomass (Figure 7.3).

7.3.2 Transitions in dynamic states

The dynamics of the system is a function of productivity. Over a wide range of productivity the return-times of the system rapidly decay with increased productivity, reaching an inflection point, and then gradually approached a limit (see Figure 7.2b). The inflection point ($r_{\lambda 2}$) represents the level of productivity where we see the onset of oscillatory dynamics during the return to equilibrium. We can estimate $r_{\lambda 2}$ by determining the level of productivity that marks the point at which point $\lambda_{(max)}$ includes both real and complex parts.

Equation 7.3 presented the dominant eigenvalue for the two species primary producer-based food chain in terms of the α_{ij} elements of the Jacobian matrix. The real part of λ_{max} describes the direction and rate of decay of the perturbation; as $r_1 \to \infty$, RT$\to 2/\alpha_{11}$. The imaginary part of λ_{max} describes the oscillations that the system undergoes during the return to steady state. The term that is responsible for the oscillations with increased productivity is α_{21} within the discriminant, μ:

$$\mu = \alpha_{11}^2 + 4\alpha_{12}\alpha_{21} \tag{7.14}$$

Recall that the α_{ij} terms are functions of the steady-state biomass (X_i or X_j) of the species involved, and for the simple two species food chain the α_{21} term which describes the influence of the consumer on its prey, is a function of the specific rate of increase (r_1) of the prey.

These results tie into our earlier discussion on the importance of trophic loops and loop weight to stability. Loop weight has been defined as either the $\alpha_{12}\alpha_{21}$ or its geometric mean. In either event, NPP increases as X_i, $X_j \to \infty$, or $r \to \infty$, and loop weight increases. Moreover, as $r \to \infty$, μ transitions from positive values, to zero, to negative values. Hence, in the simplest of model formulations, dynamic stability is a function of loop weight:

$$\alpha_{11} > 2(\alpha_{12}\alpha_{21})^{1/2} \quad \text{overdamped (monotonic damping)} \tag{7.15a}$$

$$\alpha_{11} = 2(\alpha_{12}\alpha_{21})^{1/2} \quad \text{critically damped} \tag{7.15b}$$

$$\alpha_{11} < 2(\alpha_{12}\alpha_{21})^{1/2} \quad \text{under damped (damped oscillations)} \tag{7.15c}$$

The overdamped recovery occurs when $\mu > 0$. In this situation the decay of the disturbance is governed by the real part of $\lambda_{\max}$ and the disturbance decays without oscillation to the original steady state. At some level of productivity, defined here as r_λ, $u = 0$, the overdamped recovery gives way to the critically damped recovery, where the disturbance decays in a monotonic manner but overshoots the steady state once and then returns to the steady state without oscillations. At still greater levels of productivity, when $\mu < 0$ the system develops damped oscillations with a quasi-period of $T_d \approx 2\pi/(\mu)^{1/2}$.

For the two-species primary-producer-based food chain, $r_{\lambda 2}$ can be estimated as the level of productivity at which the discriminant (μ) of Equation 7.14 becomes negative. We obtained the following relationship:

$$r_{\lambda 2} = r_{C2} + \frac{c_{11} r_{F2}}{4 c_{12} a_2 p_2} - \frac{d_2}{a_2 p_2} \tag{7.16}$$

The placement of this inflection point ($r_{\lambda 2}$) relative to the points of feasibility (r_{F2}) and cascade (r_{C2}) is not readily apparent. It is clear from Equation 7.16 that whether the onset of oscillations occurs before or after the cascade depends on the death rate and ecological efficiency of the predator and the self-regulation of the prey. We find the following three possibilities:

1) if $c_{11}/c_{12}(a_2 p_2)^{1/2} = 2$ then $r_{\lambda 2} = r_{C2}$ (7.17a)
2) if $c_{11}/c_{12}(a_2 p_2)^{1/2} > 2$ then $r_{\lambda 2} < r_{C2}$ (7.17b)
3) if $c_{11}/c_{12}(a_2 p_2)^{1/2} < 2$ then $r_{\lambda 2} > r_{C2}$ (7.17c)

As ecological efficiencies range from 0.03 for large mammals to a theoretical maxima for 0.80 for bacteria, and the self-limiting terms are often less than 0.01, we can deduce that the following would occur with increased productivity: (1) the oscillatory behavior in the dynamics that would follow a minor disturbance initiates within the window of productivity between feasibility (r_{F2}) and cascade (r_{C2}), i.e., $r_{F2} < r_{\lambda 2} < r_{C2}$, (2) that the frequency of the oscillations increases and the period decreases, and (3) that the cascade of biomass for the two-species system is more likely to experience wider fluctuations in dynamics in response to minor disturbances when compared with the pyramid of biomass.

7.3.3 Energetic efficiencies

The analysis of our simple primary-producer-based and detritus-based food chains revealed that both the feasibility and resilience of the food chains are sensitive to changes in productivity, but also to our parameter selections. Equations 7.10–7.13 indicate clear thresholds in the level of productivity required to support additional trophic levels and in the distribution of biomass. From Equations 7.17a–7.17c, we can see that the positioning of the thresholds for changes in structure (r_{Fi} and r_{Ci}) and dynamics ($r_{\lambda i}$) for a given level of productivity along a productivity gradient are dependent on the assimilation efficiencies (a_i) and production efficiencies (p_i), as the denominators of the equations for each threshold included the energetic efficiencies of the consumers. This indicates that the onset of the structural and dynamic changes hinges on the level of productivity and the energetic efficiencies of the organisms in question.

To illustrate this point, we repeated the simulations above in a way that emulated the ecological efficiencies of endothermic and exothermic organisms. The analysis compared the feasibility (F_{NPP}) and return times of food chains of length two, three, and four species with low and high energetic efficiencies. We accomplished this by sampling the efficiencies from Beta distributions, with Beta (2,9) used to sample for organisms with relatively low efficiencies such as endotherms, and Beta (9,2) for organisms with relatively high energetic efficiencies such as ectotherms. The productivity gradient was created by increasing the intrinsic rate of increase of the primary producer, r_1, for the primary-producer-based food chains, and by increasing the input rate of detritus, R_D, for the detritus-based food chains over the range of 10^{-2}–10^5 units for each. All other parameters were sampled from a uniform distribution, of Uniform (0,1) as before. The results of the analysis are presented in Figures 7.4 and 7.5.

As expected we do not see a departure in a qualitative sense in the responses of feasibility and resilience (as measured by return time) of either type of food chain or with increasing food chain length along our productivity gradient from the results presented earlier (Figures 7.4 and 7.5). The results follow the same patterns observed in the previous analyses and confirm what we had suspected, that the underlying distribution(s) of the variables used in the models affect the feasibility and resilience of the food chains. Interestingly, the gulf between the feasibility and resilience of our ectothermic and endothermic organisms grew with increased food chain length. Our sampling for this exercise generated extremes within the means of the parameters for the physiological types, with an average energetic efficiency of

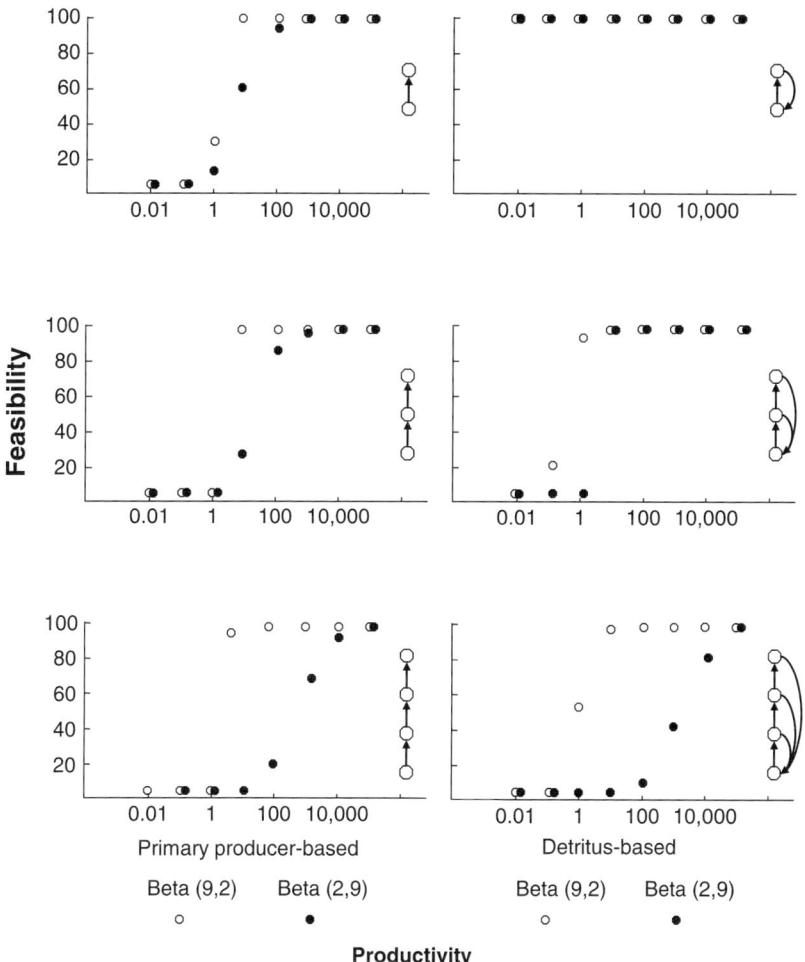

Figure 7.4 The feasibility (sensu Roberts, 1974) of primary-producer-based and detritus-based food chains of length 2, 3, and 4 using the equations presented in Figure 7.1 parameterized for ectothermic (open symbols) and endothermic species (closed symbols). The food chains were evaluated at fixed levels of productivity r_1 for the primary-producer-based food chains and R_D for the detritus-based food chains beginning at 10^{-2} units and increasing by orders of magnitude to 10^5 units. For the remaining parameters, the ectotherms followed a Beta (9,2) and the endotherms followed a Beta (2,9) (redrawn from Moore and de Ruiter, 2000).

~67% for the ectotherms and ~3% for the endotherms, thus straddling the 50% used in the previous analysis. Nonetheless, the differences in feasibility and resilience are consistent with current theory and empirical observations.

All the two-species detritus-based food chains were feasible over the range of R_D that we selected. In all other cases, the feasibility of the primary-producer-based

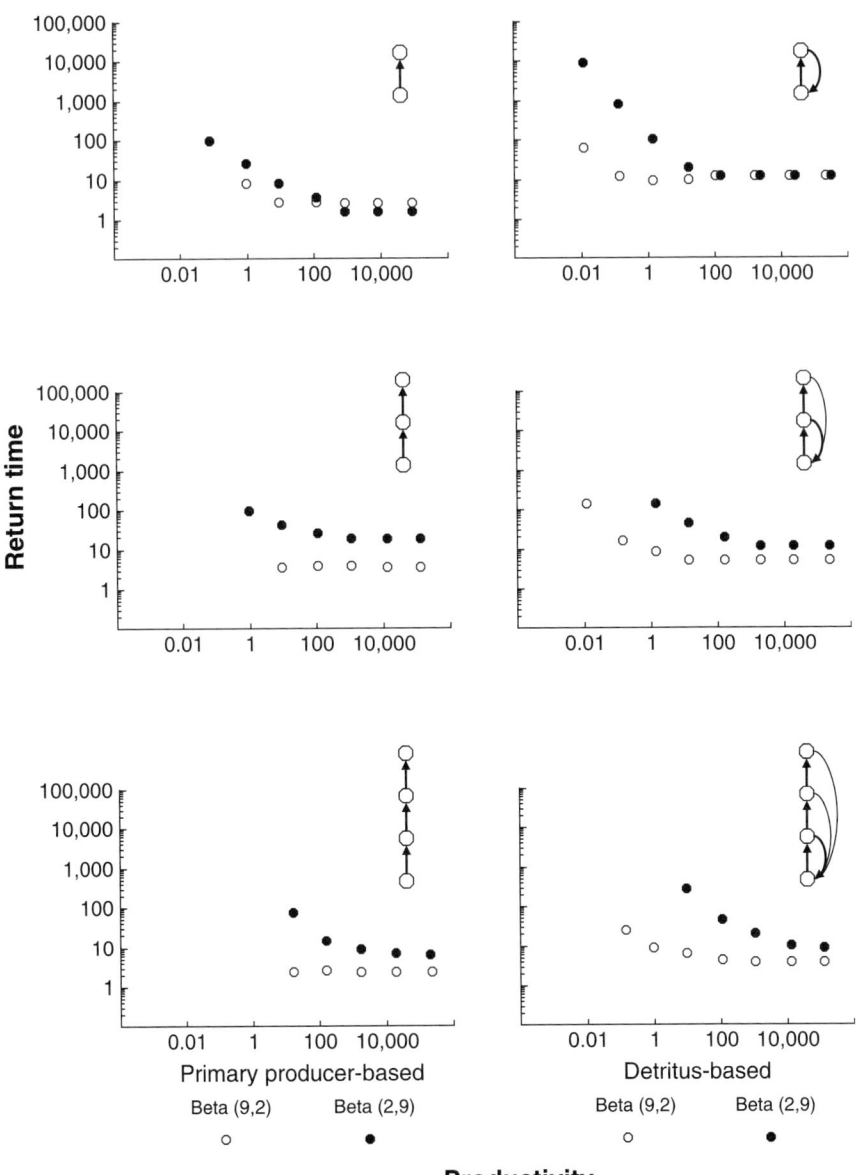

Figure 7.5 The resilience, as measured by return time (RT), of primary-producer-based and detritus-based food chains of length 2, 3, and 4 using the equations presented in Figure 7.1 parameterized for ectothermic (open symbols) and endothermic species (closed symbols). The food chains were evaluated at fixed levels of productivity (r_1) for the primary-producer-based food chains and R_D for the detritus-based food chains beginning at 10^{-2} units and increasing by orders of magnitude to 10^5 units. For the remaining parameters, the ectotherms followed a Beta (9,2) and the endotherms followed a Beta (2,9) (redrawn from Moore and de Ruiter, 2000).

and detritus-based food chains increased with increased values of r_1 and R_D, respectively. The ordering of the points along the gradient where sills of the curves for food chains of length 2, 3, and then 4 reach 100% feasibility follow the predictions of the traditional thermodynamic arguments of EEH (Elton, 1927, Odum, 1953, Hutchinson, 1959, Oksanen et al., 1981). The thermodynamic arguments are further advanced as the food chains modeled with the higher energetic efficiencies approached the sills (100% feasible) at lower levels of productivity than their less-efficient counterparts (Figure 7.4). This would be expected if a higher percentage of consumption were directed toward growth and reproduction rather than metabolic maintenance—then, all else being equal, lower rates of production would be needed to maintain an individual or population of individuals.

The resilience of the food chains was influenced by source of production, rate of production, length of food chain, and energetic efficiency of the consumers (Figure 7.5). Return times decreased with increased rates of production. Detritus-based food chains possessed longer return times than primary-producer-based food chains. Longer food chains exhibited longer return times regardless of source or the energetic efficiency of their consumers. Food chains dominated by consumers with higher energetic efficiencies had faster return times than those dominated by organisms with lower energetic efficiencies along much of the productivity gradient. The differences were much greater for the detritus-based food chains than for the primary-producer-based food chains. Many of these results are evident from the relationships presented in Equations 7.10–7.13. For a given rate of production, and all else being equal, increased energetic efficiency translates to increased energy flow through the chain, and shorter return time. Also evident from our models, particularly Equation 7.14 and Equation 7.16, are the transitions in the dynamic states.

7.3.4 Body size and home range

The observed pattern of increased feasibility within increased productivity supports the corollary to the productivity hypothesis that food chains dominated by ectotherms should be able to persist in lower productivity environments than those dominated by endotherms. However, this gulf may be blurred if body size and home range are factored into the argument. Ectotherms tend to have smaller body size than endotherms. If we couple this with a minimum population density for each physiological and body size type, the feasibility curves shift to the right along the productivity gradient, but more so for the endotherms (Figure 7.4). When home range is factored into the argument by allowing larger organisms to have larger home ranges than smaller organisms, the larger organisms encounter more units of production than smaller organisms at a given point along the productivity gradient. This would shift the curves to the left along the productivity gradient, with less production per unit of area being required for the endotherms, given their larger home ranges (Figure 7.4).

156 • *Energetic Food Webs*

The consumption coefficient of the consumers (c_{ij}) is an important parameter when considering the placement of the structural and dynamic thresholds with increased productivity. We forego a presentation of the analyses or lengthy discussion, as the results are similar to that for the energetic efficiencies. What is clear from Equations 7.10–7.17 is that the structural and dynamic thresholds depend in part on the consumption coefficient of the consumer on prey. All else being equal, high consumption rates require greater rates of productivity to maintain the populations. Having said this, the foraging range of a consumer is an aspect of consumption that needs to be considered. Recall that the dimensions of the populations (X_i) in our models are expressed in terms of mass per unit of space (e.g., g C m^{-2}). Given this, when the foraging range of a consumer exceeds the scale on which productivity is measured, the spatial extent that it encompasses should be included in the estimate of r_{Fi}. One way to visualize this would be to scale productivity in terms of the foraging range of the consumer rather than an arbitrary area (i.e., a m^2 in our case). This in effect would redefine the thresholds in terms of the area of a given rate of production that would be required to support a consumer with a given physiology and foraging area (Figure 7.6).

In addition to increasing the realized productivity through a larger spatial extent, increased home range affects the spatial distribution of consumers and the degree of

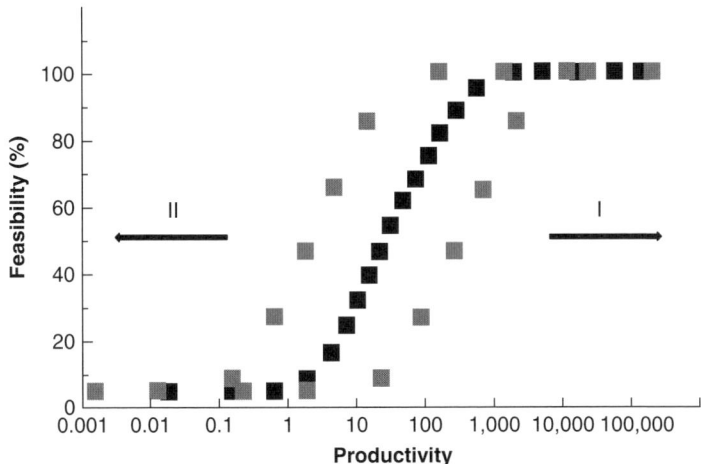

Figure 7.6 The feasibility of a food chain of a given length (■) as affected by physiology, body size, and home range (■) along a productivity gradient under different scenarios. I. If the home ranges of the organisms are fixed, body size and physiology impose constraints of trophic structure. Communities dominated by endotherms would require greater rates of production than those dominated by ectotherms. Communities dominated by large organisms would require greater rates of production than those dominated by smaller organisms. II. If the home ranges are allowed to vary with all else being equal, then communities dominated by organisms with larger home ranges would require lower rates of production then those dominated by organisms with smaller home ranges.

connectivity of patches of production, prey, and consumers. We will discuss these points in greater depth in later chapters.

7.4 More complex models

The transitions in structural and dynamic states along a productivity gradient that we discussed above using our simple models are telling but not unique to them. More complex deterministic formulations exhibit behaviors that bring us to similar conclusions, but with an important difference (Rosenzweig, 1971, May, 1976, Hastings and Powell, 1991, Cushing et al., 2003, McCann et al., 2005, Carpenter et al., 2007, Hastings et al., 2008). Studies of more complex formulations demonstrate an interesting array of dynamic changes to productivity following disturbances and in the intrinsic nature of their trajectories, be they simple repeating oscillations or complex chaotic trajectories. Here we explore the impact of enrichment on variations of our models. Below we will discuss three ways of layering in complexities: (1) altering the functional responses of the trophic interactions, (2) applying different model formulations, and (3) focusing on chaos and the changes in exogenous factors.

7.4.1 Attack rates and dynamics

Alternative models that possess more complex predation terms and functional responses are capable of generating a suite of dynamic states dependent on the choice of parameters. May (1973) reviewed several models of this sort, noting that functional responses tended to be destabilizing if the rate of the predators' consumption increased at a less than linear rate with increasing prey densities, and stabilizing if it were to increase at a more than linear rate. Our aim here is to focus on how models of systems with different functional responses respond to changes in productivity (enrichment).

Recall that a functional response characterizes how a predator adapts or adjusts its attack rate to changes in prey density. We can rewrite the predation rate $c_{12}X_1X_2$ that we have used up until now in a more general form as $F(X_1)X_1X_2$, where $F(X_1)$ is the predation term, and $F(X_1)X_1$ represents the functional response of the predator to changes in prey density. The form of the functional response affects the dynamic properties of the interaction and the response of the system to enrichment. For our two-species model with the type I functional response, the level of enrichment did not affect its local stability, but it did influence the return time and dynamics of the system as it returned to equilibrium following a disturbance (Figure 7.2). Type II and type III functional responses offer a richer set of possibilities. We will focus on the two-species primary-producer-based food chains used above, but substitute the type I functional response with the type II response to make this point. The equations are as follows:

$$\frac{dX_1}{dt} = r_1 X_1 - \frac{r_1}{K_1} X_1^2 - \frac{m_{12} X_1}{b_{12} + X_1} X_2 \quad (7.18a)$$

$$\frac{dX_2}{dt} = -d_2 X_2 + e_2 \frac{m_{12} X_1}{b_{12} + X_1} X_2 \quad (7.18b)$$

where K_1 is the carrying capacity of X_1 ($K_1 = r_1/c_{11}$ when $X_2 = 0$), and e_2 is the energetic efficiency of the predator ($e_2 = a_2 p_2$). The functional response expressed in Equation 7.18a and Equation 7.18b is as follows:

$$F(X_1) = \frac{m_{12} X_1}{b_{12} + X_1} \quad (7.19)$$

where m_{12} represents the consumption coefficient of the predator and b_{12} represents the population level of the prey where the predation rate per unit of prey is at half of its maximum, i.e., the half-saturation constant.

The dynamic properties of this model can be explained by evaluating the relationship between the isoclines for the prey and predator in a graph of prey density against predator density (Figure 7.7). The prey isocline is a collection of points defined as $dX_1/dt = 0$ evaluated for X_2, whereas the predator isocline is the collection of points defined by $dX_2/dt = 0$ evaluated for X_1. Hence the prey isocline is given by:

$$X_2 = \frac{r_1 b_{12}}{m_{12}} + \left(\frac{r_1}{m_{12}} - \frac{r_1 b_{12}}{m_{12}}\right) X_1 - \frac{r_1}{K_1 m_{12}} X_1^2 \quad (7.20)$$

and the predator isocline is given by:

$$X_1 = \frac{d_2 b_{12}}{e_2 m_{12} - d_2} \quad (7.21)$$

The prey isocline generates a hump-shaped curve that intersects the y-axis at $r_1 b_{12}/m_{12}$, intersects the x-axis at K_1, and an apogee at $\partial X_2/\partial X_1 = 0$, the nontrivial case in which the effect of the predator on the dynamics of the prey is zero. The predator isocline is a vertical line. The intersection of these isoclines defines the equilibrium. Given the shape of the prey isocline, the positioning of the predator isocline relative to the prey isocline determines whether the equilibrium is stable (Figure 7.7). If the predator isocline is positioned to the right of the apogee the equilibrium is stable, and if it is positioned to the left of the apogee the equilibrium in unstable, leading to oscillatory dynamics or extinction.

It is clear from the relationships presented in Figures 7.2 and Figure 7.7 that the dynamics and stability of our two-species system, modeled with either a type I or a type II functional response, is affected by the level of enrichment. If we assume that the predators experience neither intraspecific competition nor cannibalism, then the predator isocline was defined solely by the phenotype of the predator and its prey. The prey isocline was also defined by the phenotypes, but was subject to changes in enrichment through r_1. Increasing r_1 increases the height of the apogee and K_1 of the prey isocline, whereas the predator isocline remains fixed (Figure 7.8). If enrichment

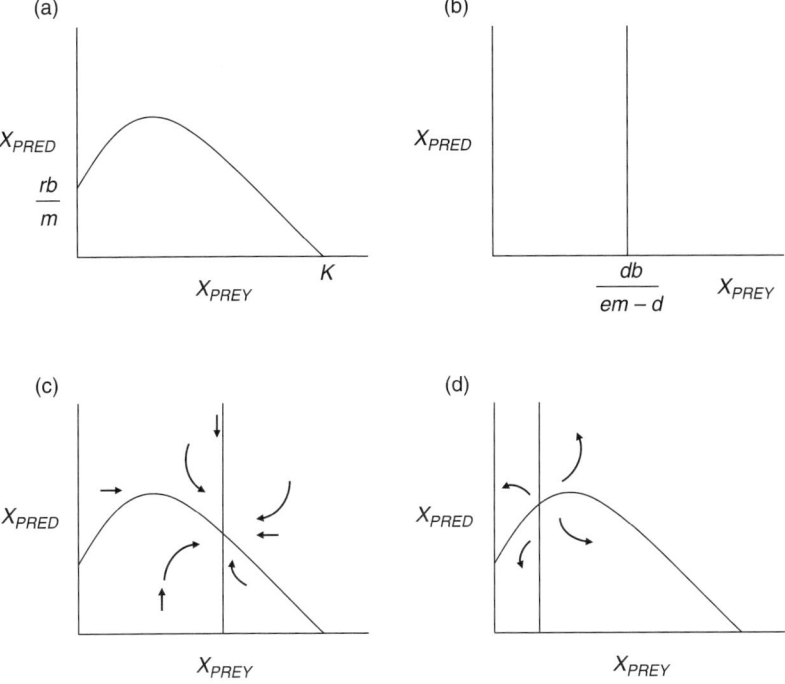

Figure 7.7 Isoclines for (a) prey and (b) predators modeled with type II functional responses. (c) The system is stable if the predator intersects to the right of the prey isocline apogee. (d) The system is unstable if the predator intersects to the left of the prey isocline apogee.

is to proceed to the point that the equilibrium falls to the left of the apogee, the system oscillates, and the equilibrium does not appear stable.

7.4.2 Paradox of enrichment

Rosenzweig (1971) dubbed the phenomenon illustrated in Figure 7.8, in which increased productivity leads to oscillations and possible instability, as the "paradox of enrichment." He assessed the dynamics of six models that shared the characteristics of the model presented above with the type II functional response at different levels of enrichment. One of the models was structured similarly to the primary-producer-based model we presented above, for a producer X_1 and predator X_2. Rosenzweig (1971) modeled the dynamics of the producer and consumer as we did in Figure 7.1 using the Pearl–Verhulst logistic form $r_1 X_1 (1 - X_1/K_1)$ for the producer where the carrying capacity is $K_1 = r_1/c_{11}$. The difference in the equations used by Rosenzweig (1971) was in the functional response of the consumer on the prey. Rosenzweig (1971) modified the form to $c_{12} X_2 X_1^g$, where $1 \geq g > 0$, providing

160 • Energetic Food Webs

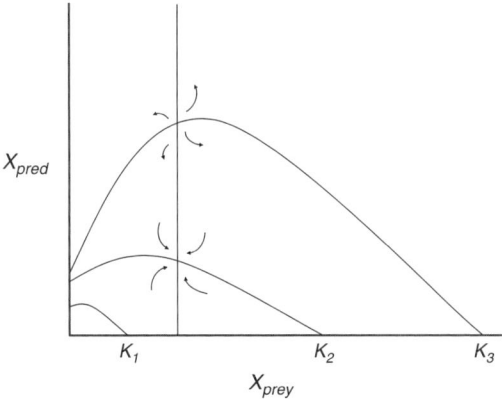

Figure 7.8 The relative positioning of the isoclines is a function of productivity and energy flow rates. For feasible (F_{NPP}) systems the predator isocline must intersect at or to the left of K for prey isocline. With increased productivity the prey isocline moves to the right. If the predator isocline is positioned to the right of the apogee of the prey isocline, the system is locally stable. If productivity is too high the predator isocline is positioned to the left of the apogee of the prey isocline, and the system is not locally stable and is prone to oscillations.

the prey a refuge from the predator. The primary-producer-based model we presented (Figure 7.1) with a type I functional response is the extreme case in which $g = 1$. Once again, the key results of the analyses can be gleaned from the positioning of the predator isocline relative to the prey isocline (Figure 7.8). The prey isocline is a vertical line up to the apogee defined by the refuge, beyond which it is hump-shaped. The predator isocline is a vertical line. Enrichment affects the height, extent, and positioning of the apogee of the hump in the prey isocline, but has no effect on the positioning of the predator isocline. If the predator isocline is positioned to the right of the apogee of the prey isocline the system is unstable, and to the left it is stable. Hence, enrichment has the potential to destabilize a system.

7.4.3 Enrichment and donor control

Enrichment of the detritus-based models like those presented above does not appear to induce the instabilities we observed for their primary-producer-based counterparts. To illustrate this point we provided detritus a refuge following the approach of Rosenzweig (1971) by substituting the density of detritus, X_i, in the predation terms with X_i^g, g, where $1 \geq g > 0$. This makes sense biologically, as at a given point in time detritus may be in a form that is inaccessible or highly recalcitrant. With this change the equilibrium values for detritus (X_i) and the consumer (X_j) are as follows:

$$X_i^* = \left(\frac{d_j}{a_j p_j c_{ij}}\right)^{\frac{1}{g}} \quad (7.22a)$$

$$X_j^* = \frac{R_D p_j}{d_j(1 - p_j)} \quad (7.22b)$$

The equilibrium for the detritus remains fixed for the same reasons as the formulations without the refuge, but at a higher density given the refuge. The equilibrium of the consumer remains the same. The inclusion of a refuge for detritus leads to the following modification of the Jacobian matrix:

$$A = \begin{bmatrix} -a_j c_{ij} X_j^{*g-1} & d_j - a_j c_{ij} X_i^{*g} \\ a_j p_j c_{ij} X_j^{*g-1} & 0 \end{bmatrix} \quad (7.23)$$

Providing a refuge for detritus does not affect the dynamic properties of the system in a qualitative sense. The following conclusions can be drawn when comparing the system with the refuge to one that does not include the refuge: (1) both systems are stable, (2) for a given level of input, R_D, the system with the refuge possesses longer return times, (3) the onset of oscillations following a disturbance occurs at greater level of input for the system with the refuge, and (4) the influence of the diagonal term in the Jacobian matrix on stability, i.e., diagonal dominance, is reduced by the magnitude of the refuge, g.

7.4.4 Chaos and complex dynamics

Studies of chaos have greatly expanded our understanding of the possible structural and dynamic states and the responses of structure and dynamics to enrichment. May (1976) presented the seminal work on chaos for biological systems using the following one-dimensional discrete logistic equation by increasing the intrinsic rate of increase (r) of the model organis

$$x_{t+1} = r x_t (1 - x_t) \quad (7.24)$$

At low levels of r the system possessed a single point attractor, or stable equilibrium, but as r increased the population density (X) increased to a point after which a bifurcation occurred, giving rise to two-cycle attractor consisting of a maximum and alternative minimum (Figure 7.9). At still large values of r the system transitioned to a four-cycle attractor, then an eight-cycle attractor, to an attractor that possessed multiple (possibly infinite) points.

Later work yielded similar transitions in dynamic states by exploring chaos using models of simple food chains and the development of a staged life cycle (Hastings and Powell, 1991, Cushing et al., 2003). In these studies, parameters other than the intrinsic rate of increase were used to induce transitions in dynamic states, as alterations in the energetic efficiencies, cannibalism and consumer consumption rates, and death-rate-induced bifurcations. The parameters share the property of affecting the rate of the transfer of energy through the system. To illustrate this point we turn to the seminal

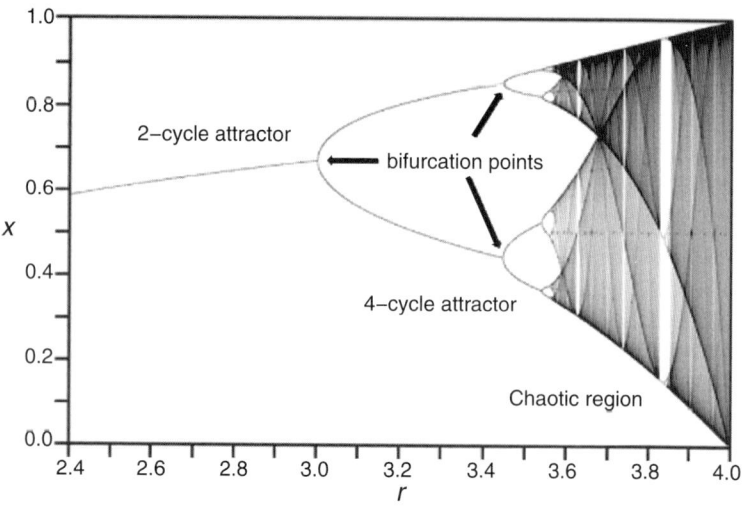

Figure 7.9 The logistic map, also known as the bifurcation diagram for the simple logistic difference equation for a single population (Equation 7.24; May, 1976). The graph depicts the behavior of the population density, X_t, at different values of the intrinsic rate of increase, r, as time $t \to \infty$. For low values of r the population densities converge to an equilibrium. At $r = 3$ a shift in the dynamic state occurs as the population density does not converge on a single value, but undergoes a bifurcation, after which it oscillates between two values. Between $r = 3.0$ and $r = 3.57$ additional bifurcations occur. At $r = 5.57$ and beyond, chaos ensues followed by oscillations of period 3, 6, 12, and 24, and then back to chaos.

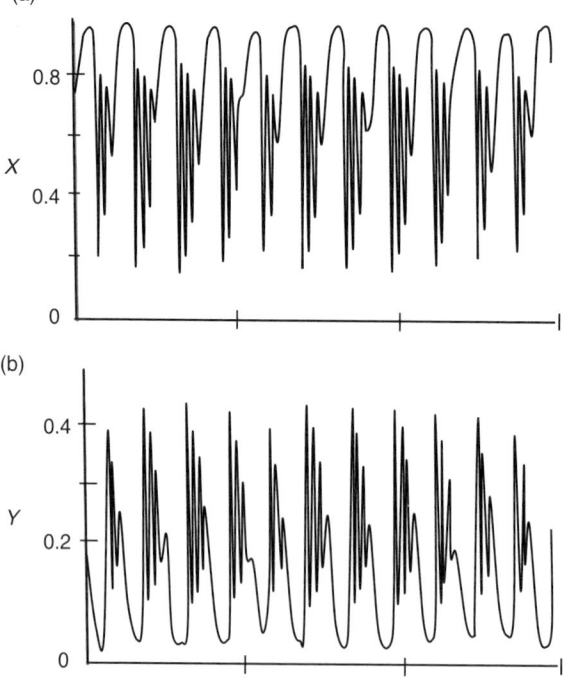

Enrichment, trophic structure, and stability • 163

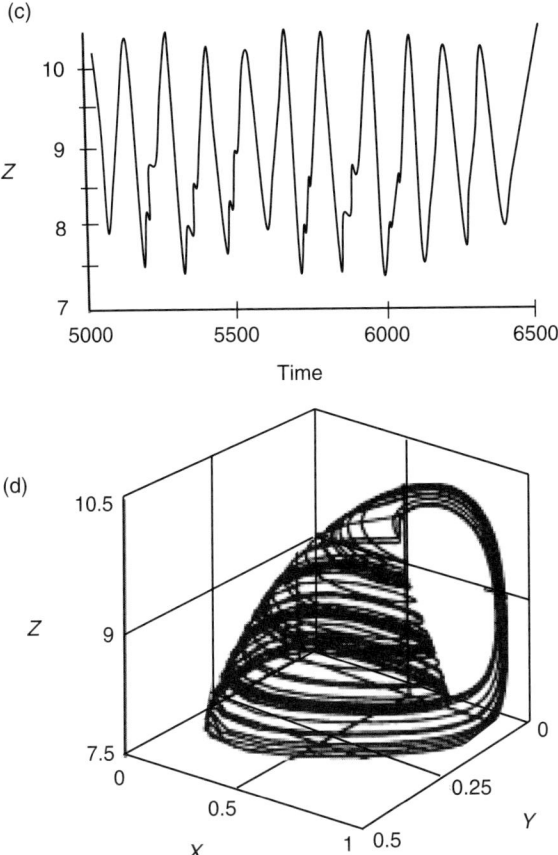

Figure 7.10 The population dynamics of (a) a basal species, (b) a consumer of the basal species, and (c) a predator of the consumer, and the phase-space representations for the three species food chain model as presented by Hastings and Powell (1991). The model included type II functional responses for all trophic interactions.

work by Hastings and Powell (1991) and their study of chaos using a model of a simple three-species food chain (Figure 7.10), not unlike the ones presented above, which incorporated predation terms that lead to type II functional responses (Holling, 1959).

The formulations used by Hastings and Powell (1991) were as follows:

$$\frac{dX}{dt} = rX(1 - X/K) - e_1 F_1(X)Y$$

$$\frac{dY}{dt} = F_1(X)Y - F_2(Y)Z - d_1 Y \quad (7.25)$$

$$\frac{dZ}{dt} = e_2 F_2(Y)Z - d_2 Z$$

with the functional response as represented as:

$$F_i(U) = A_i U/(B_i + U) \tag{7.26}$$

for $i = 1,2$ and species $U = X, Y,$ or Z

where X is the basal species with carrying capacity K, Y is a consumer of X, and Z is a consumer of Y. The constant r is the specific growth rate of species X_1, d_2 and d_3 are the specific death rates, and e_1 and e_2 are the conversion efficiencies of the prey to predators for species X_2 and X_3, respectively. After simplifying the equations to a nondimensional form, the equations and type II functional responses are as follows:

$$\begin{aligned} dx/dt &= x(1-x) - f_1(x)y \\ dy/dt &= f_1(x)y - f_2(y)z - d_1 y \\ dz/dt &= f_2(y)z - d_2 z \end{aligned} \tag{7.27}$$

and

$$f_i(u) = \alpha_i u/(1 + b_i u) \tag{7.28}$$

where x, y, and z are the nondimensional measures of population size, t is the nondimensional variable for time, α_i is the new constant that represents the consumption coefficient of the predator, and b_i represents the new constant for the population level of the prey where the predation rate per unit of prey is at half of its maximum, i.e., the half-saturation constant. A model of this form is capable of: (1) stable equilibrium dynamics, (2) limit cycle behavior, and under the right choice of parameters, (3) chaos. As shown in Figure 7.10, the model indeed generated chaotic dynamics with a parameter set that was biologically reasonable, if not extreme (see McCann and Yodzis, 1994).

Hastings and Powell (1991) were able to control the dynamics by varying a parameter, b, which scaled the carrying capacity K_1 to the half-saturation constant B_1 from the functional response of the plant to herbivore interaction ($b = K_1/B_1$). By manipulating b in this manner, one of two extremes or all options in between were taking place. On the one hand, the carrying capacity of the basal species was held constant and the half-saturation term was decreased. This scenario emulates changing the efficiency by which the consumer captures prey and converts the prey to biomass. As the efficiency of the consumer increases, i.e., the half-saturation term decreases, the dynamics of the model transition toward chaos. On the other hand, the half-saturation term B_1 was held constant and the carrying capacity K_1 of the basal species increased. Falling back on the logic presented by Rosenzweig (1971), if K_1 represents the standing crop, the basal species at equilibrium, in the absence of consumers ($Y = 0$), then K_1 is proportional to the influx of a limiting resource or energy (i.e., $K_1 = r_1/c_{11}$). Hence, an increase in K_1 equates to enrichment.

The model provides a way to connect nutrients and nutrient dynamics, as proposed by Odum (1969), for developing and mature communities to our study of dynamics and stability. With this model we can induce transitions in dynamic states by manipulating exogenous factors such as water and nutrients that affect growth and reproduction. To illustrate these points we reconfigured the models of Hastings and Powell (1991) so that the intrinsic rate of growth rate of the basal species was dependent on a resource pool in a type II manner. The resource pool was replenished by an allochthonous source. The new model in some regards was akin to adding a new basal resource and transforming the models from a three-species exploitative model to a four-species model that includes a donor-control component. We could increase the productivity of the original basal resource by either varying the half-saturation term for the uptake of resources by the basal species or by increasing the allochthonous input of the resource. As shown in Figure 7.11, we were able to duplicate the results of Hastings and Powell (1991). However, if we decreased the rate of enrichment the dynamics transitioned from chaos to what appeared to be a limit cycle and then a stable equilibrium point. If we increased enrichment beyond the levels that generate chaos, the system transitioned to yet another state.

Several implications can be drawn from this analysis. First, the exercise illustrates what Rosenzweig (1971) discussed in the paradox of enrichment but with more complex models. Enrichment leads to a change in dynamic states, some of which are not locally stable. Second, a review of the distribution of biomass within the food chain along the gradient of enrichment shows a pyramidal structure at low levels of enrichment when the system possesses a stable equilibrium, to a transition toward an inverted pyramid of biomass as the system moves toward chaos. In other words, enrichment moves biomass to higher trophic levels, and this movement is associated with oscillatory dynamics and instabilities. This is the conclusion we drew from our analyses of the Central Plains Experimental Range (CPER) food web presented in Chapter 6, and study of the simple food chains in this chapter (Figures 7.2 and 7.3). Third, given the first two conclusions it may be that resource limitations have a stabilizing affect on systems. Nitrogen limitations are common in terrestrial ecosystems, whereas phosphorous limitations are seen in many aquatic systems. Enrichment of these nutrients where they are limited often leads to the types of instabilities exhibited in the models. Fourth, our model system that was open to external nutrient sources was prone to oscillations and chaos, whereas those models that were closed or less reliant on the external sources and more reliant on internal nutrient sources exhibited attenuated dynamics. These results mirror the observations made by Odum (1969) for systems in early developmental stages and later mature stages, respectively. Finally, if systems were to exist in a resource-limited state, there are clear advantages to the species within that system having the capacity to survive in a resource-limited environment or at a low level of productivity, and yet have the capacity to respond to enrichment.

7.5 Connections to real-world productivity

We started our analyses in Chapter 2 by randomly selecting all parameters from a uniform distribution, I (0,1). In this chapter we stretched the analysis along a productivity gradient over several orders of magnitude, ranging from 10^{-2} to 10^5

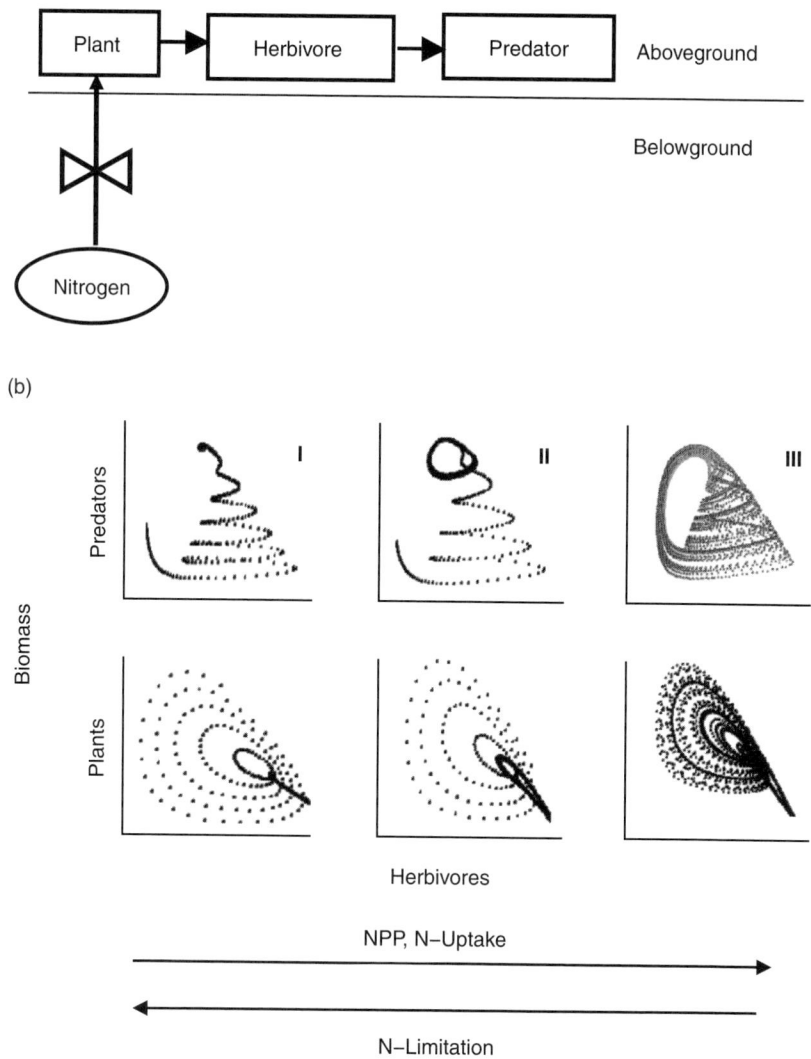

Figure 7.11 Phase-space of the biomass for plants vs. herbivores (lower panels) and predators vs. herbivores (upper panels) for a plant–herbivore–predator food chain coupled to a decomposer subsystem linked through the rhizosphere. (a) The models included type II functional responses for all the trophic interactions, including the uptake of nitrogen from the decomposer subsystem through the rhizosphere to the plant. (b) Net primary productivity (NPP) was regulated by nitrogen availability through rhizosphere activity by altering the saturation constant, within the type II functional response or the size of the nitrogen pool. The plant–herbivore–predator portion was initialized using the parameter selection from Hastings and Powell (1991). At low levels of the NPP the system approaches a stable equilibrium (I). At increased levels of NPP the system exhibits limit cycle behavior (II) and ultimately chaos to produce the "teacup" phase-space (III).

units of productivity. We then skewed the parameter selection in a way that created ectotherm-like and endotherm-like organisms. We finished by exploring how enrichment might affect more complex models. How do these analyses scale to world physiologies and observed levels of productivity? This is an important question given the nature of the responses, particularly for those along our productivity gradient. For example, if real-world productivities were to scale to within the 10^3–10^5 range of productivity units for the simple models shown in Figure 7.2, then we might conclude that productivity sets a lower limit to community structure but would have little effect on food chain length and resilience. However, if the real-world levels of productivity in g C m^{-2} yr^{-1} were to scale to the 10–10^3 range of productivity in the models, our models would align well with real-world levels of productivity.

We turn to parameters (a_i, p_i, d_i, and c_{ij}) and biomass estimates (B_i) for the soil food webs we presented to scale the units of productivity used in the model to estimates of real-world productivity (Table 7.1). By definition the assimilation and production efficiencies range from 0 to 1, and as explained above, correspond to energetic efficiencies that range from 0.03 to 0.67. These values span the observed efficiencies of large vertebrates to bacteria, respectively. Next we aligned the time step in the model to years in the real world. The nominal death rates of the microorganisms, protozoa, and invertebrates, d_i, ranged from 0.5 to 6.0 yr^{-1}. Given that we sampled death rates in the model from 0 to 1, the observed range in nominal death rates maps one time unit in the model to approximately 0.17 years. We then turned to the consumption coefficients, c_{ij}, which ranged from 0 to 1.08 (gm^{-2})$^{-1}$ yr^{-1}, not too far from the 0–1 interval in the model. We obtained these estimates by dividing the feeding rates, $c_{ij}B_iB_j$, by the appropriate estimates of steady-state biomasses, B_i. However, if we want our real-world estimates to correspond to our new time step of 0.17 years, the mass unit scales to 6.0 g m^{-2} to maintain the variation from 0 to 1. Hence in the extreme, one unit of productivity in the model corresponds to $\sim$36.0 g C m^{-2} yr^{-1}. As most of the death rates were < 2.0 yr^{-1}, a more conservative estimate would scale one unit of model productivity to 4.0 g C m^{-2} yr^{-1}.

The proposed alignment of the model units of productivity to real-world levels of productivity suggest that a high percentage of real-world levels of productivity are well beyond the observed thresholds in feasibility suggested by the models. Most studies place real-world levels of productivity from 25 g C m^{-2} yr^{-1} to an extreme of 4000 g C m^{-2} yr^{-1} (Table 7.1), but local inputs or throughputs of detritus can far exceed the upper estimates, particularly in freshwater streams. Although this range does correspond to the steep increase that we observed in our modeling studies, the studies did reveal a high percentage of feasible configurations regardless of energetic efficiency or energy source. If we further consider how ubiquitous detritus is in real ecosystems, and the high percentage of these models that were feasible, these results may explain why analyses of food web descriptions have failed to show definitive effects of productivity on food chain length (Pimm, 1982, Post, 2002). Even at low extremes in productivity a considerable number of configurations were found to be feasible. In short, the physiologies of organisms have evolved in concert

Table 7.1. Estimates of the range and mean net primary productivity (NPP) of the world's ecosystems (Whittacker, 1975, Brewer, 1994, Moore et al., 1996). These estimates indicate that between 50.3 and 99.96% (depending on scaling) global NPP occurs within ecosystems with levels of productivity within the range in which feasibility and return times are relatively insensitive to small changes (see Figure 7.2).

Productivity (g C m^{-2} yr^{-1}) Model Range*			Ecosystem Type	NPP (g dry mass m^{-2} yr^{-1})		World NPP (Tg C yr^{-1})	Total World NPP (%)	Cumulative World NPP (%)
A	B			Range	Mean			
0–10	0–10		Extreme deserts: rock, sand, ice	0–10	3	0.07	0.04	0.04
10–100			Desert and semi-desert scrub	10–250	90	1.6	0.94	0.98
			Open ocean	2–400	125	41.5	24.39	25.37
			Tundra and alpine	10–400	140	1.1	0.65	26.02
			Lake and stream	100–1500	250	0.5	0.29	26.31
			Continental shelf	200–600	360	9.6	5.64	31.95
			Upwelling ocean zones	400–1000	500	0.2	0.12	32.07
			Temperate grassland	200–1500	600	5.4	3.17	35.24
			Cultivated land	100–3500	650	9.1	5.35	40.56
	10–100		Woodland and shrubland	250–1200	700	6.0	3.53	44.12
			Boreal forest	400–2000	800	9.6	5.64	49.77
			Savanna	200–2000	900	13.5	7.93	57.71
			Temperate deciduous forest	600–2500	1200	8.4	4.94	62.65
			Temperate evergreen forest	600–2500	1300	6.5	3.82	66.47
			Estuaries	200–3500	1500	2.1	1.23	67.70
		>100	Tropical seasonal forest	1000–2500	1600	12.0	7.05	74.75
			Swamp marsh	800–3500	2000	4.0	2.35	77.10
			Tropical rain forest	1000–3500	2200	37.4	21.98	99.08
			Algal beds and reefs	500–4000	2500	1.6	0.94	100.00
Total					333	170.00		

* Model ranges correspond to the productivity increments of the x-axes in Figure 7.3. If 50% of dry mass is carbon, then one unit of productivity on the model scales to either 4 (column A) or 36 (column B) g C m^{-2} yr^{-1}, depending on initial assumptions.

with real-world levels of productivity (a physiological attribute) and have adapted to them.

7.6 Summary and conclusions

The results presented in this chapter raise some interesting points that have implications for real-world systems and the approaches taken to study them. We will address four aspects of systems: (1) comparisons of the structural and dynamic responses of primary-producer-based and detritus-based food chains to enrichment, (2) a model for community development along a gradient of productivity that combines structural and dynamic considerations (e.g., lower limits of productivity, pyramids of biomass, and dynamics), (3) comparisons of ectothermic and endothermic physiologies, life histories, and body sizes suggest some of the "devious strategies" that have evolved to cope with variation in productivity, and (4) the structure and dynamics of food webs are inextricably linked through energetics as simple and complex systems undergo abrupt changes in dynamics with subtle changes in energy inputs. We will end with a discussion of the implications of these aspects on the approaches taken to study them.

We discussed how primary-producer-based models and detritus-based models respond to increases in the level of productivity (enrichment). Monte Carlo simulations demonstrated that both classes of models yield qualitatively similar results with some important differences. The length of food chains in both classes of model increased with increases in productivity. However the two-species detritus-based models were feasible at extremely low levels of productivity. The dynamic properties of primary-producer-based and detritus-based models shared similarities at the upper levels as well, with both classes exhibiting decreases in return times. Although we modeled and studied them separately, within real food webs primary-producer-based and detritus-based systems are intertwined at several levels.

We then compared models constructed with parameters selected from distributions designed to emulate the dynamics of ecologically efficient organisms to those that were distributed in a manner that emulated less-efficient organisms. In general, the responses followed what many had predicted (Yodzis and Innes, 1992), that the models dominated by more efficient organisms were more likely to be feasible and had shorter return times at a given level of productivity than less-efficient organisms. The nature of the interaction between structure and dynamics depended on the efficiencies of the organisms within the food chains and whether the food chains were based on plants or on detritus.

Detritus and nutrients as resources provided interesting twists to how we might interpret the response of systems along a real productivity gradient. The theoretical lower limit in productivity of the detritus-based model appears to be limited only by the energy requirements needed to maintain a single cell. If this is not an artifact of our construction, the implications of the result are that for certain types of organisms (e.g., unicellular ectotherms) extremely low levels of input are required for survival.

At more nominal and extreme levels of productivity, the detritus-based models dominated by endotherms tended to perform less well than their ectotherm-dominated counterparts. Although detritivory is ubiquitous across all heterotrophic taxa and physiological types, it clearly appears to be dominated by ectotherms.

The exercises also revealed that, all things being equal, enrichment of systems that are closed to new top predators tend to lead to movement of biomass to the upper levels at steady state. Coupled with the upward movement of biomass are changes in the dynamic states. In the case of the simple models, return times decreased but oscillations ensued with increased enrichment. In the case of the more complex models, the upward movement of biomass with enrichment resulted in transitions from dynamics near equilibrium to oscillatory states (limit cycles and chaos), then to eventual collapse. These findings parallel those of the analysis of our empirically based soil food webs, in that stability was closely coupled to the distribution of biomass.

The lengths of food chains within ecosystems are thought to be limited either by the productivity of ecosystems or by the resilience of a system following a minor disturbance. Moreover, for a given level of productivity, ecosystems that are dominated by organisms that efficiently utilize energy for growth and reproduction are expected to possess longer food chains than those dominated by organisms that utilize energy less efficiently. These conclusions were based on simple models of ecosystems that did not explicitly consider the energetic efficiencies of organisms, nor the internal cycling of matter, nor the role of detritus. Our results suggest that both productivity and dynamics are in play. A final observation was made when attempting to align the gradient of productivity studied in our models to the observed gradients of productivity across the globe. The real-world levels of productivity appear to be aligned with the levels of productivity of the models in such a way that food webs have positioned themselves within the optimum regions of space defined by the interplay between the constraints imposed by productivity and those imposed by dynamics.

An assessment of more complex models with nonlinear functional responses provided a richer array of dynamic possibilities and has revealed how small changes in parameter values, particularly those associated with the predation rates and functional responses, might influence responses to enrichment. On balance the results were qualitatively similar to those of the simpler systems. The transitions in dynamic states that we presented above share some important features. In each case they were induced by decreasing the resident time of energy—hence increasing the flow rate of energy through the systems. The result of this increase in flow rate that preceded a transition in dynamic state was a redistribution of biomass within the system, which on average resulted in movement of biomass to higher trophic positions.

As for nutrients, limitations in supply can influence the reproductive rates of primary producers, limiting their potential and thereby limiting the availability of production to higher trophic levels. The implications here are that nutrient limitations can serve to attenuate the dynamics of systems that have a built-in capacity for high levels of production. On the one hand, in the long term systems might evolve to develop at the lowest level of productivity that is physiologically possible, yet in the short term have the capacity to respond to increases in productivity.

Numerous studies have established clear relationships between productivity and the dynamic stability (Rosenzweig, 1971, May, 1976, DeAngelis et al., 1978, Hastings and Powell, 1991), and productivity and trophic structure (Oksanen et al., 1981, Moore et al., 1993). The reported relationship between loop weights and pyramids reflects a richer relationship inasmuch as both are functions of productivity and energy flow. That is, the pyramid of biomass and loop weight are not the basis for stability, but rather were respective co-occurring features in trophic structure and trophic dynamics that occur with changes in productivity or the flow of energy through a system. Hence, any factor that increases productivity or accelerates the flow of energy through the whole web or within a segment of a web also increases loop weight, alters trophic structure by moving biomass to higher trophic levels (i.e., pyramid of biomass to a nonpyramidal structure of biomass) and alters the dynamics (i.e., increases the likelihood of instability). This reinforces the conclusion that energetics, trophic structure, and trophic dynamics are inextricably interrelated.

Linking these simple constructs at higher trophic positions to form more diverse and compartmentalized food webs is a topic we will discuss at length in Chapter 8. For this discussion, the relationship between architecture and stability suggests that communities with a higher diversity of production at the base of the food web that give rise to subassemblages of species (compartments) support a greater diversity of consumers (Moore and Hunt, 1988), and for both aquatic and terrestrial systems are more productive—and stable, in that they are less prone to oscillations (Naeem et al., 1994, Tilman et al., 1996, McCann et al., 2005, Rooney et al., 2006, Rooney et al., 2008).

8

Modeling compartments

8.1 Introduction

In this chapter we explore a special type of community organization: the compartmentalization and hierarchical arrangement of trophic interactions, energy flow, and biomass in space and time. The interest in compartments arose from studies on the relationships among diversity, community complexity, and community stability. The concept that diverse and complex ecosystems were more stable then depauperate simple ecosystems emerged during the early and mid-twentieth century from decades of observations of real ecosystems. Many had argued that diverse complex systems were more stable than simpler ones, given their perceived greater ability to resist and respond to disturbance (Elton, 1927, MacArthur, 1955, Hutchinson, 1959). Empirical field observations on complexity and stability aside, mathematics, systems thinking, and hierarchy theory offered an important twist to the debate (Simon, 1962, Gardner and Ashby, 1970, May, 1972): namely, that for a given number of interacting entities, the stability of the system depended on the number of interactions among the entities, and more importantly the arrangement of interactions. The hierarchical organizations of interacting species into subsystems, or compartments, separated either spatially or temporally, or coexisting but operating at different temporal or spatial scales, is one such arrangement (Allen and Starr, 1982, Salthe, 1985, O'Neill et al., 1986, Ahl and Allen, 1996).

Our connectedness, energy flux, and functional descriptions of the soil food webs possess many of the attributes of a hierarchically organized system. This was in part by design, as the connectedness description was constructed in terms of functional groups of species that were defined in part in terms of two important boundary-defining features of a hierarchical organized system—space (habitat and habitat selection) and time (life histories and life cycles), connected to one another by food and mode of feeding. These initial criteria influenced the energy flux descriptions as well, as the life histories defined rates of turnover of populations, and with it consumption rates and energetic efficiencies, all of which contribute to the formulation of flux rate. The functional food web description provides a means of relating structural arrangement and fluxes through the estimates of interaction strengths and their placement in the Jacobian matrix.

We begin with a brief background on the complexity and diversity debate. We then define compartments, in a food web context, using information from our connectedness, energy flux, and functional description. In doing so, we present different approaches that have been used to study compartments, review the evidence for and against the presence of compartments of species within food webs, and discuss the putative reasons behind the formation and persistence of compartments.

8.2 Complexity, diversity, compartments, and stability

The eigenvalue-based analyses of randomly constructed networks conducted by Gardner and Ashby (1970) and May (1972) revealed that more complex systems were not necessarily more stable than simpler systems. We introduced the results and conclusions of these analyses back in Chapter 3 (see Figure 3.3). An important outcome was the relationship that May established among diversity (S), connectance (C), and the average interaction strength (i) within the interaction matrix and the likelihood of the system being stable. Recall that the probability $P(S, C, i)$ that a community matrix constructed in such a manner was stable was near certain ($P \to 1$) if

$$i\sqrt{SC} < 1, \tag{8.1}$$

and unstable (P→0) if

$$i\sqrt{SC} > 1. \tag{8.2}$$

Given the nature of Equations 8.1 and 8.2, the implications are clear. The transition between stable to unstable states became more abrupt with increased diversity (S), connectance (C), and average interaction strength (i). May (1972) ended the paper with an interesting caveat. If the species were arranged into "blocks," or compartments, then the likelihood of stability increased. To make this point, he compared 12-species systems with a given connectance, ones constructed at random (reticulate), and ones in which the species were arranged into three interacting blocks (compartmented) of four species. Nearly all of the reticulate webs were unstable, whereas 35% of the compartmented webs were stable.

Two studies provided important follow-up and nuance to May's work on random versus patterned food web architectures and the relationships between diversity, complexity, and stability presented in Equations 8.1 and 8.2 (Box 8.1). McMurtrie (1975) compared the stability of systems that had a specified size, connectance, and interaction strength but that differed in the compartmentalized arrangement (described as crude hierarchies) of the nonzero interactions within the matrix A. Contemporary to this work was a study by Austin and Cook (1974) that focused on the effects on community stability of predator attack rates that link compartments and feedback loops that result from the cycling of nutrients, detritus and the consumption of decomposers. The stability of the systems was dependent on the degree of compartmentalization and the ways in which consumers linked the compartments.

Box 8.1 Patterns within Jacobian matrices and their impacts on dynamic stability

May (1972) proposed that for diverse systems ($S \gg 1$) the stability of matrix A hinged on the diversity (S), connectance (C), and average interaction strength (i), such that if $i\sqrt{SC} < 1$, the system is stable, and it is unstable otherwise.

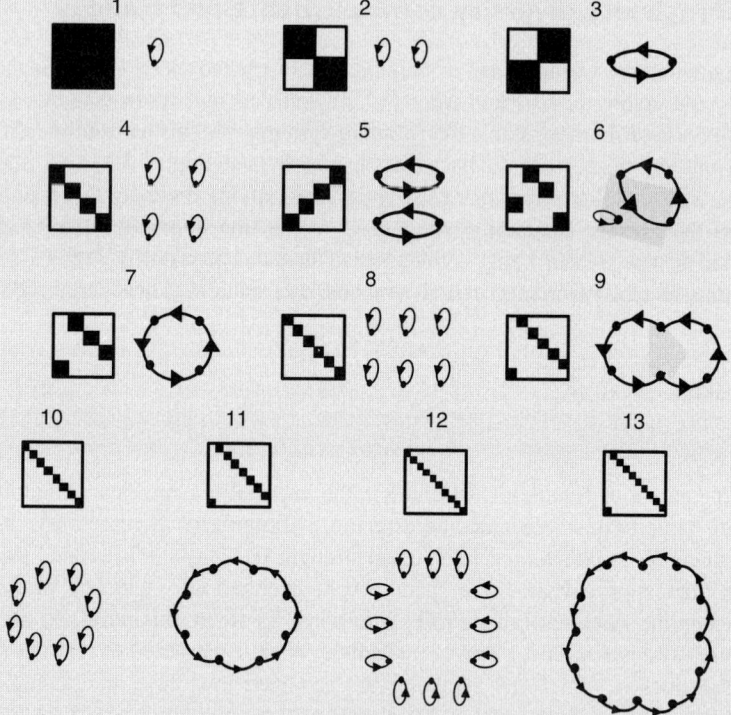

Figure B8.1 Schematic representations of the compartmentalized patterns and linkages studied by McMurtie (1975).

The compartmentalized pattern and linkages within Jacobian matrices that McMurtie (1975) studied were found to promote stability (Figure B8.1). These structures bear a remarkable resemblance to the patterns we observed in Jacobian matrices for our soil food web (see Figure 5.1 in Chapter 5). The crude hierarchies have the nonzero terms arranged in blocks, but they vary in the way these blocks are positioned over or outside the diagonal. The 13 different arrangements create cycles that differ in length, ranging from 1 to 12. Monte Carlo modeling shows that the matrices generating the longer cycles are much more likely to be stable than the matrices generating only short cycles (Table B8.1; Figure B8.1). Hence, matrices with long cycles could overcome the constraints of diversity and connectance and can have much larger connectance values, as would be indicated by the inequality of May (1972). Although these hierarchical structures do not represent food webs that we might expect to occur in nature (see Warren, 1990), the results of McMurtie (1975) clearly show that well-developed structures can be much more stable than random ones. Moreover, the partition of interactions and the role of cycles in the stability of the constructed matrices by McMurtie (1975) bear a significant resemblance to cycles observed in real food webs. For example, organisms will mostly feed on organisms that are smaller than themselves, which creates an ordering of trophic interactions along trophic level (Cohen and Newman, 1985), cycles that are created by predators that link trophic compartments by feeding on both (Rooney et al., 2006, Rooney et al., 2008), and cycles created by flows of materials back to detritus, e.g., in the form of dead bodies and unassimilated fractions of consumption (see Figure 5.1 in Chapter 5) that subsequently may serve as the resource for primary decomposers in food webs (DeAngelis, 1992, Moore et al., 2005).

Table B8.I. The relationship between stability and the structure of the system ($S = 24$, $C = 2/24$, $i^2 = 0.5$, see McMurtie (1975) for definition of interaction strength)

Regime	Fraction Stable	Average Cycle Length (S. E.)	Fraction of Vertices Involved in Cycles
1	16/25	3.40 (0.07)	0.65
2	10/25	2.47 (0.07)	0.69
3	19/25	3.55 (0.07)	0.69
4	9/25	1.83 (0.06)	0.76
5	13/25	2.74 (0.08)	0.76
6	16/25	3.12 (0.08)	0.76
7	26/35	4.33 (0.06)	0.76
8	4/25	1.51 (0.05)	0.86
9	23/25	6.02 (0.02)	0.84
10	3/25	1.33 (0.04)	0.92
11	20/25	8 (0)	0.89
12	4/50	1 (0)	1.0
13	23/25	12 (0)	1.0

The results presented by May (1972) and others (Austin and Cook, 1974, McMurtrie, 1975) made it clear that the diversity and complexity of communities are not related to stability in a simple direct way, but rather that diversity cast within certain types of complexity promote stability. This conclusion, that blocks of interactions promote stability, aligned the broader community ecology concept of the ecological niche (Hutchinson, 1959), in which many interactions occur within the clusters and few occur between the clusters, and the then contemporary, more narrow notions, of feeding guilds and interactive clusters of species (Paine, 1966, Root, 1967, Schoener, 1974). The conclusion had clear implications from the energy-based ecosystem perspective as well. Lindemann (1942) was one of the first to describe and partition trophic interactions in energetic terms and to compartmentalize ecosystems into subsystems that relied on living and dead organic material. MacArthur (1955) linked these concepts to mathematical ideas of stability when he proposed that systems that possess multiple pathways of energy flow possess a greater capacity to recover from disturbances and that they are more likely to persist than systems with fewer pathways. In other words, trophic interactions and dynamic stability studied through energetics provides the theoretical underpinning that explains the ubiquitous patterns and ideas about community organization.

8.3 Defining compartments

A compartment within a food web is a subgroup of species in which the species within the subgroup interact more intensely among themselves than with other species within the web. All species or groups within the food web need to be connected in some way. For example, if the scenario using the 12-species food web presented by May (1972) included three four-species compartments that were not linked in some fashion to one another, then arguably we are dealing with three independent systems. Moreover, if the counterexamples to May's supposition used by Pimm (1982) were considered (Figure 8.1), then the unstable system deemed to be compartmentalized represents two independent food chains, one stable and one not, while the stable system deemed reticulate might be better characterized as compartmentalized. Our view is that the interactions among compartments are integral to compartmentalized architectures, and their presence alone should not negate the integrity of a compartmentalized architecture (Figure 8.1). The key is to establish objective criteria to establish the boundaries of the compartments.

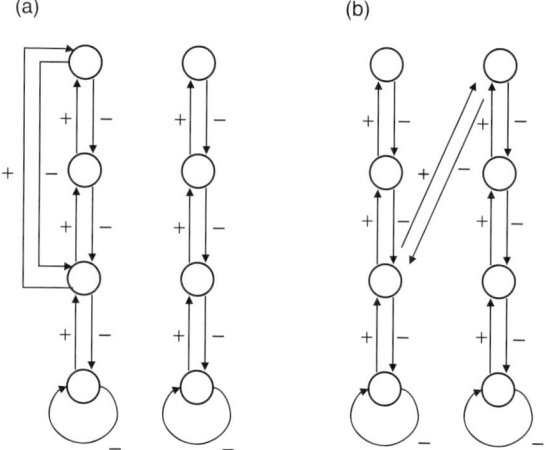

Figure 8.1 Different representations of systems (a and b) that possess parallel food chains with a single omnivore. The systems have the same diversity (S) and connectance (C). (a) The food chains are not linked. (b) The food chains are linked by the omnivore interaction. Pimm (1982) found high percentage of models in System (b) were stable (81%) while none of the System (a) models were stable. System a is compartmentalized but is arguably two systems. System (b) is arguably compartmentalized and is more representative of patterns we find in published descriptions. The cross-linkages between compartments will be an important recurring theme of this chapter.

For food webs, the boundaries of compartments should reflect patterns in the utilization of resources by species. Three perspectives help define the boundaries of compartments within food webs and provide insights into their role in food web dynamics and stability. First, the boundaries of compartments are defined by the adaptive attributes of individual species. Specifically, differences in species physiologies, energetic efficiencies, life histories, body sizes, and foraging behaviors contribute to the formation of the boundaries that shape and link compartments within ecosystems. The adaptive attributes listed above frame a second perspective as they define the concept of ecological niche for individual species. Schoener (1974) concluded that the dominant dimensions within niche space are food, habitat, and time. From this point of view, a compartment represents a grouping of species linked through trophic interactions (food axis) that occupy similar regions or threads along the habitat and time axes.

Lastly, hierarchy theory (Allen and Starr, 1982, O'Neill et al., 1986) and the related principles of spatial and temporal scaling offer ways to identify an organizational structure that is compartmentalized into coexisting subunits of interacting species. Not unlike the niche-based approach discussed above, an approach based on hierarchy theory establishes the boundaries between compartments using differences in the rates at which components (organisms and matter) within a compartment operate, their spatial arrangement, and their responses to external drivers. An important tenet in hierarchy theory is that the organizing principles and rules that govern the system apply to its subsystems as well. Put another way, the structural and dynamic properties of the compartments within a food web should be governed and constrained in the same ways as the whole food web. In the sections that follow we will elaborate on these features, but we will start with a review of the different approaches that have been undertaken to identify and study compartments within food webs.

8.4 Approaches to studying compartments

Two broad approaches have been used to study compartments. One approach seeks to identify clusters of species based on the frequency of interactions and then compares the observed pattern to that generated by a null model. A second approach establishes a set of rules or guidelines from which a compartment might be based and then uses the limit presented in Equation 8.1 to see if the parameters within the compartments relate to one another in a manner consistent with the limit.

Regardless of the approach, trying to identify compartments within published webs is problematic for several reasons, given the ways in which they are described and the level of information used in the description. Most published webs are connectedness food web descriptions that simply indicate the presence or absence of a feeding relationship among a relatively small set of species. Weak linkages carry the same weight as strong linkages, thereby obscuring the ability of simple statistical tests based on binary data to distinguish compartments. Many published food web descriptions are either source webs that establish feeding relationships originating from a defined set of basal resources, sink webs that identify predators and trace their prey down to basal resources, or community webs that describe the interactions within a given habitat. By design, each type of description may in and of itself represent a compartment. Nearly all published food webs average interactions that occur over time and space. The extent to which temporal and spatial averaging obscures the presence of compartments is often unclear. What follows are presentations of approaches that are based on the information that can be gleaned from connectedness, energy flux, and functional food web descriptions.

8.4.1 Evidence of compartments—binary data I

The simplest approach to studying compartments focuses on the number of species and the presence or absence (binary data) of interactions among the species. Pimm and Lawton (1980) developed a statistic, $\bar{C}_1$, that described the degree to which a compartment based on the off-diagonal elements of the interaction matrix, A, describes trophic interactions. Nonzero elements of the matrix were reassigned the following index of similarity, s_{ij}, which indicated the degree to which a species share predators and prey:

$$s_{ij} = \frac{\text{The number of species that both i and j interact}}{\text{The number of species that either i or j interacts}} \qquad (8.3)$$

The compartment statistic, $\bar{C}_1$, is derived as follows:

$$\bar{C}_1 = \frac{1}{n(n-1)} \sum_{i=1}^{n} \sum_{\substack{j=1 \\ j \neq i}}^{N} s_{ij} \qquad (8.4)$$

The statistic can take values over the interval (0,1) and has the property that as $\bar{C}_1 \rightarrow 1$ the systems are more compartmentalized, and as $\bar{C}_1 \rightarrow 0$ they are more reticulate. The analysis starts by first calculating $\bar{C}_1$ for the description (observed web). Next, all possible permutations of the interactions within the web are generated, with $\bar{C}_1$ calculated for each (random webs). Let P represent the proportion of random webs that generate values of $\bar{C}_1$ greater (more compartmentalized) than the observed web. Small values of P indicate that the food web is compartmentalized beyond that expected by chance alone, while large values indicate that the web is more reticulate than would be expected by chance alone.

Pimm (1982) applied this analysis to several of the food webs compiled by Cohen (1978) and found little evidence that these observed food web structures are more compartmentalized than would be expected by chance (Table 8.1). These results may say more about the effectiveness of using binary data and sensitivity of the analysis, than about the presence or absence of compartments.

Table 8.1. Degree of compartmentalization ($\overline{C}_1$, Equation 8.4) of selected food webs (from Pimm, 1982, Pimm and Lawton, 1980). The value P is the proportion of random webs with values of $\overline{C}_1$ greater than the observed web.

$\overline{C}_1$	P	Ecosystem	Source
0.170	0.556	Prairie	Bird (1930)
0.151	0.400	Willow forest	Bird (1930)
0.144	0.727	Rocky intertidal: gastropods	Paine (1966)
0.264	1.00	Rocky intertidal: starfish	Paine (1966)
0.337	0.402	Freshwater stream	Minshall (1967)
0.175	0.015	Mudflat	Milne and Dunnett (1972)
0.137	0.115	Mussel bed	Milne and Dunnett (1972)
0.224	0.330	Spring	Tilly (1968)
0.253	0.732	Freshwater stream	Jones (1949)
0.140	0.800	Lake fish	Zaret and Paine (1973)
0.135	0.333	Marine algae	Jansson (1967)

8.4.2 Evidence of compartments—binary data II

The approach outlined above determines the degree to which the observed patterning within a food web is compartmentalized. An alternative approach pivots off the relationships developed by May (1972), presented in Equations 8.1 and 8.2, when applied to multiple food web descriptions. Given that diversity varies across ecosystems, and if we assume that the systems are stable, Equation 8.1 would indicate that either connectance (C) or the maximum average interaction strength (i) are constant, or that they covary in a manner that satisfies the condition for the level of observed diversity (S).

From Equations 8.1 and 8.2 we can see that if the maximum average interaction strength (i) were constant across webs, the connectance (C) should decrease with increased diversity in a hyperbolic manner (Figure 8.2). Several studies have confirmed this relationship and have used it to produce general laws about trophic interactions, diversity, and connectance. As connectance is a function of both the number of trophic links (L) and the total number of trophic species (S), the relationship depicted in Figure 8.2 could rise if the number of links that each species engaged in were constant, or if each species engaged in a fixed fraction of links in the web. Both extremes have been proffered based on the relationship between the number of links and species from published webs. Pimm (1982) argued that if the number of species a species feeds on were independent of the number of species in the food web, then the hyperbolic relationship in Figure 8.2 would result. Cohen and Newman (1985) formalized the link–species scaling law, which states that the number of links per species in a food web is constant on average and scale invariant at roughly two (i.e., $L/S \cong 2$ or $L \approx 2S$). Martinez (1992) proposed the constant connectance hypothesis as an alternative to the link-species scaling law. Martinez (1992) argued that the number of links should increase in proportion to the number of species in the community (i.e., $L/S \approx kS$ or $L/S^{\approx 2} = k$, where k is a constant).

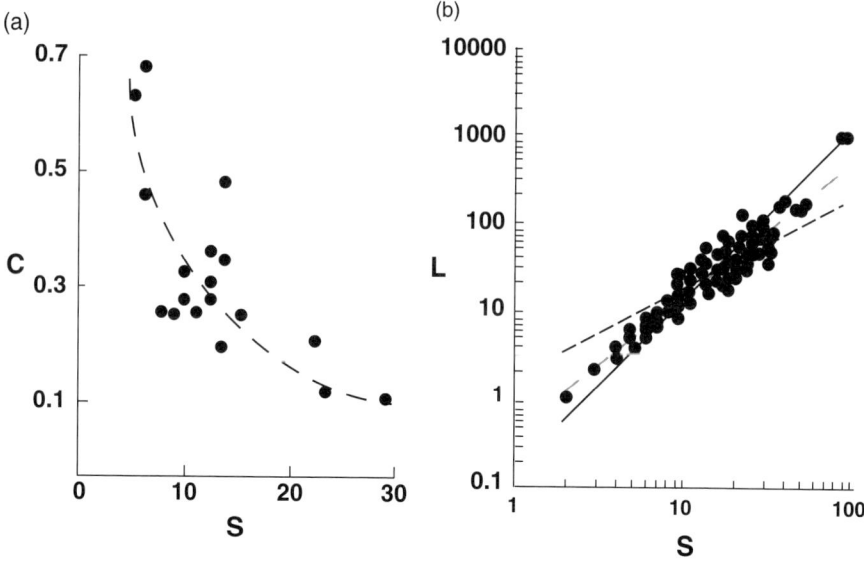

Figure 8.2 (a) The connectance (C) versus diversity (S) of several published food webs as presented by Pimm (1982). (b) Different underlying mechanisms could generate the observed decline in connectance with increased diversity ($L = 0.41S^{1.54}$;- - -). The link-species scaling relationship ($L = 1.9S$; - - -) presented by Cohen and Newman (1985), where links (L) among species increase linearly with increased diversity. The constant connectance hypothesis ($L = 0.14S^2$; —) proposed that links increase with the square of the number of species (from Martinez, 1992).

8.4.3 Evidence of compartments—carbon and nitrogen flux

The organization of communities into assemblages of species based on the types of resources was perhaps one of the least contentious arguments for compartmentalization. Odum (1962, 1963) made a distinction between the grazing food chains, which are assemblages of species based on primary producers, and the detritus food chains, which include consumers of decaying organic material and associated microorganisms (Figure 8.3). Research at that time used isotopes to trace trophic interactions and energy flow through systems to reveal compartments. Marples (1966) used P^{32}-labeled plants and detritus in a *Spartina* salt marsh to demonstrate that indeed the diets of the primary consumers of these resources do not overlap, but that the consumers are preyed upon by common predators.

The same pattern is evident in published connectedness food webs (Moore and Hunt, 1988, Polis and Strong, 1996, Moore et al., 2004). To illustrate this point, one needs only to deconstruct a connectedness description into its set of source webs and then categorize the basal resources as based on either primary producers or detritus. Moore and Hunt (1988) applied this approach to the set of 40 community food webs

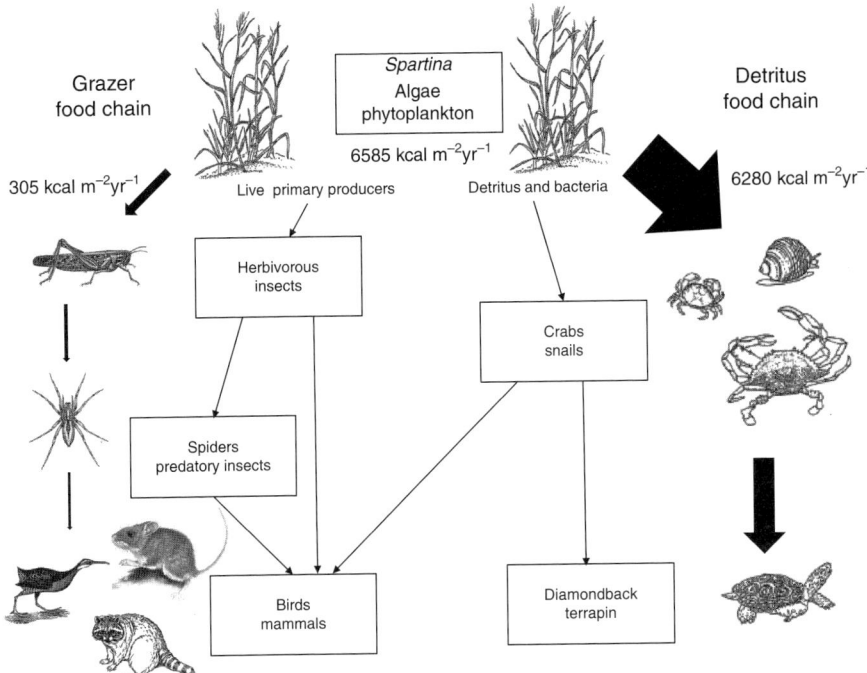

Figure 8.3 The grazer and detritus food chains based on a simple connectedness and energy flux (kcal m^{-2} yr^{-1}) description of a salt marsh ecosystem (after Odum, 1961). Of the total net primary productivity of 6585 kcal m^{-2} yr^{-1}, only 4.6% was consumed by herbivores within the grazer food chain, and 95.4% became detritus to be utilized by bacteria and detritivores (estimates from Teal, 1962).

compiled by Briand (1983) yielding 138 source webs, which they dubbed "energy channels" (Table 8.2). The majority (72.5%) of the community webs possessed both primary producers and detritus. Of the 138 energy channels, 63% originated with a primary producer, 20% with detritus, and 17% with a consumer. Predators linked the channels (Table 8.2).

The approach can be taken a step further using the flux estimates from an energy flux description to estimate the fraction of a group's biomass that originates from a primary producer or detritus. Moore and Hunt (1988) used estimates of nitrogen flow (g N m^{-2} yr^{-1}) obtained from the Central Plains Experimental Range (CPER) soil food web model of the North American shortgrass steppe to estimate the fraction of each functional group's diet that originated from the different basal resources of roots or detritus (labile and resistant substrates; Table 8.3). The exercise could have used carbon flow as well. The estimates are derived from the fluxes, starting with each basal resource to their consumers up through the web to the top predator (Figure 8.4). Each flux estimate is then partitioned into subfluxes originating from each basal resource.

Table 8.2. Summary of the number of energy channels based on basal resources from the food webs compiled by Briand (1983).

Food Web	Type and Number of Basal Resources				Reference
	Producers	Detritus	Consumers	Total	
Cochin estuary	1	1		2	Qazim (1970)
Knysna estuary	2	1		3	Day (1967)
Long Island estuary	3	1		4	Woodewell (1967)
California salt marsh	2			2	Johnston (1956)
Georgia salt marsh	2		1	3	Teal (1962)
California tidal flat	1	1		2	Macginitie (1935)
Narragansett Bay	2	1		3	Kremer and Nixon (1978)
Bissel Cove marsh	3	1		4	Nixon and Oviatt (1973)
Lough Ine rapids	2			2	Kitching and Ebling (1967)
Exposed intertidal (New England)	1	1		2	Menge and Sutherland (1976)
Protected intertidal (New England)	2	1		3	Menge and Sutherland (1976)
Exposed intertidal (Washington)	2	1		3	Menge and Sutherland (1976)
Protected intertidal (Washington)	2	1		3	Menge and Sutherland (1976)
Mangrove swamp (station 1)		1		1	Walsh (1967)
Mangrove swamp (station 3)		1		1	Walsh (1967)
Pamlico River	3	1		4	Copeland et al. (1974)
Marshallese reefs	2	1		3	Hiatt and Strasburg (1960)
Kapingamarangi atoll	8			8	Niering (1963)
Moosehead Lake	1		1	2	Brooks and Deevey (1963)
Antarctic pack ice zone	2	1		3	Knox (1970)
Ross Sea	2	1		3	Patten and Finn (1979)
Bear Island	5	1		6	Summerhayes and Elton (1923)
Canadian prairie	1			1	Bird (1930)
Canadian willow forest	3		1	4	Bird (1930)
Canadian aspen communities	2	1	1	3	Bird (1930)
Aspen parkland	5	4	4	9	Bird (1930)
Wytham Wood	3	1		4	Varley (1970)
New Zealand salt meadow	5	1	1	7	Paviour-smith (1956)

Arctic seas	1	1		2	Dunbar (1954)
Antarctic seas	1			1	Mackintosh (1964)
Black Sea epiplankton	1	1		2	Petipa et al. (1970)
Black Sea bathyplankton	1	1		2	Petipa et al. (1970)
Crocodile Creek	1	2	3	5	Fryer (1959)
River Clydach	2	2		4	Jones (1949)
Morgan's Creek	1	1		2	Minshall (1967)
Mangrove swamp (station 6)	3	1	4	8	Walsh (1967)
California sublittoral			6	6	Clarke et al. (1967)
Lake Nyasa rocky shore	2	1		3	Fryer (1959)
Lake Nyasa sandy shore	3	1	1	5	Fryer (1959)
Malaysian rain forest	2	1		3	Harrison (1962)
Summary totals	87	28	23	138	
Percentages	63%	20%	17%		

Table 8.3. Estimates of the proportion of energy (indexed by carbon) that the different functional groups potentially derive from within the bacterial, fungal, and/or plant root energy channels within the soil food web of the CPER.

	Energy Channel		
Functional Group	Bacteria	Fungi	Roots
Protozoa			
Flagellates	100	0	0
Amoebae	100	0	0
Ciliates	100	0	0
Nematodes			
Phytophagous nematodes	0	0	100
Mycophagous nematodes	0	90	10
Omnivorous nematodes	100	0	0
Bacteriophagous nematodes	100	0	0
Predatory nematodes	68.67	3.50	27.83
Microarthropods			
Collembola	0	90	10
Cryptostigmatid mites	0	90	10
Non-cryptostigmatid mites	0	90	10
Nematophagous mites	66.70	3.78	29.52
Predatory mites	39.54	38.56	21.91

The pathway of the subfluxes through the web from basal resource to top predator constitutes an "energy channel," a concept that we will formalize below.

When applied to the CPER food web, the exercise revealed that the web was compartmentalized along resource types and habitat to form different energy channels. The web possessed primary-producer-based and detritus-based energy channels (Table 8.3). The compartment originating from detritus was divided into two pathways—a bacterial pathway composed of bacteria and their consumers, and a

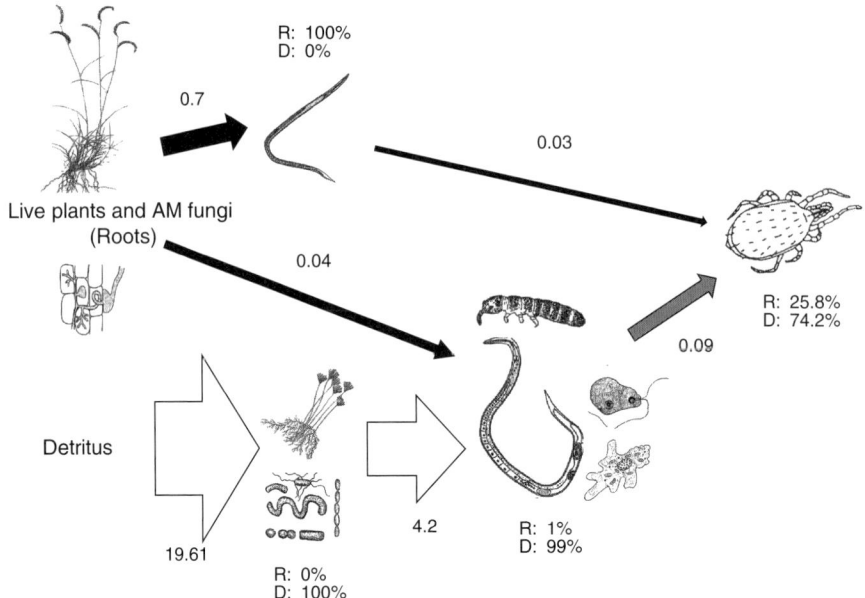

Figure 8.4 The root (R) and detritus (D) energy channels for a simplified version of the CPER soil food web based on the nitrogen flux rates provided by Hunt et al. (1987). The values over the arrows represent the N flux rates (g N m^{-2} yr^{-1}). The percentages represent the fraction of N flux derived from roots (R) or detritus (D). The detritus energy channel can be further subdivided into a bacterial energy channel and a fungal energy channel (see Table 8.3).

fungal pathway composed of fungi and their consumers—based on previous observations (Coleman et al., 1983, Hunt et al., 1987). The pathways were joined by a suite of nematode and arthropod predators, and terminated with predatory mites. About 25% of the predatory mites' diet originated from plant roots, and 75% from detritus, from which nearly equal proportions originated from the bacterial and fungal pathway.

8.4.4 Evidence of compartments—carbon flux and interaction strengths

Krause et al. (2003) used a statistical approached developed in the social sciences to identify social networks (Frank, 1995, Frank, 1996, Frank and Yasumoto, 1996). Like the approach used by Pimm (1982), this analysis attempts to determine the extent to which interactions are grouped into compartments more than might occur by chance alone. Frank (1996) put it best by saying, "...[A]re there really subgroups in the data, or have subgroups been imposed on a fluid pattern of interactions?" The idea is based on the following logit model, in which $x_{ij} = 1$ if species i and j interact, and zero otherwise:

Table 8.4. The relationship between membership within a compartment and the interactions between taxa (from Krause et al., 2003).

		Interaction occurring		
		No	Yes	
Membership	Different	A	B	$n(n-1)-\sum_g n_g(n_g-1)$
	Same	C	D	$\sum_g n_g(n_g-1)$
		$n(n-1)-\sum_i\sum_j X_j$	$\sum_i\sum_j X_j$	$n(n-1)$

The terms are defined as follows: X_j represents the presence (1) or absence (0) of an interaction between taxon i and taxon j, n is the number of taxa (S) in the food web, and n_g the number of taxa in compartment g, A = unrealized interactions between compartments, B = realized interactions between compartments, C = unrealized interactions within compartments, and D = realized interactions within compartments.

$$log\left|\frac{P[x_{ij}-1]}{1-P[x_{ij}-1]}\right| - \theta_0 + \theta_1 \text{same compartment}_{ij} \qquad (8.5)$$

where θ_i same compartment$_{ij}$ = 1 if species i and j are members of the same compartment, and 0 otherwise. The term θ_1 represents the increase in the likelihood of a relationship being present. Maximizing θ_1 in Equation 8.5 is equivalent to maximizing the odds ratio (AD/CB) as presented in Table 8.4. The odds ratio can be interpreted as the likelihood that species within a compartment will interact relative to the likelihood that species in different compartments will interact. The analysis begins by seeding species to subgroups and estimating θ_1. Unaffiliated species are assigned to groups, and then species are iteratively reassigned to groups seeking to maximize θ_1. Once the sill for θ_1, is reached through the iterative process, Monte Carlo simulations are used to generate a sampling distribution for θ_1, and the statistical significance is determined.

An interesting feature of this analysis is that the odds ratios can be computed using the unweighted interactions determined by the presence of the interaction, or (if the data were available) weighted interactions by predation frequency (events ha^{-1} d^{-1}), carbon flow (g C m^{-2} yr^{-1}), or interaction strength (geometric mean of pairwise interaction strengths). The carbon flow and interactions strength metrics are calculated using the same approaches we presented for the CPER food web in Chapter 4 and Chapter 5, respectively.

Krause et al. (2003) applied this approach to 17 variations of five food webs (Table 8.5) and found evidence of one to six compartments within the webs. Only seven of the variations of webs generated odds ratios that were significantly higher than would be expected by chance alone. The analyses based on interactions weighted by carbon flow or interaction strength were more likely to reveal compartments than those based on unweighted interactions. In fact, all the analyses in this set of studies that were based on carbon and carbon fluxes revealed evidence of compartments beyond chance alone.

Table 8.5. Analysis for compartments of five food webs (Krause et al., 2003).

Name	n	Weight of Interaction	Odds Ratio	P	Number of Compartments	Overall IC	Within IC	Between IC
Ythan Estuary	134	None	4.19	≥0.999	3	0.033	—	—
Little Rock Lake	92	None	3.14	≥0.999	2	0.12	—	—
St Martin Island	181	None	10.04	≤0.001*	4	0.072	0.17	0.020
	44	None	4.20	≤0.907	5	0.11	—	—
	44	Frequency	28.93	≤0.803	6	0.0065	—	—
Chesapeake Bay	33	None	8.61	≤0.751	3	0.067	—	—
	33	Strength	642.55	≤0.001*	2	0.0029	0.0059	0.0000093
	33	Carbon	618.75	≤0.012*	2	0.0035	0.0071	0.000012
	45	None	9.63	≤0.200	4	0.069	—	—
	45	Strength	114.92	≤0.001*	2	0.0052	0.0099	0.000087
	45	Carbon	165.84	≤0.001*	3	0.0018	0.0044	0.000026
Cypress wetland								
Dry season	64	None	5.52	≤0.285	2	0.11	—	—
	64	Strength	11.93	≤0.918	3	0.0021	—	—
	64	Carbon	228.18	≤0.001*	5	0.00057	0.0020	0.0000088
Wet season	64	None	7.81	≤0.001	1	0.11	—	—
	64	Strength	12.34	≤0.910	4	0.0026	—	—
	64	Carbon	384.81	≤0.001*	3	0.00044	0.0013	0.0000033

The text and Table 8.4 provide definitions and calculations for the odds ratio, overall IC, and the definitions of A, B, C, and D. Overall IC = [(B + D)/(A + B + C + D)], Within IC = D/(C + D), Between IC = B/(A + B). Asterisks indicates significance.

8.5 The energy channel

We advance the energy channel as an energetics-based compartment within an ecosystem. An energy channel is a collection of energy or material fluxes that start with a basal resource and end with a top predator top predator in a food web. The concept of an energy channel melds the organism-based community ecological food web approach with the nutrient-based ecosystem ecological approach to studying food webs. In this regard, an energy channel is the energy flux and functional equivalents (sensu Paine, 1980) of a source web within a community web (sensu Cohen, 1978). The energy channel offers a way of exploring compartments beyond the presence or absence of trophic interactions by including the following considerations.

1. Food webs may contain multiple energy channels, defined by trophic interactions and nutrient transfers among species that are based on different basal resources. Primary producers and detritus are two basal resources common to most ecosystems, but the observer can parse the system at any scale or resolution.
2. Energy channels within food webs are linked to one another through consumers and the cycling of dead organic material. This property shows that there is overlap between compartments within communities, or—put another way—that compartments do not exist in isolation.
3. The assemblages of species within an energy channel transfer and turn over matter at different rates. This property arises from the physiological and life history attributes of the species within the energy channel, and the predisposition to be in temporal synchrony with one another.
4. Habitats and time play a role in defining and shaping the boundaries of energy channels. There are two caveats to this: a) habitat boundaries may be obscured if the food web is a community food web description (sensu Cohen, 1978) that singles out a particular type of habitat or does not have fine enough resolution to distinguish subhabitats; b) temporal averaging a description over extended times periods will obscure compartments that operate at different times (i.e., diurnal or seasonal).
5. The structural and dynamic properties of energy channels within a food web are governed by the same constraints as the whole food web.
6. Given the above, energy channels may respond to disturbances differently in a "quasi-independent" manner.

In the sections that follow we will elaborate on these features. We will rely heavily on our studies of soil systems, but will provide connections to other systems.

8.5.1 Time—flux rates and turnover

The boundaries of an energy channel are defined by the flux rates of matter among groups and turnover rates of biomass of groups, as defined by the production: biomass (P:B) ratio of the groups (O'Neill et al., 1986). For the soil food web, Coleman et al. (1983) proposed that nutrient dynamics in soils could be partitioned

Table 8.6. A reformulation of Table 3.2 to characterize the boundaries that form compartments within food webs. Habitat use, life-history characteristics, and energetic efficiencies of broad classifications of organisms encountered in belowground food webs (Hunt et al., 1987, Coleman, 1996, and Moore and de Ruiter, 1997). These attributes along with the food preferences serve as the basis of the compartmentalization of trophic interactions along the principal niche axes of food, habitat and time (Schoener, 1974).

Taxon Habitat	Bacteria Water/ Surfaces	Fungi Free Air/ Surfaces	Protozoa Water/ Surfaces	Nematodes Water Films/ Surfaces	Collembola Free Air Spaces	Mites Free Air Spaces
Minimum generation time (h)	0.5	4–8	2–4	120	720	720
Turnover time (season 1)	2–3	0.75	10	2–4	2–3	2–3
Assimilation efficiency (%)	100[†]	100[†]	0.95	0.38–0.60	0.5	0.3–0.9
Production efficiency (%)	0.4–0.5	0.4–0.5	0.4	0.37	0.35	0.35–0.40

[†]The assimilation efficiencies of bacteria and fungi are 100%, given that microorganisms absorb materials across their membranes as opposed to ingesting or engulfing prey or materials.

into pathways that process matter at fast and slow rates. The fast cycling pathway and the slow cycling pathway correspond to the bacterial and fungal channels presented above (Table 8.3). The differences in cycling rates were attributed to the physiologies of species within each energy channel. Head-to-head comparisons of the taxa at each trophic position within the channels reveal that organisms within the bacterial energy channel have shorter generation times and turn over their biomass at faster rates (higher P:B) than those in the fungal pathway (Table 8.6). To put this in perspective, a single bacterium can complete its life cycle in 20 minutes; field data indicate that a gram of bacterial biomass turns over two to three times during a growing season. Protozoa (the primary consumers of bacteria) can complete their life cycles in 4 hours, and turn over as many as ten times in a growing season. Under optimal conditions, fungi require 4–8 hours to complete their life cycles, and achieve only 75% turnover during a growing season. The arthropod consumers of fungi require 2–3 months to complete their life cycle and might turn over two to three times per season.

Rooney et al. (2006) used this approach to propose that many communities are compartmentalized into fast and slow energy channels based on the flux rates among organisms and their turnover rates. The differences in rates within the channels result in asynchronous dynamics between the channels that is important to stability. The analysis revealed remarkable similarity in the structure of the terrestrial soil food webs and marine food webs (Figure 8.5; Table 8.7). In both cases, the fast and slow energy channels are separate at lower trophic positions and merge into one another by predators at upper trophic positions. We will discuss the importance of the asynchronous dynamics to the stability below.

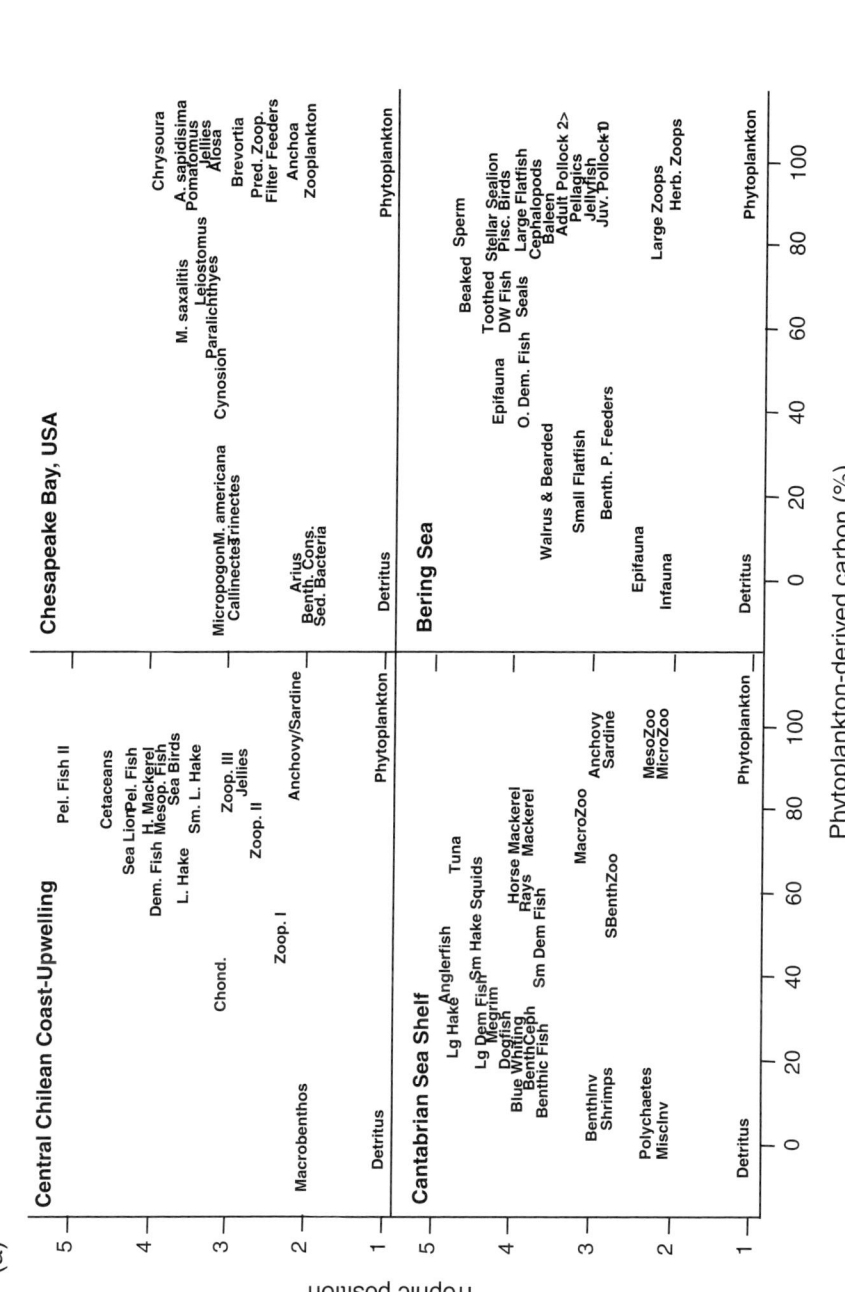

Figure 8.5 *Continues overleaf.*

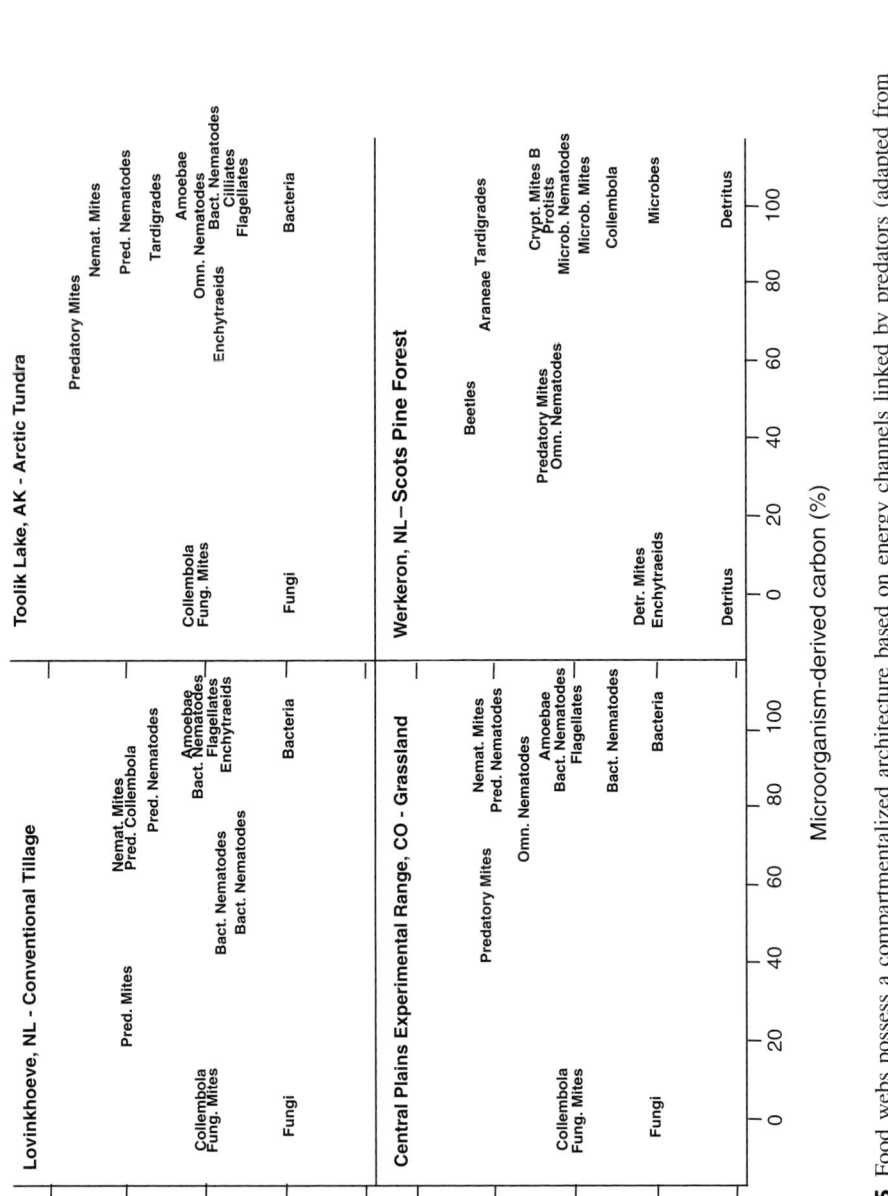

Figure 8.5 Food webs possess a compartmentalized architecture based on energy channels linked by predators (adapted from Rooney et al., 2006). Representations of energy channels within marine (a) and soil (b) food webs based on the percentage of carbon derived from basal resources (x-axis). Trophic positions of groups based on feeding interactions are presented on the y-axis.

Table 8.7. Total production (g C m^{-2} yr^{-1}), biomass (g C m^{-2}) and production: biomass ratios of species grouped by trophic position and basal energy resource for aquatic and terrestrial ecosystems (from Rooney et al., 2006).

Aquatic Food Webs

Food Web	Trophic Level	Benthic Channel			Pelagic Channel		
		Production	Biomass	P:B Ratio	Production	Biomass	P:B Ratio
Chilean upwelling	2+	7.17	2.01	**3.57**	712.65	78.74	**9.05**
	3+	0.16	0.44	**0.36**	379.25	51.77	**7.33**
	4+	0	0	—	17.97	28.38	**0.63**
Cantabrian shelf	2+	130.38	35.15	**3.71**	537.30	22.68	**23.69**
	3+	14.76	20.47	**0.72**	72.79	29.76	**2.45**
	4+	1.10	2.49	**0.44**	0.31	0.38	**0.82**
Chesapeake Bay	2+	5359.32	3873.09	**1.38**	18448.9	1909.14	**9.66**
	3+	103.83	214.90	**0.48**	2827.69	690.70	**4.09**
Bering Sea	2+	12.95	18.42	**1.43**	397.11	69.67	**5.61**
	3+	8.45	2.98	**0.39**	45.81	43.60	**1.05**
	4+	0	0	—	0.32	1.09	**0.30**

Terrestrial Food Webs

Food Web	Trophic Level	Fungal Channel			Bacterial Channel		
		Production	Biomass	P:B Ratio	Production	Biomass	P:B Ratio
Lovinkhoeve conventional	2+	1.36	3.28	**0.41**	61.73	245.75	**0.25**
	3+	0.16	0.55	**0.28**	29.92	19.96	**1.50**
	4+	0.02	0.08	**0.21**	0.004	0.01	**0.26**
Arctic tundra	2+	38.56	19.00	**2.03**	261.18	8.55	**30.55**
	3+	2.35	0.57	**4.13**	70.86	4.44	**15.96**
	4+				6.16	1.97	**3.13**
CPER	2+	380.60	63.00	**6.04**	1817.67	304.00	**5.98**
	3+	42.87	3.91	**10.95**	180.43	17.36	**17.36**
	4+				1.01	0.16	**6.33**

Food Web	Trophic Level	Detrital Channel			Microbial Channel		
		Production	Biomass	P:B Ratio	Production	Biomass	P:B Ratio
Wekerom scots pine forest	2+	20.94	1.85	**11.35**	99.52	14.45	**6.89**
	3+	4.84	0.86	**5.63**	12.20	1.56	**7.80**
	4+				3.52	1.82	**1.93**

8.5.2 Habitats—spatial arrangement

Habitats form natural boundaries for compartments. Not surprisingly, compartmentalization of species is common in food web descriptions that include multiple habitats (Pimm and Lawton, 1980). The prevalence of compartments within habitats is another matter, as the spatial scale of the description becomes an issue. Whether the food web description is a source, sink, or community type as defined by Cohen (1978), the spatial extent of a food web is defined by the observer. For source webs the distribution of the basal resource (the source) broadly defines the spatial extent of the food web. Basal resources tend to define habitat boundaries or reside within a habitat. If the basal resources are too coarsely defined, trophic interactions that emerge from them will not reveal compartments. For sink webs, the distribution or home range of the top predator (the sink) defines the spatial extent of the food web. Many top predators move from one habitat type to another. Hence, habitat-based compartments may readily emerge. For community food webs the distribution of a habitat or set of habitats (the community) defines the spatial extent of the food web. The likelihood of uncovering compartments is part and parcel of the habitat boundaries that are defined. In other words, the likelihood that a description will reveal habitat-based compartments depends on the choices the observer makes when formulating the description.

Linking information from the source webs and energy flux descriptions helps to clarify habitat boundaries. Referring back to Figure 8.5, we see that the fast and slow pathways are based not only on their basal resources and transfer rates, but on habitat as well. For the Cantabrian Sea shelf, the fast energy channel based on phytoplankton operates within the pelagic zone, whereas the slow energy channel based on detritus operates largely in the benthic zone, with the two merging in the water column. Likewise, for our soil food webs, the organisms within the fast bacterial energy channel and slow fungal energy channel occupy different regions of the soil habitat. For the bacterial energy channel, bacteria and their consumers (protozoa and nematodes) are largely aquatic organisms or require water films for survival. For the fungal energy channel, fungi and their consumers (arthropods and nematodes) live within the water films or within air-filled pore spaces. The two energy channels merge through the action of predators that roam within the soil matrix.

8.5.3 Temporal arrangements and spatial interactions

Time and space often interact in ways that define the boundaries between compartments. We see this in the case of migratory populations, particularly higher-order consumers and predators that move from one habitat to another on a seasonal basis. There are several case studies to illustrate this point. Two examples include the patterns of resource use of the nomadic people of the South Turkana region of East Africa, and the impact of migratory birds on ecosystems within North America (Coughenour et al., 1985, Jefferies, 1999, Jefferies et al., 2004).

The studies of the nomadic people of South Turkana illustrate the importance of seasonal migration to alternative habitats (compartments) on the persistence of the higher-order consumers (Coughenour et al., 1985, Ellis and Swift, 1988, Ellis et al., 1993, Swift et al., 1996). These studies involved detailed analyses of the pastoralists' diet, which included the caloric content of the food, the quantities consumed, how it was partitioned within the tribe, and when and where it was consumed (Figure 8.6). The studies revealed that the Turkana people compartmentalized the utilization of their food resources both spatially and temporally in ways that were important to their survival (Coughenour et al., 1985). Not unlike the top predators in the CPER food web, the diet of the Turkana people was spread out across distinct energy channels, which in their case originated from several species of plants, with over 50% derived from trees and shrubs, about 30% from grasses and forbs, and 20% from nonpastoral sources (wild animals, plants, and purchased goods). Coughenour et al. (1985) proposed that by compartmentalizing their resources in this manner and moving to different regions during dry and rainy periods the Turkana people were able to subsist in a harsh and seasonal environment.

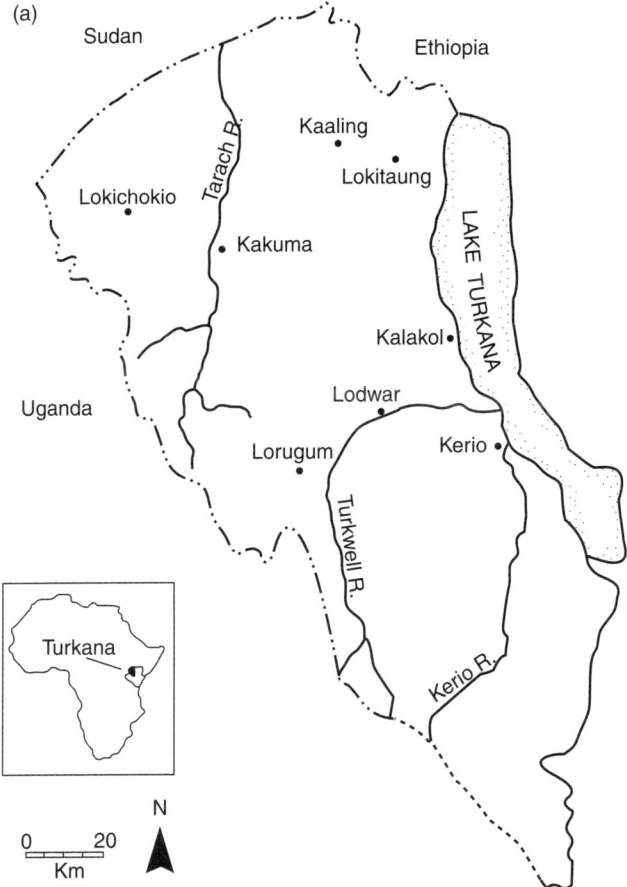

Figure 8.6 *Continues overleaf.*

(b)

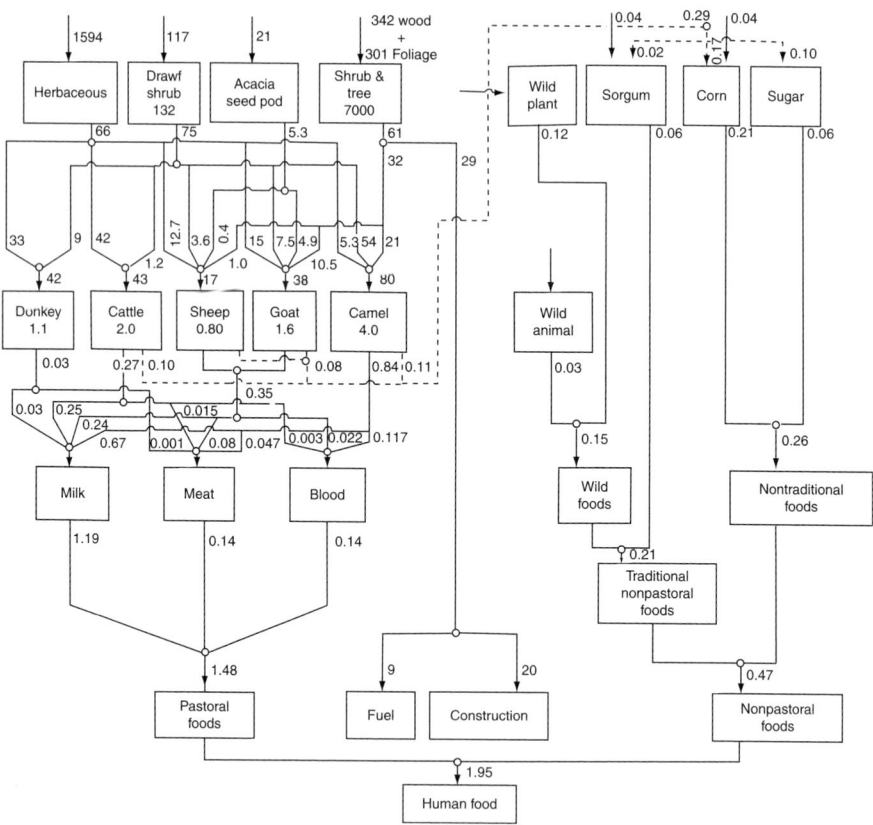

Figure 8.6 Energy flux description of a human sink web for the Ngisonyoka Turkana ecosystem (Coughenour et al., 1985, McCabe, 1990). (a) The Ngisonyoka Turkana ecosystem is located in northwest Kenya in the region west of Lake Turkana. The Ngisonyoka are pastoralists whose annual migration patterns follow seasonal rains and growing seasons to optimize the use of native pastoral and traditional nonpastoral and cultivated nontraditional nonpastoral food sources. (b) The energy flux description is expressed in gigajoules per person per year with the numbers in the boxes referring to standing stocks (dashed lines indicate livestock sales or barter). The food web is compartmentalized into multiple pathways of energy flow (31 distinct flows) within spatial and temporal networks that are essential to the stability and persistence of the system. Coughenour et al. (1985) identified three complementary energy channels or modes of resource use: (1) dependence on a reliable energy pathway (woody plants to camel milk), (2) opportunistic use of a seasonal pathway (herbaceous plants to cattle milk), and (3) contingency conversion of biomass (meat, blood) to energy for humans as needed. Seasonal migrations and the maintenance of a diverse array of livestock are essential.

Studies of migratory bird populations in North America illustrate how changes in the energetic properties of one compartment can influence another through the linkages provided by higher-order consumers. Geese and other migratory fowl use the Canadian high arctic wetlands during the summer and temperate wetlands along the Atlantic coast of North America during the winter and on their spring and fall migrations (Jefferies, 1999, Jefferies et al., 2004). Agricultural production in the temperate regions has increased production that has led to increased survival and in the winter has resulted in larger populations of the migratory fowl. Upon return to the northern reaches during the summer, the larger populations of fowl exceed their carrying capacity to the point that the northern boreal and arctic ecosystems are negatively impacted by increased grazing and fecal deposition.

8.5.4 Structural and dynamic properties

The structural and dynamic properties of the energy channels within a food web should be governed and constrained in the same ways as the whole food web. This idea emerges from a tenet of hierarchy theory about the similarities in the properties of subsystems (read "energy channels") to the system (read "food web"). To illustrate this point we will work with the 40 community food web compiled by Briand (1983) and a few assumptions.

We will start with the assumptions that basal resources by definition are consumed and do not consume other species or resources. We will further assume that each basal resource forms at least one energy channel. Couple these assumptions with the knowledge that basal resources often occupy different habitats or subhabitats and may possess different morphologies, or different dynamic properties (Table 8.8). Given these assumptions and caveats, we end up with no interactions among basal resources, few consumers that feed on multiple basal resources, and few interactions among the consumers of basal resources. This leads to the prediction that connectance of food webs should decrease with increased numbers of basal resources.

We further assume that each energy channel may be subdivided into multiple energy channels that are cross-linked by predators. At this time we will not concern ourselves with cross-linkages or common predators among the energy channels. We will discuss the importance of these linkages below in section 8.6. Empirical evidence suggests that food chains on average possess three to four species (Pimm and Lawton, 1977). A conclusion that we took from Chapter 7 is that for a given level of energy input, the lengths of food chains are constrained by dynamic stability for the simplest to the more complex set of interactions, involving both primary producers and detritus. From these results we would expect that the diversity of consumers should increase with an increase in the diversity of basal resources.

The conclusions we drew from the first and second assumptions above seem to bear out. In line with the first assumption, the connectance of the descriptions declined with increased diversity of basal resources (Figure 8.7a). Moreover, if we parse the 138 energy channels within the 40 food webs compiled by Briand (1983)

Table 8.8. List of attributes of basal resources (primary producers and detritus) that contribute to the formation of species assemblages and compartmentalization of energy flow within food webs.

Basal Resource Type	Niche Separation of Basal Resource (Food, Habitat, Time)
Primary Producer	
Phenology	Seasonality of growth and production
Morphology	Single cells to multicellular
	Structural versus nonstructural
Habitat	Aquatic versus terrestrial
	Freshwater versus marine
	Climate: temperature and precipitation
	Elevation and depth
	Slope and aspect
	Microhabitats on and within primary producers
Dynamics	Life-history strategy: r-selection versus K-selection
	Photosynthetic rate
	Generation time
	Life-history type: continuous versus discrete
Detritus	
Phenology	Seasonality of input rates
Morphology	Size and structure (molecules to particles to macro scale), chemical composition (e.g., C:N, lignin, 2° compounds)
Habitat	Aquatic versus terrestrial
	Surface versus soil or benthos
	Spatial distribution of detritus
	Microhabitats on and within detritus
Dynamics	Donor-controlled

into subwebs based on primary producers and subwebs based on detritus, and compare the scatter, assumptions of independence notwithstanding, of the connectance and maximum average interaction strength with increased diversity, both declined for both types of subwebs in a similar manner (Figure 8.7b,c). Interestingly, the maximum average interaction strength was constant (average i_{max} was 0.39) across the food webs regardless of the number of basal resources. Consistent with the second assumption, the diversity of consumers within described food webs increases with the diversity of basal resources (Figure 8.7d).

8.5.5 Quasi-independence of dynamics

The results presented above suggest that energy channels form a structural foundation of food webs and that the interactions within one channel are invariant to some degree to the interactions within another. Here we show that differences in life history and physiological traits of organisms within energy channels and the temporal and spatial arrangements of the channels leads to an asymmetry and shifts in

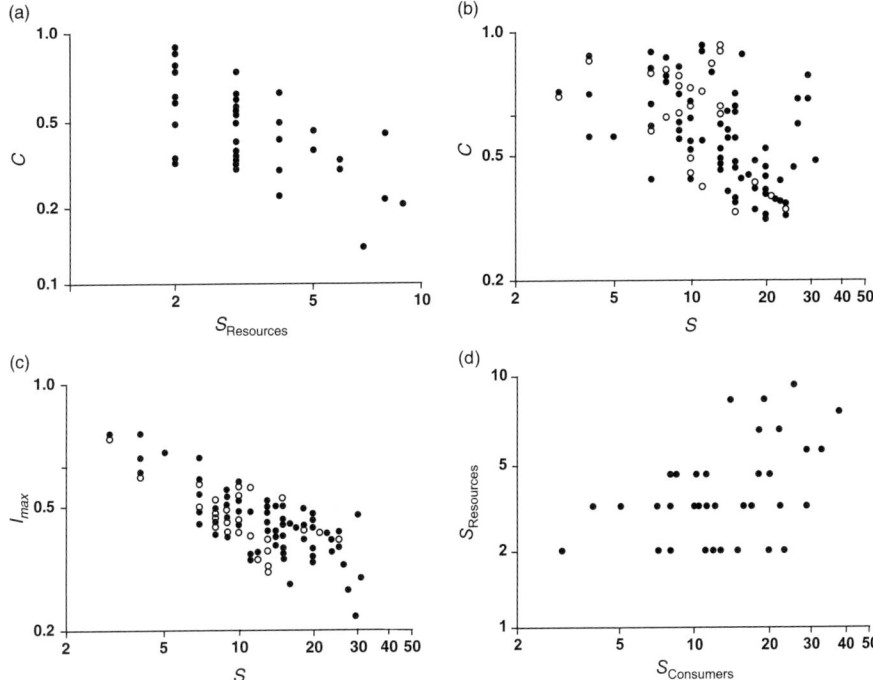

Figure 8.7 Patterns in food web metrics that support a compartmentalized architecture based on energy channels (from Moore and Hunt, 1988). (a) The relationship between the connectance (C) and the number of energy channels ($S_{Resources}$) in the food web ($r = -0.66$, $p < 0.001$). (b) The relationship between the connectance (C) and diversity (S) of energy channels based on primary producers (●, $r = -0.66$, $p < 0.001$), detritus (○, $r = -0.57$, $p = 0.0017$), and combined ($r = -0.64$, $p < 0.001$). (c) The relationship between the maximum interaction strength (i_{max}) estimated from Equation 8.1, and the diversity (S) of energy channels based on primary producers (●, $r = -0.86$, $p < 0.001$), detritus (○, $r = -0.71$, $p = 0.001$), and combined ($r = -0.83$, $p < 0.001$). A simple likelihood ratio test revealed that for the relationships presented in b and c, that the regression equations for the primary producer and detritus did not differ significantly at the $p < 0.10$ level. (d) The relationship between the number of energy channels ($S_{Resources}$) and the number of consumers ($S_{Consumers}$) in the food web ($r = 0.42$, $p = 0.0065$).

flows among energy channels, and differences in the responses of energy channels to disturbance. These features, coupled with the results from the analysis presented in the previous section, suggest that within a food web energy channels exhibit quasi-independent dynamics. The best examples come from the soil ecology literature.

Several studies have subjected soil communities to disturbances, e.g., freeze-thaw cycles, wet-dry cycles, or methyl bromide fumigation, and then tracked the initial response and eventual recovery (Allen-Morley and Coleman, 1989, Hunt et al., 1988, Wall and Moore, 1999). In each instance, bacteria and their consumers returned to their predisturbance steady-state densities more rapidly than did fungi and their consumers. Monte Carlo simulations using mathematical models of trophic interactions that incorporated the physiologies and life history characteristics of the bacterial and fungal pathways yielded results that were similar to those of the experiments. Simulations indicate that the bacterial pathway returned to initial conditions more rapidly than those of the fungal pathway (Moore et al., 2004). In these cases it was assumed that the communities, both in terms of structure and dynamics, possessed a steady state, or, in the case of the models, a strict equilibrium, and that the disturbances simply perturbed the system.

Certain disturbances change conditions to such a degree that the underlying structure of the soil community and key processes change. For example, field studies in agricultural settings have shown that agricultural management practices change the distribution of biomass within the web and the flow of energy through the bacterial and fungal pathways (Moore, 1986, Hendrix et al., 1986, Andrén et al., 1990, Moore and de Ruiter, 1991, Doles, 2000). Four patterns have emerged from studies that compared conventional and intensive management practices to reduced tillage practices, and tracked the biomasses of soil biota, nitrogen mineralization rates, and soil respiration over several growing seasons. First, the densities and activities of biota within the bacterial energy channel were enhanced relative to the biota within the fungal energy channel under the more intensive tillage practices. Second, the temporal separation in the activities of organisms within the fungal and bacterial energy channels within a growing season had diminished under the conventional practices (Figure 8.8). Third, the organic matter content of the soil decreased more rapidly and to lower levels under conventional management practices compared with less intrusive practices (Brussaard et al., 1988, Elliott and Coleman, 1988). Fourth, soils that have been subjected to intensive tillage practices compared with those subjected to reduced tillage practices often exhibited higher rates of nitrogen mineralization during the shoulders of the growing season, when the nitrogen is less likely to be used by the crops and more likely to be lost from the system through leaching.

Several mechanistic explanations have been offered to explain the patterns. The conventional management practices employ tillage methods that mix soils and disrupt soil aggregates, thereby releasing labile (low C:N) carbon (Camberdella and Elliott, 1994, Six et al., 1999, Wander and Bidart, 2000, Simpson et al., 2004), rely heavily on inorganic nitrogen fertilizers, and manage soil water in ways that favor the biota within the bacterial pathway over those within the fungal pathway. The reduced tillage management practices are less detrimental to the existing soil aggregate structure. These practices incorporate resistant (high C:N) organic carbon in the form of crop residues or manures into the soils, resulting in a stratification of organic residues by depth and a more heterogeneous mix of water- and air-filled pore spaces that favor a balance between the bacterial and fungal pathways (Figure 8.9).

Changes of the types mentioned above have not been restricted to agricultural settings. Studies addressing the impacts of increased temperatures and concentrations

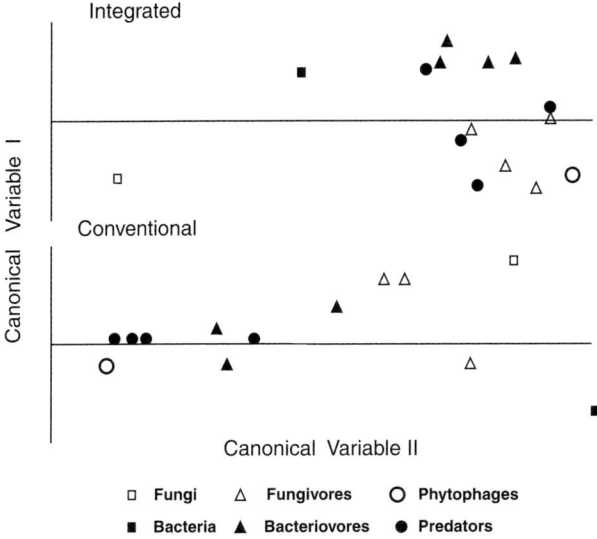

Figure 8.8 Comparison of the seasonal relationships between functional groups within the bacterial, fungal, and root energy channels and their predators from soils from the integrated and conventional management practices at the Lovinkhoeve Farm (Marknesse, The Netherlands) using canonical discriminant analysis. Significant temporal separation (compartmentation) in the dynamics of groups within the fungal and bacterial energy channels was found in the integrated management practice (Wilk's λ, p > F = 0.026). The temporal separation in dynamics was not found in the conventional practice. Adapted from Moore and de Ruiter (1991).

of atmospheric CO_2 point to similar shifts in activity within and between the bacterial and fungal pathways (Mack et al., 2004, Phillips et al., 2006). For example, ten years of nitrogen addition to the arctic tundra at Toolik Lake, Alaska has resulted in the loss of mosses, an increase in woody plants, a loss in soil organic matter, and a shift in activity below ground from the fungal channel to the bacterial channel (Doles, 2000, Mack et al., 2004). Johnson et al. (2002) focused on the microbial community of the arctic tundra in northern Sweden, and found that elevated CO_2 and ultraviolet B radiation affect the C:N ratios of substrates and the densities and diversity of the bacterial community. Phillips et al. (2006) reported that after ten years of CO_2 enrichment of pasture in Switzerland, soil protozoa and nematodes responded in a manner that is consistent with shifts predicted for the availability of soil bacteria and fungi. The results suggest an increase in fungal density (mycorrhizal and saprobic) and their consumers, whereas enrichment result in a decline in soil protozoa. Higher concentrations of atmospheric CO_2 result in increased plant productivity and root turnover, thereby decreasing litter quality (higher C:N) and favoring the fungal energy channel over the bacterial energy channel (Figure 8.9).

In each of the cases discussed above, disturbances affected the structure and dynamics of food webs as a whole but in a patterned way as the energy channels

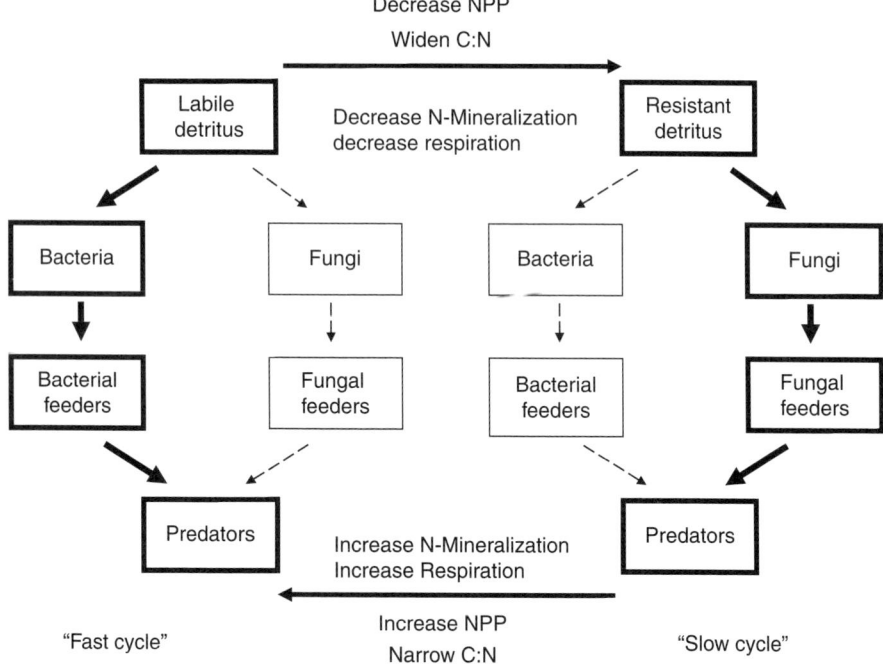

Figure 8.9 Simplified representation of coupled bacterial and fungal energy channels of a rhizosphere food web. When the bacterial energy channel is dominant the system is operating as the 'fast cycle' owing to the higher turnover rates of bacteria and their consumers relative to the fungi and their consumers (Coleman et al., 1983). When the fungal energy channel is dominant the system operates as the 'slow cycle.' Natural and anthropogenic disturbances that affect changes in the C:N ratio of the detritus or net primary productivity (NPP) also affect rates of nitrogen mineralization and decomposition (respiration or C-mineralization) have been associated with shifts in the relative dominance of one channel to the other.

(read "compartments") within the food webs respond to the disturbances to different degrees. These differences in the responses of the compartments include differential responses of populations among energy channels, movement of biomass to higher trophic positions within and among energy channels, rates of the flux of matter and energy among energy channels, and dynamic states among energy channels.

8.6 Energy channels—structure and stability

In the previous chapters we discussed the ways in which these types of alterations in structure, rates of flux, and dynamics to simple and complex systems are related to stability. So far in this chapter we have defined ways to characterize and study compartments within food webs. Here we discuss why compartments form, and ways in which they can interact to promote stability. We have borrowed from Pimm (1982) and distilled the argument down to a set of biological and dynamical reasons, concluding that some combination of the two is at work. The basis of the biological argument is that the presence of compartments within food webs is a consequence of the adaptation species make to their environment and to one another. Compartments are generally separate at lower trophic positions and are linked by predators at higher trophic positions. The dynamic stability argument originated from May's (1972) observation that compartmentalized food webs are more likely to be stable on average than their noncompartmentalized reticulate counterparts. We present several examples in which the linking of compartments by consumers and predators serves as a stabilizing force.

8.6.1 Coupled pathways and weak links

McCann et al. (1998) presented an important result that demonstrated that weak interactions among species can serve to stabilize otherwise unstable configurations. The dynamics of simple food web models described with type II functional responses were compared. The simplest systems included two coupled resource-consumer interactions (compartments), one of which included a top predator (Figure 8.10). The model was initiated with the top-predator-consumer-resource (P–C_1–R) food chain, with the second consumer (C_2) being allowed to invade. The dynamics of the model were explored by plotting the local minima and maxima of the top predator against the ratio of the interaction strengths of mid-level consumers and the basal resource. If the interaction strength of the invading consumer on the resource (I_{c_1}–R) was small relative to the interaction strength of the original consumer on the resource (I_{c_2}–R) the dynamics of the system shifted from complex chaotic dynamics to a simpler periodic signal. Under certain conditions, linking the second consumer to the predator tempered the dynamics to a stable equilibrium.

Post et al. (2000) took a different approach by assessing the importance of a top predator's preference for prey originating in two separate food chains, akin to the energy channels we presented above. They viewed the exercise as one that added species to an existing group of species, or one that linked two spatially distinct groups of species (i.e., compartments). The models included a type II functional response of the predator, but allowed for prey-switching by the predator to generate a sigmoid type III-like functional response. The study identified plausible ranges of predator preferences for prey between the food chains (channels) that changed the dynamics of the system from chaotic dynamics to limit cycles, and eventually a stable equilibrium.

Moore et al. (2004), following the approach presented by Post et al. (2000), explored the issue with models that included parallel energy channels, one that transferred and turned over matter at a slow rate and one that did so at a fast rate, characterized by the input rates of the basal resources and the death rates, and rates at which the consumer species processed matter (Figure 8.11). The stability of these

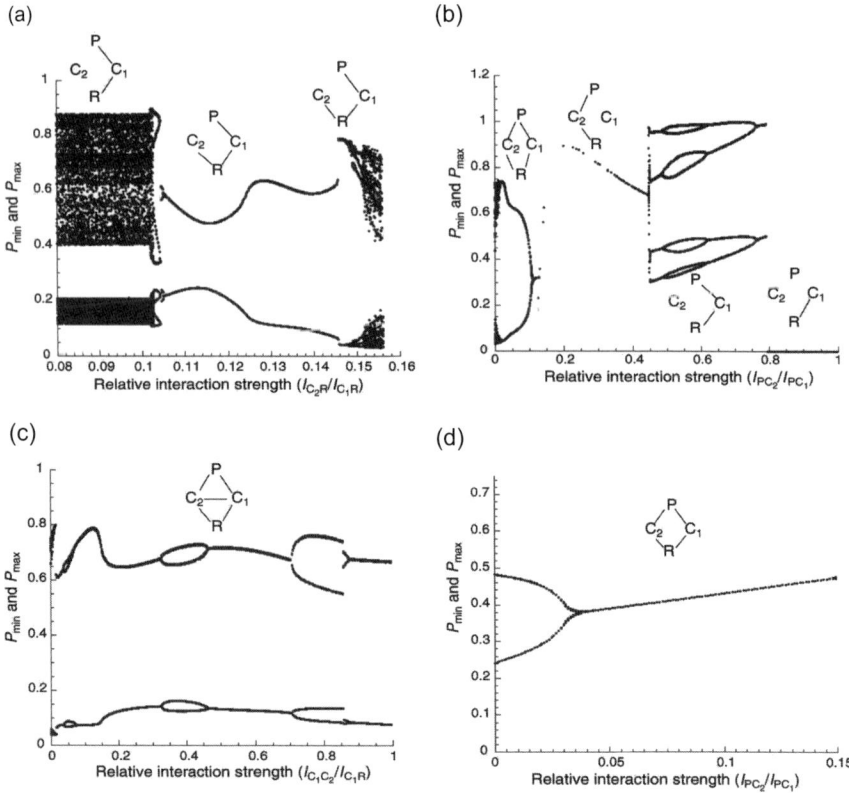

Figure 8.10 Weak linkages among between compartments and the coupling compartments by consumers and predators through (a) exploitative competition, (b) apparent competition, (c) intraguild predation, or (d) a different parameterization of apparent competition can stabilize otherwise unstable configurations (McCann et al., 1998). The figures show the local minima and maxima for top predator density, P, for a range of relative interaction strengths (I) among different combinations of resource (R), consumers (C), and predators (P) and within the food webs. The configurations are given as a function of the relative interaction strength based on the stated combinations. Species that are not connected to others by an explicit link cannot persist. Reprinted with permission from Macmillan Publishers Ltd: Nature.

model systems was dependent on the linking of the fast and slow channels by predators, and the balance of material passing through the fast and slow channels. Shifts in the relative flux rates of the fast and slow channels affected both resilience and stability.

Rooney et al. (2006) found that the organization of systems into fast and slow pathways was ubiquitous across terrestrial and aquatic ecosystems. A compartmentalized food web architecture based on multiple energy channels that process and turn over matter at different rates provides ecosystems with a potent mechanism for

Modeling compartments • 203

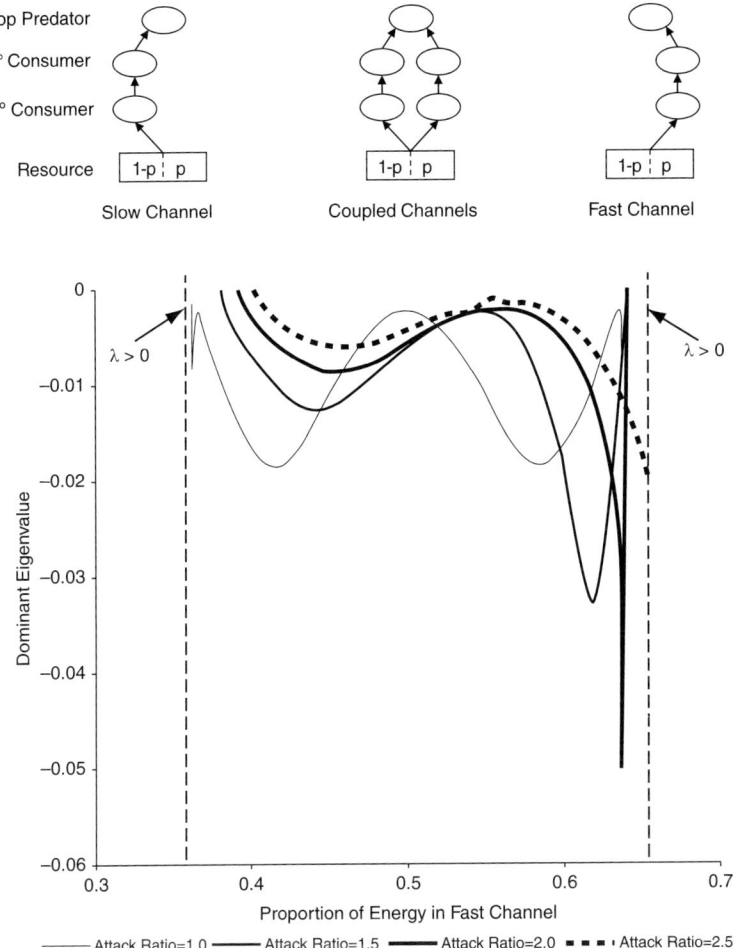

Figure 8.11 The relationship of the dominant eigenvalues of a series of models possessing two parallel food chains, or energy channels, drawing energy from a heterogeneous source and linked by a common predator. The food chains differ in the rates in which that process and turnover energy, i.e. fast (right) and slow (left) channels, and in the preference for the fast channel (attack ratio) of the top predator. The x-axis represents the proportion of resource passing through the fast channel (p), while the y-axis represents the dominant eigenvalue (λ) for the system for each partitioning of resource. The total amount of energy passing through each system is the same for each partitioning of resource. The dashed vertical lines represent unstable transitions. Note that the most stable configuration occurs when the two channels are coupled, and unstable if most of the energy passes through the fast channels ($\lambda > 0$).

responding to perturbations, both small and large (i.e., dynamics near and far from the equilibrium, respectively). They further concluded that the asymmetry in the energy flux rates in the coupled fast and slow energy channels provided the top predators with complementary dynamics that work to stabilize the system. Following a disturbance, the fast energy channel provides the predator with a prey base that allows for the rapid recovery of predator populations. If the fast energy channel were not coupled to the slow energy channel, and production were sufficiently high, it would almost certainly result in the oscillations and runaway consumption dynamics observed in the earlier enrichment studies, i.e., the paradox of enrichment (sensu Rosenzweig, 1971). Rooney et al. (2006) concluded that the presence of a slow channel enhanced compensatory responses near the system attractor. The complementary functions of the fast and slow energy channels produce a rapid and stable recovery from a perturbation.

Each of the examples presented above illustrates how linkages between subgroups of species can influence the dynamics of the overall system in a positive way. Changes in the linkages or the relative flow of matter within the subgroups affect system resilience and stability. In the sections that follow we offer explanations that draw from both the biological and dynamical arguments, in that although biology certainly plays a role, the responses of species to disturbances are central to demonstrating the existence of compartments and to explaining their importance.

8.6.2 Resistance and resilience

Some have argued that food webs that possess a compartmentalized architecture are more resistant to disturbance than their reticulate counterparts. MacArthur (1955) proposed that systems that are configured in such a way that they possess multiple pathways of energy flow possess a greater capacity to recover from disturbances and are more likely to persist than systems with fewer pathways. Krause et al. (2003) surmised that compartments serve to confine disturbances, thereby minimizing the impacts of a disturbance on the entire food web and stabilizing the system. There is an appeal to these explanations, as they draw from analogous situations in engineering dealing with redundancy and structural buffering, but there is phenomenological circularity to the arguments as well. We will focus on the second example to make this point.

Compartments are revealed in part through the coincident responses of a subset species to disturbance. We have argued that the similarity of responses to disturbance arises from species sharing similar habitats and turnover rates, and by virtue of their feeding relationships, i.e., the effect of a disturbance is limited to within a compartment. The problem with taking this approach is twofold. First, if disturbances were to propagate through a whole food web or a compartment and any species were to drop out as a result, by definition the system is unstable. Second, the argument asserts that the dynamic stability of the food web is affected by the compartments themselves rather than both the compartmentalized structure and the way in which compartments interact with one another.

A simpler explanation can be formulated if we assume that the rules governing the structure and dynamics of the configurations of a compartment of species are the same rules that apply to the food web as a whole. Assume that different basal resources attract consumers that operate at different temporal and spatial scales, which in turn attract predators, resulting in a series of simple food chains. The individual food chains are constrained by all of the factors that we discussed earlier in Chapter 7, to the point that there is sufficient biomass to support an additional trophic level, to the point at which biomass begins to accumulate at the upper trophic position, and finally to the point at which they reach their respective thresholds from stable to unstable dynamics. We have evidence that the latter scenario is more likely. Cross-linkages between and among compartments that possess different flow rates have a higher likelihood of persisting than embedding additional species within an existing compartment or by linking compartments with similar flow rates, given the conflicting constraints of the availability of energy and the impact of energy on dynamics.

8.6.3 Enrichment, predators, and energetic bottlenecks

Compartmentalized food webs possess energetic bottlenecks that restrict energy flow and runaway consumption and growth. Bottlenecks may come in the form of compensatory and complementary dynamics of coupled fast and slow energy channels, polyphagous predators that ameliorate unstable dynamics (Rooney et al., 2006), and higher-order consumers that are energetically less efficient than their prey. In the lexicon of hierarchy theory, higher-order consumers that link energy channels serves as filters that obstruct, delay, or attenuate the free flow of information within an energy channel. The information here is the energy flux and the eigenvalue-based information involving interaction strengths that define stability. This mechanism can be traced back to the "paradox of enrichment" and the potentially destabilizing effect of enrichment (Rosenzweig, 1971). Any interaction or organism that restricts the flow of energy through a food chain from another compartment could work in this capacity (McCann et al., 1998, Debruyn et al., 2007).

Rooney et al. (2006) offered a two-part corollary to this general result: (1) perturbations that alter the relative flow of matter through the channels either through synchronizing or removing channels may destabilize these systems relative to their heterogeneous counterparts, and (2) removal of mobile higher-order consumers that couple the energy channels eliminates the energetic bottleneck and may destabilize these systems relative to their coupled counterparts. Related to this, in line with the findings of Post et al. (2000), a compartmentalized food web possesses interactions that ameliorate competitive interactions and unstable dynamics. This mechanism is related to the previous mechanism on enrichment but focuses on the positioning and nature of the interaction within the food web and its effect on stability regardless of the level of enrichment.

8.6.4 Donor-controlled dynamics

Reviews of published webs suggest that systems are compartmentalized into food chains based on primary producers and consumers and food chains based on nonliving organic matter (detritus) and detritivores that are merged by predators after two to three trophic transfers (Hunt et al., 1987, Moore et al., 1988, Moore et al., 2004). This observation was not lost on early theoreticians (see Box 8.1), and provides a starting point to address how detritus-based compartments might stabilize whole food webs. We will draw from results presented throughout the book to make the point.

In Chapter 2 we discussed how models based on detritus possess elements of donor-controlled dynamics and mathematical properties that are inherently stable. Simple representations of a two-component detritus-based food chain were shown to be globally stable (Neutel et al., 1995), and resilience was demonstrated to be governed by the rates of allochthonous and autochthonous inputs of detritus and the efficiencies of the consumers of detritus. Our treatment here leads us to conclude that compartments based on detritus that possess donor-controlled dynamics when coupled with compartments based on primary producers have—under the right conditions—the potential to stabilize food webs.

Detritus operates in a bottom-up manner as a resource that influences productivity directly, inasmuch as it is the basal resource, and indirectly, as it regulates primary production. In Chapter 7 we demonstrated that detritus-based systems are more likely to be feasible at low rates of input than their primary-producer-based counterparts. A modicum of detritus input can support a consumer—far less than that required within a primary-producer-based system. We further demonstrated that dynamics and stability are dependent on the accessibility of detritus for a given level of input. In this case we presented a detritus-based analog to a model presented by Rosenzweig (1971), in which detritus was given a refuge from consumption. Two important results from this analysis were that refugia afforded the detritus-based systems longer return times, and greater levels of input before the onset of oscillations following a disturbance. In a related exercise, we made the growth rate of a primary producer dependent on the allochthonous and autochthonous inputs of detritus-based nutrients. Here, the availability of plant-limiting nutrients affected the growth rate of the primary producer and subsequently those of its consumers, and the dynamic state of the whole compartment.

Detritus can also affect the way in which consumers affect systems in a top-down manner. In this case, the predator feeds on organisms or resources from both the detritus-based compartment and the primary-producer-based compartment. The dynamics and stability of these systems are linked to the prey preferences and attack rates of the consumers or predators that link the detritus-based and primary-producer-based energy channels, and the relative rates of the flows through the different energy channels. The donor-controlled properties of detritus-based energy channels, the asymmetry in the flow rates from the detritus-based channel to the primary-producer-based channel, and the degree of dependence of the primary-producer-based channels on the detritus-based channel, attenuate system-level oscillations and modulate runaway consumption by predators for the reasons discussed above, e.g., bottlenecks in flow rates, weak links, and asymmetry in flow rates.

8.7 Summary and conclusions

The debate on whether the architecture of food webs is compartmentalized or reticulate arose from the broader discussions on the relationships among food web diversity, complexity, and dynamic stability. The seminal work by Gardner and Ashby (1970) about systems in general and the follow-up by May (1972) on the implications of the results to ecological systems challenged the tenet that more diverse and complex systems are more likely to be stable. Their findings reframed the debate by forcing us to ask about the types of complexity that increase stability. For simple model systems, the arrangement of components (i.e., species for food webs) into linked blocks or compartments of tightly linked interacting subgroups were more likely to be stable than their reticulate counterparts (Austin and Cook, 1974, McMurtrie, 1975).

We reviewed different approaches to studying compartments. Studies that were based on connectedness descriptions that solely relied on the binary information of the presence or absence of linkages and their arrangements provided weak evidence of compartments. The types of basal resources and habitat preferences emerged as important compartment-forming factors. Studies that incorporated information from the energy flux descriptions and functional descriptions provided greater discrimination.

We viewed food webs as being organized in a hierarchical manner, consisting of coupled subsystems (i.e., compartments) of organisms that processed matter acquired via trophic interactions for their growth and maintenance. From this perspective we identified the energy channel as a fundamental organizing unit within a food web. An energy channel consists of a basal resource and all the trophic interactions and accompanying energy fluxes that radiate from it to the top predator.

Food webs may contain multiple energy channels, but primary producers and detritus are two basal resources common to most ecosystems. The energy channels transfer matter and turn over matter at different rates: a property that arises from the physiological and life history attributes of the species within the channel. Habitats and time play a role in defining and shaping the boundaries of energy channels. Energy channels tend to be distinct from their base and linked by a common suite of predators. The structural and dynamic properties of the energy channels within a food web are governed by the same constrains as the food web as a whole. Although coupled, energy channels operate in a "quasi-independent" manner.

We concluded by demonstrating that the stability of complex ecosystems depends critically on the maintenance of the heterogeneity of distinct energy channels, their differential dynamic properties (i.e., differential productivity and turnover), and the consumers that couple these distinct channels.

9

Productivity, dynamic stability, and species richness

9.1 Introduction

Understanding the determinants of biological diversity at local, regional, and global scales has been an important topic in ecology for some time. Several hypotheses have been proposed to explain the relationship between the species richness and latitude along a gradient from the tropics to the poles (Rohde, 1992, Rosenzweig, 1995). Typically, studies at the continental scale reveal a pattern of decreased diversity with increased latitude (Box 9.1). At first glance this pattern appears to parallel the well-established and similar latitudinal pattern of global net primary productivity (NPP). However, closer examination of these trends reveals that the relationship between species diversity and NPP is dependent on the physiologies of the organisms under study and the spatial scale at which the observations are made (Hutchinson, 1959, Mittlebach et al., 1998). At the regional scale the highest numbers of species occur not within the regions of highest NPP, but rather within regions of intermediate levels of NPP. At the regional scale, the relationship between species richness and productivity is often not linear, but "hump-shaped" (Rosenzweig and Abramsky, 1993, Gross et al., 1998, Mittlebach et al., 1998, Waide et al., 1999, Mittlebach et al., 2001). At the local scale the relationship appears to be either positive or nonsignificant (Alder et al., 2011).

Box 9.1 Diversity in a food web context

Examples of hump-shaped species diversity curves for various groups of organisms and from various kinds of ecosystems and ecosystem drivers are presented in Figures B9.1A–B9.1C. The original premise for the relationships was based on latitude, with the apogees in diversity occurring at intermediate latitudes, corresponding to intermediate levels of productivity. These relationships have been challenged for being too simplistic, and often fail to materialize for terrestrial plants (Figure B9.1D). For example, Mittlebach et al. (1998) argue that the relationships were dependent on the spatial scale at which the observations were made, while others have discounted the relationships altogether for plant species (Adler et al., 2011). Our approach has been to view diversity from a food web perspective, taking into account the rate and availability of production and the effects that these have on the dynamic stability of the system. The theory that we develop generates hump-shaped curves with the apogees at intermediate levels of productivity and reconciles the discrepancies between theories presented in the literature.

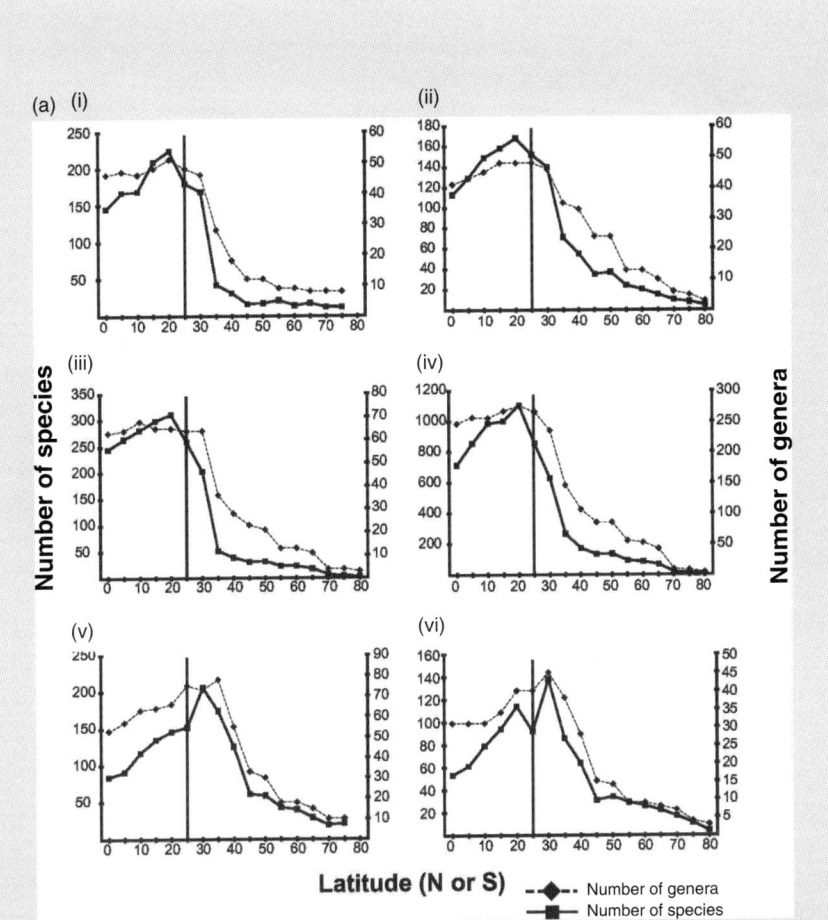

Figure B9.1A Bivalve (species and genera) diversity along altitude (degrees north or south). (i) Arcoida, (ii) Mytiloida, (iii) Pteriomorphia, (iv) Veneroida, (v) Anomalodesmata, (vi) Carditoida. The vertical line marks 25°, the general position of the tropical–temperate boundary. Reprinted from Krug et al. (2007), © 2007 National Academy of Science, U.S.A.

Box 9.1 (*Cont.*)

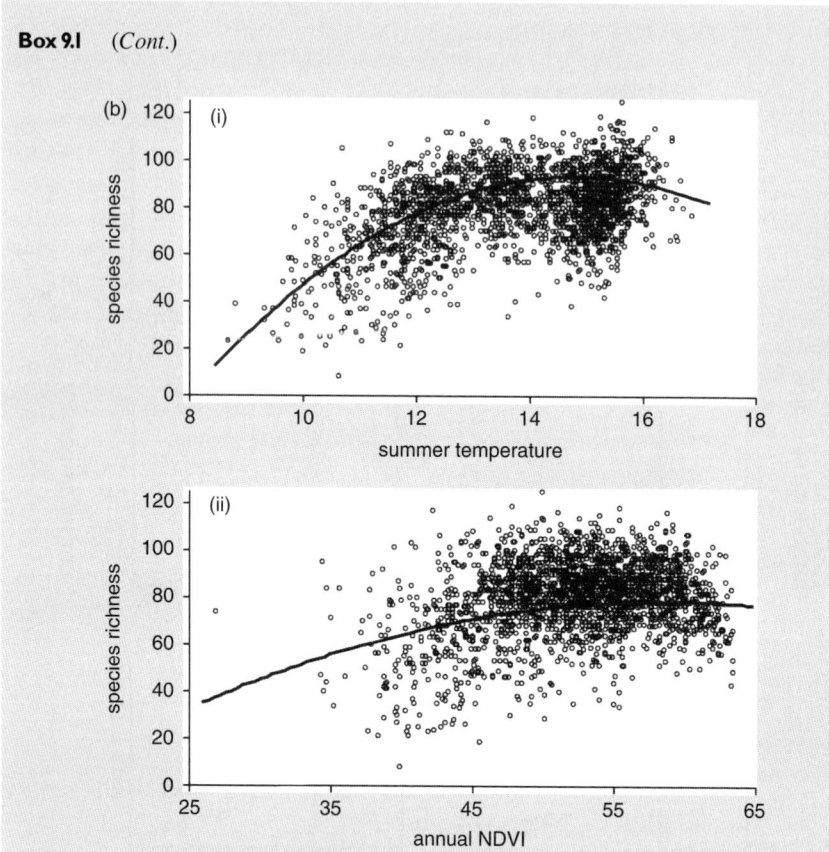

Figure B9.1B Relationships between species richness and (i) temperature and (ii) normalized difference vegetation index (NDVI) across the entire assemblage of British breeding birds. Open circles represent raw data and lines represent the predicted relationship from models that take into account spatial autocorrelation and use a single measure of energy, the seasonal measure that gives the best fit to the data. From Evans et al. (2005), with permission from The Royal Society.

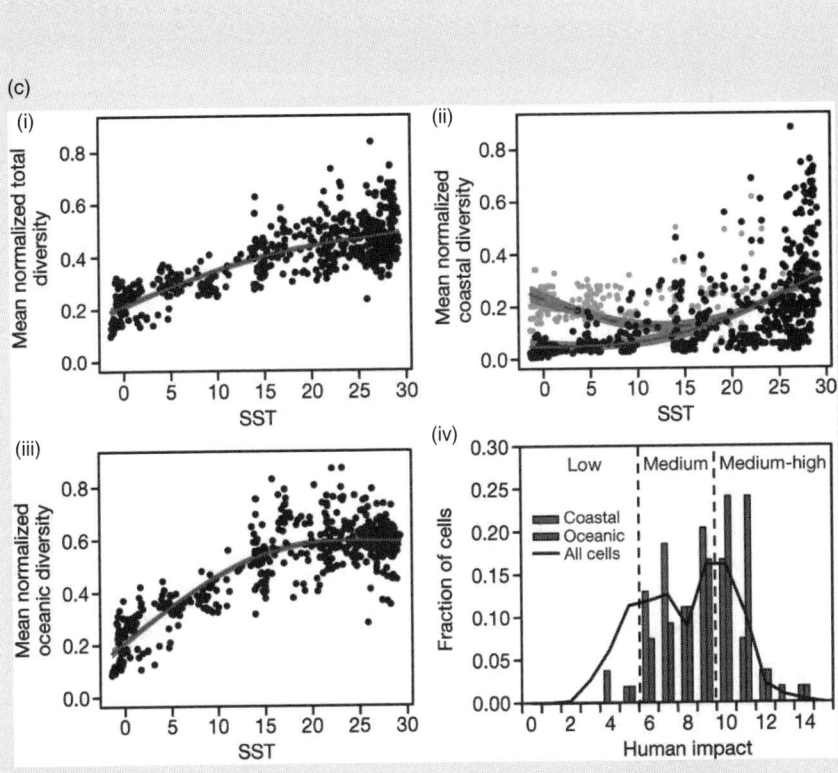

Figure B9.1C Relationship between oceanic diversity (ranging from zooplankton to mammals) and sea-surface temperature (SST) for (i) all taxa, (ii) coastal taxa without pinnipeds (solid line and black points) and coastal taxa with pinnipeds (dashed line and gray points) and (iii) oceanic taxa. Trends (red lines with gray 95% confidence limits) indicated by generalized additive model fit with basis dimension 3. (iv) Histogram of diversity hotspots (10% of cells with highest mean richness) by human impact. From Tittensor et al. (2010), reprinted with permission from Macmillan Publishers Ltd: Nature.

Box 9.1 (*Cont.*)

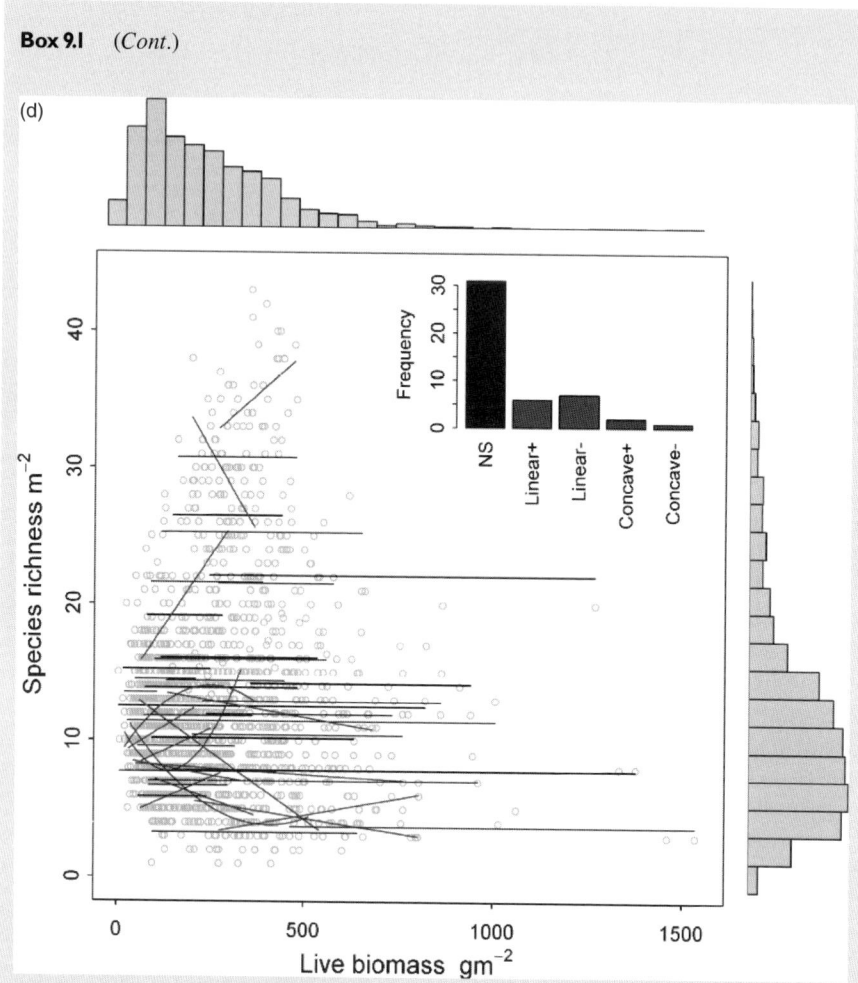

Figure B9.1D The within-site relationships between productivity (indexed as peak biomass) and plant species richness for 47 sites across North America (n =35) Europe (n =4), Africa (n =4), Asia (n =1), and Australia (n =3) (from Alder et al., 2011). The inset shows the frequency of relationships that were nonsignificant (NS, black lines), positive or negative (gray lines), and concave up (+) or down (–) (gray curves). The histograms on the margins show the frequency of species richness and the frequency of peak biomass across all sites. Details of the statistical analyses and separate figures for each site are presented in Alder et al. (2011).

In this chapter we describe energetically based food web processes that can explain the observed relationship between NPP and species diversity. Our argument draws on the interrelationship between NPP and dynamic stability presented in Chapters 6 and 7 (Rosenzweig, 1971, Pimm and Lawton, 1977, Moore et al., 1993), and the compartmentalized nature of ecosystem structure that we presented in Chapter 8 (May, 1973, Hunt et al., 1988, Krause et al., 2003, Rooney et al., 2006). We start with independent models of trophic dynamics that can be applied to any spatial or temporal scale. The models show that the likelihood of sufficient energy to sustain a simple food chain of a given length, and the likelihood of such food chains exhibiting stable dynamics following a disturbance, interact with and vary with productivity so that the potential for more food chains exists at intermediate levels of productivity. Next, we present both theory and empirical evidence that the number of consumer species increases as the number of food chains within a community increases. Species diversity is therefore influenced by both the rate and diversity of production. The interaction of these two factors generates a unimodal pattern in species diversity along a productivity gradient. We adjust the models to study how different physiologies (compare ectotherms to endotherms), life histories, and home ranges (i.e., we alter the temporal and spatial scale within the models) might influence the shape of the relationship between NPP and species diversity. Here we demonstrate that the wide range of scale-dependent observations can result from the interaction between the rate and diversity of production and dynamic stability. We present a framework for community development that integrates these ideas with those about compartments and their impacts on structure and stability. We conclude with a review of the hypotheses presented over the past decades on the relationship between NPP and diversity through the lens of our proposed framework.

9.2 Trophic structure, dynamics, and productivity

In earlier chapters we examined the effects of productivity on both the structure of communities and their dynamics. For models of simple primary-producer-based and detritus-based food chains, the feasibility (all species maintaining positive steady-state densities) and the time required for food chains to return to steady state (the return time, or RT) following a minor disturbance (a small deviation from steady state) were a function of productivity (Figure 9.1).

The more complex models with nonlinear functional responses yielded qualitatively similar results. The feasibility of these models was governed by the influence of productivity on the steady-state densities of species and their dynamics. At low rates of input the complex models behaved similarly to the simpler ones, as minimum levels of input were required to sustain all species. The models differed in the range of responses that were possible at higher rates of enrichment. Unlike their counterparts with linear functional responses, the complex models exhibited a range of dynamic states, from unstable equilibria that gave rise to extinction to those that generated stable limit cycles and chaotic dynamics.

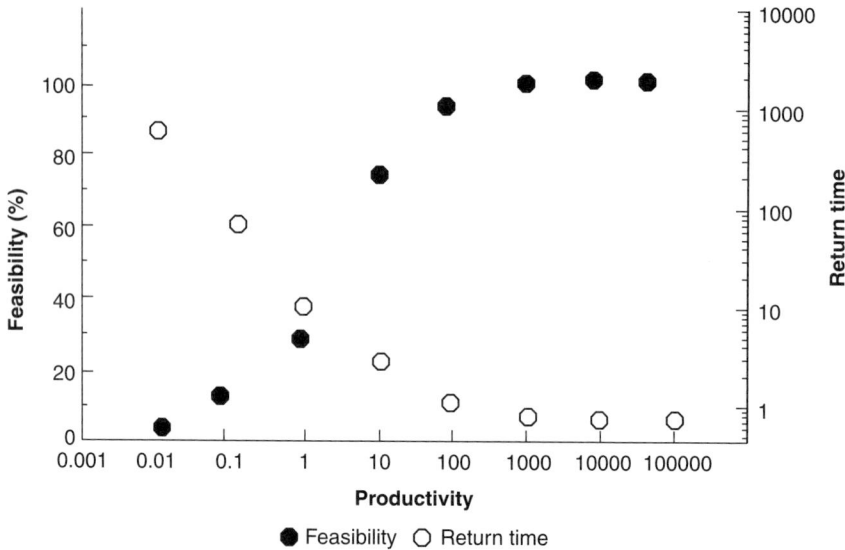

Figure 9.1 Feasibility and return time as a function of productivity for two-species Lotka–Volterra primary-producer-based food chains. Food chains were evaluated at levels of productivity beginning with 0.01 units, increasing by orders of magnitude to 100,000 units. The analysis included 1000 iterations for the chains at each level of productivity. All other parameters were selected at random from a uniform distribution over the interval [0,1]. Depending on initial assumptions, one unit of productivity equals 4 to 36 g C m^{-2} yr^{-1} (adapted from Moore et al., 1993).

To address how productivity and the subsequent interplay feasibility and dynamics might operate to affect species richness requires that we revisit the analysis presented previously. Different levels of productivity were achieved by changing the specific rate of increase of primary producers for models depicting primary-producer-based food chains or by changing the input rate of detritus from outside sources for models representing the detritus-based food chains. Each trial was akin to sampling organisms at random, assembling food chains, assessing their feasibility, and estimating return times after a disturbance. The results presented in Figure 9.1 could explain lower species richness in low-productivity habitats, but not the decline in diversity in high-productivity habitats. Declines in diversity at high productivity are possible if the concept of feasibility includes a dynamic component (Rosenzweig, 1971).

9.3 Feasibility revisited

The discussion above leads us to expand our notion of feasibility. We indentify the following aspects of the feasibility of a system ($F_{[sys]}$): feasibility based on net primary production ($F_{[NPP]}$), and feasibility based on dynamics ($F_{[\lambda]}$). These factors interact so to define the feasibility of a system as follows:

$$F_{[SYS]} = F_{[NPP]} \cap F_{[\lambda]} \qquad (9.1)$$

In working through this relationship we should satisfy the following conditions.

1. X_i and $X_j > 0$ for all i, j species.
2. The system possesses a persistent dynamic state, e.g., is globally stable, locally stable, exhibits a stable limit cycle, or chaos.
3. The parameter choices used to construct the model generate dynamics that are biologically reasonable.

Conditions 1 and 2 bring in the thresholds or transitions in structural and dynamic state of systems that occur along a gradient of productivity. Condition 1 addresses feasibility based on primary production ($F_{[NPP]}$), which hinges on the familiar thermodynamic-based hypotheses proposed to explain trophic structure (Hutchinson, 1959, Oksanen et al., 1981). Condition 2 addresses feasibility based on dynamics $F_{[\lambda]}$, which can be traced to "the paradox of enrichment" (Rosenzweig, 1971) and studies of chaos (May, 1976). Models of simple predator–prey interactions with nonlinear functional responses exhibited either unstable dynamics, extreme oscillations at high levels of productivity, or trajectories that bring populations close to zero. Condition 2 questions the feasibility of systems that exhibit these types of dynamic behaviors.

Condition 3 requires that we introduce some additional biological realism to the discussion. In doing so, we can see that the thresholds defined by $F_{[NPP]}$ and $F_{[\lambda]}$ are sensitive to the parameters selected. We will consider three ideas to make this point. First, we know that for some models stability is not certain given their architecture, but depends on the choice of parameters used to initiate them. Second, we also know that models can transition to states that possess an unstable equilibrium but nonetheless satisfy condition 1, as in the cases of stable limit cycles and chaotic states. Some forms of oscillation are feasible, and all things being equal, these types of systems will persist. Third, even though all our mathematical notions of stability are met, the parameter choices should be reasonable. We know, for example, that energetic efficiencies ($e = ap$) cannot equal or be less than 0%, or exceed 100%. Beyond the obvious, seemingly reasonable choices can come into question as well, as some aspects of physiology or life history are contingent on others.

To further evaluate condition 3 we will start with the simple food chain models that include linear functional responses (Figure 9.1). Although stable, these models demonstrate different dynamics along an NPP gradient, as resilience increases with increased NPP (Figure 9.1). Greater resilience has been linked to greater stability, in the sense that the longer a system takes to return to its original steady state the more likely additional perturbations will intercede and push the system toward an alternative steady state (May, 1973, Pimm, 1982). This implies that higher rates of productivity promote greater resilience, a conclusion that clearly appears at odds with the paradox of enrichment. However, closer examination of these results reveals their similarity when condition 3 is taken into consideration.

The resilience of the food chains presented in Figure 9.1, whether primary-producer-based or detritus-based, is estimated as the inverse of the return time (RT) of the food chain, RT = $-1/\lambda_{max}$, where λ_{max} is the real part of the dominant (least negative) eigenvalue. To illustrate this, we will rely once again on the solution for the eigenvalues for the system of equations for the two-species primary-producer-based food chain. From our discussions in Chapter 7 we know that the dominant eigenvalue for this food chain is:

$$\lambda_{max} = \frac{\alpha_{11} + \sqrt{\alpha_{11}^2 + 4\alpha_{12}\alpha_{21}}}{2} \quad (9.2)$$

The transient dynamic behavior (see Neubert and Caswell, 1997, Verdy and Caswell, 2008, Hastings, 2001, Hastings, 2010) of the decline and eventual decay of RT with increased NPP presented in Figure 9.1 can be explained by a decrease in the value of the discriminant (μ) of λ_{max}, where $\mu = \alpha_{11}^2 + 4\alpha_{12}\alpha_{21}$. Although the system is locally stable along all levels of NPP > 0, enrichment will induce transitions through different transient dynamic states, some of which are arguably tenuous.

Within the discriminant (μ) the α_{21} term depends in part on productivity. When NPP $\rightarrow \infty$, μ transitions from positive to negative, giving rise to the imaginary part of the critical eigenvalue and the onset of oscillations on return to equilibrium. Hence, as NPP $\rightarrow \infty$, μ undergoes transitions, from positive, to zero, to negative, leading to three distinct types of recovery: (1) overdamped, (2) critically damped, and (3) underdamped (see Equation 7.15). For this discussion we will focus on the oscillations that occur in the underdamped case, as the overdamped and critically damped recoveries do not possess oscillations and are feasible dynamics from a biological standpoint.

It is at the greater levels of NPP that dynamic feasibility ($F_{[\lambda]}$) of the system becomes questionable for biological and practical reasons. At the high levels of NPP, when $\mu < 0$, the system becomes reactive and develops damped oscillations with a quasi-period of $T \approx 2\pi/\mu^{-1/2}$. Some populations would not able to survive large and frequent oscillations in their food supply, despite the fact that on average, the supply would suffice (see Dodson et al., 2000). For the models with the linear type 1 functional response, the frequency of the oscillations ($\approx 1/T$) increases with increases in productivity beyond the level of the onset of oscillations. In the absence of a compensatory mechanism, the oscillations eventually reach a pitch that is beyond the growth capabilities of some species: in such cases, the systems are feasible from a theoretical point of view but not feasible for biological reasons (Figure 9.2). Even if organisms were capable of responding to such oscillations, there are practical reasons to suggest that the system would not persist.

The argument presented above is not restricted to simple models. The dynamics for related models with nonlinear functional responses possess a wider array of possible dynamics, depending on the particulars of the species involved. In many cases, after the onset of oscillations the amplitude of the oscillations increases as the

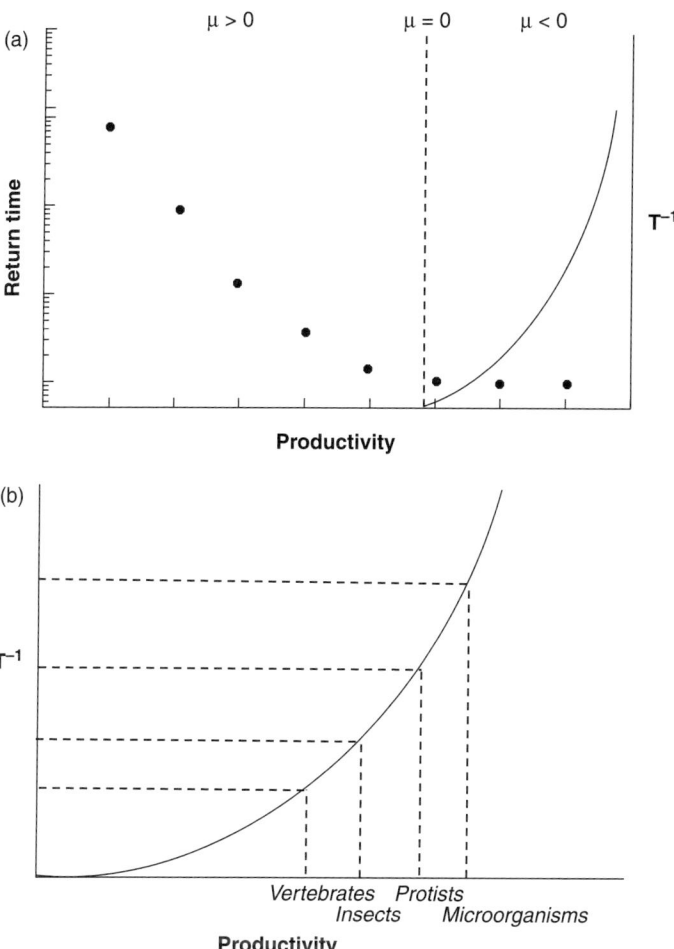

Figure 9.2 The nature of the transient dynamic of stable systems is a function of the level of productivity, whereas the capacity of a system to endure the transient is a function of the turnover rates (life histories) and processing rates (production:biomass ratios) of the organisms within the system. (a) The return time of the two-species food chain decreases with increased productivity. The transient dynamic behavior of the system, expressed here by the frequency of oscillations (T^{-1}), is also a function of productivity. For low to intermediate levels of productivity in this example, the transient behavior does not possess oscillations ($\mu \geq 0$, see Equation 9.2). At higher levels of productivity, the transient behavior transitions to when $\mu < 0$ and the system develops damped oscillations with a quasi-period of $T \approx 2\pi/\mu^{1/2}$ and frequency $\approx T^{-1}$. (b) Taxa differ in the reproductive capacities and processing rates, and hence will differ in the types and degree of transient behavior that they can endure.

frequency decreases (Gilpin and Rosenzweig, 1972, May, 1972). In these situations, the oscillations never reach zero, but they approach zero so frequently that the likelihood of extinction is near certain (Gilpin, 1975). Likewise, models that possess chaotic dynamics may satisfy conditions 1 and 2, but the paths species take may bring the populations perilously close to zero. Additionally, there are a large number of models in which enrichment may lead to instabilities (Rosenzweig, 1971). In many cases, as shown previously, an evaluation of the eigenvalues for the systems in question reveals a transition in dynamic states that is dependent on the level of productivity and coincident with the eigenvalues equaling or passing through some critical value (e.g., the critical eigenvalue is equal to zero; the critical eigenvalue is nonzero and purely imaginary).

A final example relates to the approach we used to assess the stability of multispecies populations (see Chapter 6), which employs a compensatory mechanism to attenuate dynamics. Recall that the stability of these systems were gauged by the magnitude of a constant, s, required to increase the magnitudes of the negative diagonal elements of the Jacobian matrix, to the point that the real parts of all eigenvalues were negative. This approach is predicated on the assumption that systems with lower values of s are inherently more stable than those with higher values of s. Arguably this is a strange assumption, as all the systems within the comparison are stable, i.e., the real part of $\lambda_{max} < 0$. However, if we view this through condition 3, we can conclude that higher values of s imply higher death rates and turnover rates, given that the diagonal terms are functions of s and death rates. In other words, one strategy employed by populations to compensate for the untenable dynamics described above would be to adopt a form of density-dependent self-limitation. This may be biologically reasonable, but at some point, the death rates and turnover rates become unrealistically large, or are not biologically feasible. In this sense, with high values of s we could begin to question the dynamic feasibility on biological grounds.

9.4 Feasibility and the hump-shaped curve

The interaction between productivity and dynamics presented previously operates in a way that the likelihood of a system being feasible ($F_{[sys]}$) at a given level of productivity is the result of sufficient energy in the system to support all species ($F_{[NPP]}$) and, given that there is enough energy, that the dynamics of the system ($F_{[\lambda]}$) are biologically reasonable (i.e., $F_{[sys]} = F_{[NPP]} \cap F_{[\lambda]}$). At low rates of productivity, the systems are limited by the availability of energy ($F_{[NPP]}$). At high rates of productivity, the systems are limited by their dynamics ($F_{[\lambda]}$). Taking both of these together, we end up with a hump-shaped distribution in which the region of intermediate productivity is more likely to support systems (Figure 9.3) that, when coupled with the second part of the argument discussed below, leads to greater diversity.

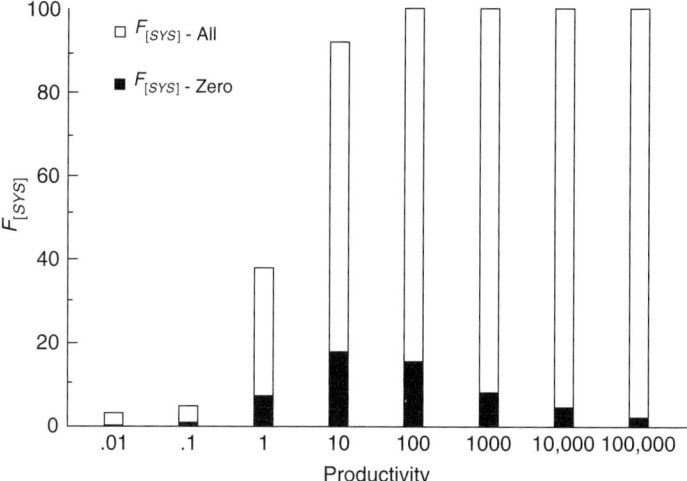

Figure 9.3 The feasibility (%) of a two-species primary-producer-based food chain ($F_{[SYS]}$ – SYS represents the system as a whole) results from the influences of productivity ($F_{[NPP]}$ – NPP represents net primary production) and dynamic stability ($F_{[\lambda]}$ – the symbol 'λ' is used to depict the eigenvalues of the Jacobian matrix). The criteria for $F_{[NPP]}$ is that both species possess positive equilibria (X_1^* and $X_2^* > 0$). The criteria for $F_{[\lambda]}$ is based on the restrictions we place on the transient behavior of the system following a disturbance (see Figure 9.2). The interaction of $F_{[NPP]}$ and $F_{[\lambda]}$ defines $F_{[SYS]}$. In this case, whether the food chains are allowed to oscillate or not. The likelihood of a food chain persisting in time is a function of there being sufficient productivity to support all the organisms in the food chain ($F_{[NPP]}$) and that the dynamics of the food chain following a disturbance are compatible with the birth and death rates of the organisms $F_{[\lambda]}$. For $F_{[SYS]-All}$ (open bars), only productivity is used as a criterion to define feasibility. For $F_{[SYS]-zero}$, (black bars) the food chains that are feasible in terms of productivity ($F_{[NPP]}$) and that oscillate following a disturbance are excluded, i.e., we use the restrictive criterion that the food chain will exhibit no oscillations following a disturbance. A hump-shaped curve results when both productivity and dynamics are considered simultaneously. The two-species food chains were constructed and evaluated in the same manner as those in Figure 9.1. At the productivity level of 0.01, we conducted 2500 iterations to increase the number of feasible runs on which to base a percentage.

A few caveats about both components of feasibility need to be discussed before going further. May's (1973) reference to the devious strategies of species or as Rosenzweig (1995) argued the clarity by which energy and dynamics interact as we posit previously '... breaks down when natural selection has enough time to do its job.' This is certainly true, but only up to a point. The shape of the distribution is affected by the life histories and physiologies of organisms. The models we have used so far are based on instantaneous growth and possess functional responses that are prey-dependent. This framework may work well in many cases but fails to take into account the time lags that are inherent in many life histories, and the physiological and behavioral adaptations that organisms have evolved to respond to and cope with changes in environmental conditions.

9.4.1 Life history strategies and adaptations

A review of our models and analytical approach reveals the importance of physiological and life history attributes to $F_{[sys]}$. Selection could operate on ecological efficiencies, birth rates, death rates, and consumption coefficients of the consumers, and the degree of intraspecific competition of the primary producers. Given basic thermodynamic constraints (entropy and the conservation of matter), combining these parameters creates a continuum of possible organisms, which in turn are assembled together to form simple systems. All else being equal, once the parameters are selected, the life histories and physiologies have been defined and $F_{[NPP]}$ and $F_{[\lambda]}$ are established. However, a criticism of the models is that they assume continuous instantaneous consumption and growth, and as such, the lower threshold of productivity established by $F_{[NPP]}$ might well exclude organisms that possess physiologies and life histories that are well suited to cope with low levels of productivity. Many organisms enter into resting stages when resources become limiting or conditions are less than optimal for growth, and then emerge from these states when conditions improve. Others simply migrate to other habitats.

We can anticipate cases in which the inputs of resources (primary production or detritus) are not sufficiently high to maintain the population densities of consumers on a continuous basis but could if the resources were available on an intermittent basis or were distributed heterogeneously over a landscape. The consumer would simply respond to cues in the environment to take advantage of the distributions and shifts in resource availability. Collins et al. (2008) provided an example of this in their review of pulse dynamics in arid grassland systems, wherein the interactions between plants, microorganisms and invertebrates, and nutrients are controlled by soil moisture and the frequency of rainfall events. Rooney et al. (2008) discussed elements of this in their discussion of the differences in the spatial arrangements of producers and resources in relation to higher-order consumers.

9.4.2 Oscillations, instabilities, and population turnover rates

Many real systems exhibit oscillations. Obviously, some systems that oscillate following a disturbance may be feasible in the sense that the populations never reach zero density and exhibit limit cycle or chaotic dynamics. Nonetheless, these systems may possess properties that are biologically unreasonable, or position the system in vulnerable states that increase the likelihood of extinction following further perturbations. Any criteria that define the likelihood of extinction or non-feasibility for a given frequency of oscillation are arbitrary.

To illustrate this point, we estimated the frequency of oscillations of our simple two-species models with linear functional responses at different productivity levels, and constructed a series of frequency of oscillation curves representing upper limits to feasibility for each frequency along the productivity gradient (Figure 9.4). Frequencies were restricted to $1/T \leq 1$ year^{-1}; $1/T \leq 1$ month^{-1}; $1/T \leq 1$ week^{-1}; and $1/T \leq 1$ hour^{-1}. By relaxing the criteria of $F_{[\lambda]}$ to include those systems that oscillate with frequencies at or below a given curve, a family of hump-shaped curves can be generated (Figure 9.4), with the $\lim_{1/T \to \infty} F_{[\lambda]} = F_{[NPP]}$. The curves approach that of $F_{[NPP]}$ as the frequency of oscillations increases. The highest proportion and number of feasible food chains is found at some intermediate level of NPP.

The models we used described instantaneous growth and possessed simple prey-dependent functional responses and similarly simple density-dependent self-limitation. We might have reached different conclusions if we had used different models. There are limits to the frequency at which the densities of any population can oscillate. In the previous paragraph we imposed these limitations on our models (Figure 9.4). Others would argue that the models themselves should include these features: for example, inclusion of predator self-limitation as we discussed in Chapter 5 and above, or the inclusion of some form of functional response that depends on the density of the

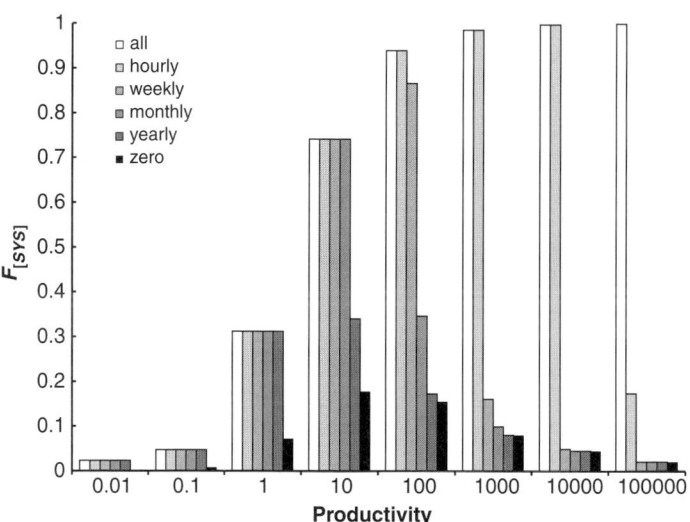

Figure 9.4 The feasibility (%) of two-species primary-producer-based food chains at fixed levels of productivity and parameter selections from the uniform [0,1] distribution as presented in Figure 9.1, but with different dynamic restrictions. The criterion for $F_{[\lambda]}$ was relaxed to include systems that oscillate with hourly, weekly, monthly, or yearly frequencies to generate a family of of $F_{[SYS]}$ hump-shaped curves.

9.4.3 Energetic properties of populations

The $F_{[sys]}$ is sensitive to the physiological and life history characteristics of the organisms within a food web. The energy needed to support consumer populations ($F_{[NPP]}$), and the onset and frequency of oscillations of a system following a disturbance ($F_{[NPP]}$), are functions of the intraspecific competition coefficient (c_{11}) of the primary producer, as well as the consumption coefficient (c_{12}), ecological efficiency ($a_2 p_2$), and specific death rate (d_2) of the consumer. Parameter selections that depict different types of organisms should change the shape of the curves for $F_{[sys]}$, given their impacts on $F_{[NPP]}$ and $F_{[\lambda]}$ that we demonstrated back in Chapter 7 (see Figures 7.4 and 7.5). Here we will refine this result. Phillipson (1981) identified three physiologies for heterotrophs: unicellular ectothermy, multicellular ectothermy, and multicellular endothermy. In the extreme, unicellular ectotherms have relatively short life spans and high energetic efficiency, whereas endotherms have longer life spans and lower energetic efficiency. The ecological efficiency of unicellular ectotherms is relatively high (0.4 – 0.5) compared with that of endotherms (< 0.03). The maximum life spans of ectotherms range from days (rotifers and many insects) to centuries (large bivalves), whereas those of mammals range from < 5 years (many rodents) to >100 years (humans). The median life spans of 274 invertebrate taxa and 170 mammals were < 5 years and ~15 years, respectively (Comfort, 1979, Eisenberg, 1981).

Using the aforementioned estimates and ranges as guides, we conducted separate Monte Carlo simulations to compare models with ectotherms and endotherms at the second trophic position, similar to the ones presented in Chapter 7. We selected coefficients for the specific death rates (d_2), and the energetic efficiencies ($a_2 p_2$) from beta distributions (beta [3,8] for ectotherms and beta [8,3] for endotherms), and the intraspecific competition coefficient (c_{11}) and the consumption coefficient (c_{12}) from the uniform distribution [0,1]. Productivity was varied in the manner presented in the previous exercises, the same random seed was used for both trials, and each trial consisted of a minimum of 1000 iterations.

The models dominated by ectotherms produced qualitatively similar but quantitatively different results from those dominated by the endotherms, indicating clear trade-offs between $F_{[NPP]}$ and $F_{[\lambda]}$ (Figure 9.5). With respect to $F_{[NPP]}$, the ectotherms were more likely to be feasible over the entire range of productivities and required lower energy inputs to reach 100% $F_{[NPP]}$ than did the endotherms (Figure 9.5). The $F_{[NPP]}$ curve for endotherms shifted to the left, whereas the curve for ectotherms shifted to the right of the curve. The opposite occurred for $F_{[\lambda]}$. Both endotherm- and ectotherm-dominated systems produced unimodal curves for $F_{[\lambda]}$; however, the apogees of these curves differed (Figure 9.5). For ectotherms, the distribution of the feasible runs along our productivity gradient possessed an apogee near 1 unit of productivity and negative kurtosis. For endotherms, the apogee was near 10 units of productivity and the distribution had a positive kurtosis. Compared with ectotherm-based systems, endotherm-based systems are less likely to be feasible for lack of sufficient production. However, endotherm-based systems are less likely to oscillate, and hence are more likely to be feasible for dynamical reasons than are the ectotherm-based systems. This results in higher $F_{[sys]}$ for endotherm-based systems when oscillations in dynamics are not allowed (Figure 9.5). This is counterbalanced by the propensity for ectotherms (particularly unicellular and small multicellular species) to possess life cycles with rapid turnover times that are conducive to oscillations.

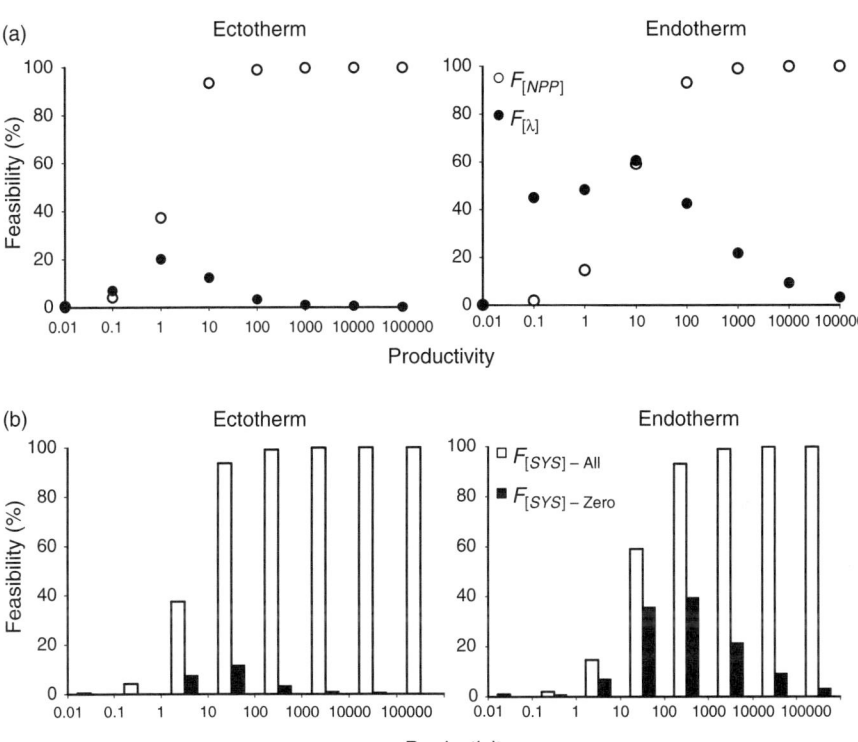

Figure 9.5 The feasibility (%) of two-species primary-producer-based food chains with either an ectotherm or an endotherm at the second trophic position. (a) $F_{[NPP]}$ (open circles) with no restrictions on the transient, and $F_{[\lambda]}$ (black circles) with the transients restricted to those without oscillations ($\mu \geq 0$). (b) $F_{[SYS]}$ without restrictions on the dynamics of the transient ($F_{[SYS] - All}$) and with restrictions (no oscillations) on the transient ($F_{[SYS] - Zero}$) ectotherms or endotherms were created by sampling a_2, p_2, and d_2 from beta [3,8] and beta [8,3] distributions, respectively. All other parameters were selected from uniform [0,1]. Note the hump-shape curves in b.

9.5 Trophic structure and the diversity of production

The second part of our argument begins with the observation that species diversity increases with the diversity of production (living or detritus) as well as the rate of production. As new food resources are added at the base of the food web (basal resource), as habitats become more heterogeneous, or when seasonality is considered, the potential exists for new assemblages of species to form within the community (Leopold, 1949, Yodzis, 1981, Coughenour et al., 1985, Lane, 1986, Moore and Hunt, 1988, Moore et al., 1989).

We have treated this subject at length in Chapter 8 and elsewhere and summarize the main points hereafter. The basis for the relationship between increased diversity and increased diversity of production can be traced to the work of Gardner and Ashby (1970) and May (1972, 1973). These studies revealed that certain types of assemblages promote stability. Random assemblages of species in model food webs are not as stable as webs that possess a blocked or compartmentalized pattern of interactions. Empirical evidence in support of compartments has been mixed (Pimm and Lawton, 1980, Yodzis, 1980).

From their work with the soil food web of the North American Shortgrass Steppe, Moore and Hunt (1988) proposed that ecosystems are organized in a compartmentalized manner based on resource use. The pattern of resource use within the soil food web is ubiquitous; species with similar growth rates form assemblages based on food resources (i.e., compartments). The pattern of energy use has been seen for communities as disparate as the soil biota of polder soils in The Netherlands to the nomadic Ngisonyoka people of the Turkana tribe of Northwest Kenya (Coughenour et al., 1985, Moore and de Ruiter, 1997). Furthermore, energy channels within food webs are structured in the same way as the entire web, but the channels independently respond to disturbances (Moore and Hunt, 1988, de Ruiter et al., 1995, Moore and de Ruiter, 1997).

The organization of communities as a cluster of quasi-independent assemblages of species is a requisite property of systems that are organized hierarchically (Simon, 1962, Koestler, 1967, Allen and Starr, 1982, O'Neill et al., 1988). First, species are organized into assemblages (subsystems) within ecosystems that are based on resource use. Second, subsystems are arranged as a series of linear food chains that operate at different efficiencies and on different time steps (a key to establishing boundaries between subsystems). Third, subsystems exhibit similarity in structure, dynamics, and responses to disturbance (i.e., the subsystems are governed by the same principles as the system as a whole). Fourth, subsystems are linked to one another yet exhibit a degree of independence.

To connect the compartmentalized nature of communities with the hump-shaped relationship between productivity and species richness, we simply need to knit together several results (Box 9.2) and modify the conclusion at the end of the modeling exercise to read "...the likelihood of feasible compartments is greatest at intermediate levels of productivity." Over a range of productivities, communities may possess compartments of species of different diversity and food chain lengths. Nonetheless, the highest incidence of compartments (species) would occur at intermediate levels of productivity.

Box 9.2 Determinants of diversity in food webs

We proposed a two-part argument on how food web energetics might explain the observed relationships between productivity and species diversity (Figure B9.1). First, the diversity of consumers in food webs is related to the rate of production (i.e., productivity) and the diversity of production. Second, food webs possess pathways, or energy channels, that propagate matter and energy at different rates and that are linked by predators. These features lead to the following: (1) variation in the basal resource and lower level consumers in the food web, (2) asymmetry in energy flow within energy channels, and (3) the coupling of energy channels by mobile predators. We discuss how these three aspects of food web structure and dynamics support greater diversity at intermediate levels of productivity.

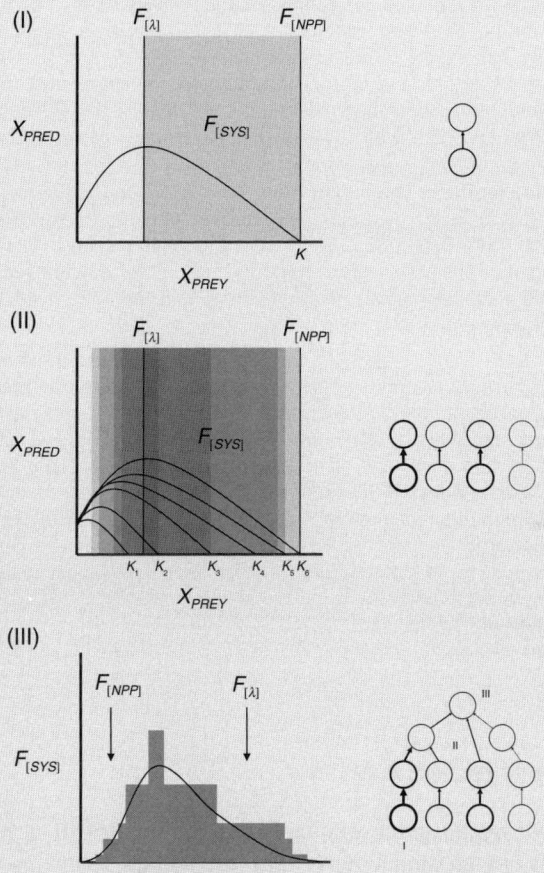

Figure B9.2 The proposed model to explain the putative relationships between productivity and species diversity in food webs is based on (I) variation in basal resources and consumers, (II) asymmetry of energy flow within energy channels, and (III) predators as couplers.

(1) **Variation in basal resources and consumers:** In Figure 9.5 we demonstrated that the $F_{[SYS]}$ is sensitive to the productivity of the basal resources (primary producers and

Box 9.2 (*Cont.*)

detritus) and the life histories and physiologies of consumers. The model food chains for the ectotherms processed higher material flux rates than those for endotherms. The predator–prey models presented by Rosenzweig (1971) illustrate this point as well. Variations in the life histories and physiologies of the prey or producer would change the shape of their isocline and fix the position of the predator relative to it. These model systems are feasible as long as the predator isocline intersects the prey isocline to the right of the apogee of the prey isocline. As with the simpler models, $F_{[SYS]}$ is bounded by productivity ($F_{[NPP]}$) and dynamics ($F_{[\lambda]}$).

(2) **Asymmetry of energy flow within energy channels:** Basal resources and lower trophic level organisms are distributed across the landscape in a heterogeneous manner. Given differences in habitat characteristics (e.g., nutrient availability, water availability, etc.) and differences in the life histories and physiologies of organisms, we can expect to see several energy channels emerge with some degree of asynchronous production and energy transfer in space and time. Over large spatial and temporal scales, the landscape would possess a collection of food chains or energy channels, each with lower and upper bounds to $F_{[SYS]}$ defined by $F_{[NPP]}$ and $F_{[\lambda]}$.

(3) **Predators as couplers:** The higher-order consumers and predators tend to operate over larger spatial scales then their prey and are capable of responding to complex environmental information (i.e., type III functional responses). These organisms couple the energy channels to form the food web. Over a heterogeneous landscape, predators will seek out regions of prey density and thereby release the low-density prey from predation pressure.

All else being equal over optimal spatial and temporal scales, the greatest number of feasible systems occurs at intermediate levels of productivity. Rooney et al. (2008) make an important point regarding the magnitude or extent of the spatial and temporal scales. If space is limited or homogeneous, or so expansive that the distance between resource patches is at the edge or beyond the movement capabilities of the predator, or if time is compressed to the point that interactions are occurring simultaneously, then the stabilizing influence of higher-order predators will be compromised. In these cases predation pressure can be excessive, leading to extinction of prey. In other words, in spatially or temporally constrained ecosystems, predators can be destabilizing (McCann et al. 2005): a topic we will explore further in Chapter 12.

9.6 A review of hypotheses

Rosenzweig (1995) evaluated several factors that could explain the unimodal pattern of species diversity with productivity, but settled on the following as being viable: (1) time, (2) dynamic stability and disturbance (combined here), (3) environmental heterogeneity, and (4) area. Although other hypotheses were discussed, they were either rejected outright or were judged to possess conceptual flaws (Rosenzweig, 1995).

9.6.1 Time

The time hypothesis asserts that productive habitats are young, with constituent species still undergoing speciation; given enough time, such habitats would become more diverse than less productive systems. Time was rejected for lack of supportive empirical evidence. Based on our results, we reject it on theoretical grounds. Monte Carlo simulations removed the age of the system as a factor, and played out the same 1000 assembly events at each level of productivity. The decrease phase of the unimodal pattern occurred independent of time.

9.6.2 Dynamic stability and disturbance

The dynamic stability and the disturbance hypotheses are related because two aspects of dynamic stability (local stability and return times) are measures of the response a system has to a disturbance. We prefer to treat disturbance as a corollary to the dynamic stability hypothesis. Rosenzweig (1995) argued that the dynamic stability hypothesis had limits. With enough time organisms would evolve compensatory mechanisms to attenuate or overcome untenable dynamics. This is certainly true, but only up to a point. In our models, selection could operate on ecological efficiencies (a_2p_2), birth rates ($a_2p_2c_{12}X_1$), death rates (d_2), and consumption coefficients (c_{12}) of the consumers, and the degree of intraspecific competition (c_{11}) of the primary producers. A combination of these parameters creates a continuum of possible organisms. There are two end points along this continuum. For a given level of productivity, one potential end point of the continuum consists of organisms whose physiologies and growth rates do not generate oscillations following a disturbance. The other extreme consists of a group of superorganisms capable of rapid changes in densities or biomass. The first end point generates the unimodal pattern along a productivity gradient (Figure 9.3). Changes in efficiencies, life spans, and birth rates (through efficiencies) may modify the curve, but they can generate the unimodal pattern as well. The second end point occurs by relaxing the restrictions on $1/T$, and is defined by $\lim_{1/T \to \infty} F_{[\lambda]} = F_{[NPP]}$ (Figure 9.5). Although this end point could be reached, it has not happened yet. There are limits to the frequency at which the densities of any population can oscillate. The unimodal pattern occurs even when selection runs its course.

9.6.3 Environmental heterogeneity

With increased environmental heterogeneity, increased productivity initially allows for increased plant diversity, which in turn allows for an increase in the number of consumers. At the high end of the productivity gradient, plant diversity declines as habitats become more uniform in limiting nutrients and as superior competitors are better able to exploit nutrients and light (Tilman, 1982). The heterogeneity hypothesis alone can explain the unimodal pattern in diversity; but it suffers from circular reasoning. Rosenzweig (1995) convincingly argues that the heterogeneity of resources (e.g., habitat, food, and time) is correlated strongly with species diversity, and in the case of habitat heterogeneity has coevolved with organisms, or: "more species means more selection for finer habitat discrimination."

Our model includes environmental heterogeneity at two levels. Both the diversity of resources and compartmentalization imply environmental heterogeneity. The important distinction, though, is that the heterogeneity was applied equally across the productivity gradient, as the same food chains were evaluated at each level of productivity. Nonetheless, a unimodal pattern of species diversity occurs with increased productivity. The model is independent of habitat, because it allows for as fine a discrimination of habitat as productivity and dynamics will allow. Initially, as the heterogeneity of the environment and productivity increase, potentially more niches can be filled, with a higher likelihood that energy is sufficient to support the species that fill them. However, at a point along the productivity gradient, dynamic stability limits the combinations of resources that can maintain consumers regardless of the number of potential niches. In the end, more species can coexist at intermediate levels of productivity.

9.6.4 Area

Area and productivity are inseparable. Larger areas would be expected to support more species than smaller areas of similar productivity. Understanding the effect of area is further complicated because the regions of the globe with the most area (the tropics) are also the regions with the highest productivity. Our analyses evaluated how productivity affected diversity within a fixed area (m^2), but the model could be applied to any spatial scale. The $F_{[NPP]}$ curve can be shifted to the left on the productivity axis by increasing the area in which a consumer can forage. Relaxing the restrictions of area by allowing for larger home range does not affect $F_{[\lambda]}$. The net result is a hump-shaped or unimodal pattern in diversity along the productivity gradient, with the maximum feasibility defined by $F_{[\lambda]}$ because $\lim_{A_{min} \to \infty} F_{[sys]} = F_{[\lambda]}$, where A_{min} is an organism's minimum home range defined in terms of a energy requirements. To reach this conclusion, we need to revisit three aspects of the analysis.

First, $F_{[NPP]}$ was defined in terms of the ability of species to maintain positive population densities at steady state within a m^2. Many organisms have home ranges that are far greater than a m^2. Second, population densities at steady state only needed to be positive without regard to the size of an individual organism or the viable population size of the species. For example, at a given level of productivity, a m^2 may be able to support a population at 1 g C year^{-1} level, whereas the minimum viable population size requires 1000 kg C year^{-1}. Third, we have modeled the transfer of energy from one trophic level to another, not from one species to another. The hidden assumptions are that the species within a trophic level are not competing with one another, and that all the production has the potential to be consumed. Obviously, production is shared with other consumers, and not all the production is available for consumption (e.g., roots are not available to most aboveground herbivores).

Do the points raised above negate our conclusions? We do not believe they do, but they do require that we reinterpret them. For example, if we were to redefine A_{min} as the smallest area needed to support a minimum viable population (P_{min}). From the models, $A_{min} = P_{min} X^{*-1}$. Hence, a population of herbivores with $P_{min} = 1000$ kg could survive in a habitat with a productivity of 1 g m^{-2} year^{-1} by establishing a minimum home range of 10^6 m^2 ($A_{min} = 1$ km^2). The same population would require a smaller home range in a more productive habitat. With competitors present, the home range would have to be even larger to compensate for the reduction in available production. If only 50% of total production was above ground, and the herbivores lived and foraged above ground, home range would have to expand even further.

At the m^2 scale, the scenarios raised above are most problematic for large organisms (predominantly mammals), as smaller organisms are more likely to acquire their energy within or close to this area. Our models predict that mammals of similar trophic position should increase home range size with increased body size, and species of similar size should have smaller home ranges in more productive habitats. Empirical evidence supports this (McNab, 1963, McNab, 1980, Kleiman and Eisenberg, 1973, Eisenberg, 1981). Marsupials, edentates, and carnivores each exhibit a positive association between home range size and body size. Moreover, for mammals of similar body size and feeding habitats, home ranges are smallest in the tropics, and increase within temperate regions through the boreal and arctic regions. We are aware of no similar data for ectotherms.

9.7 Summary and conclusions

The species richness of a community is a compromise of several factors. Earlier chapters stressed the importance of energy flow to community structure and stability. Equally importantly, the energetic organization of communities plays a role in shaping trophic structure as well as species richness. Energy flow influences trophic structure and dynamics in a way that results in a unimodal pattern of species diversity along a gradient of productivity. In this chapter we offered a simple model to explain the unimodal pattern in diversity that is based on only two factors: the amount of available energy ($F_{[NPP]}$) and the dynamics of the system following a disturbance ($F_{[\lambda]}$). The importance of energetics to both factors is inescapable. Productivity and consumption are forms of energy flow, and energy flow directly influences trophic structure, dynamics, and stability. Variation in the rate of energy input creates what May (1977) referred to as "a regime of dynamic behavior... so that continuous variation in a control variable [productivity] can produce discontinuous effects [in dynamics]." Transient dynamic behavior represents one type of a regime of dynamic behavior. We suggest that not all species have adapted to some of these effects, or if they have, they have done so by adopting a life history trait to circumvent their influences (e.g., torpor or diapause). We further posit that most endotherms and many large ectotherms are not well adapted to rapid oscillations. This type of dynamic behavior is largely the domain of unicellular ectotherms and small multicellular ectotherms.

Productivity, the energetic efficiencies of consumers, habitat heterogeneity, and the frequency of disturbance have each formed the basis of hypotheses to explain patterns in trophic structure and species richness at local, regional, and global scales. We argue that species richness is a function of how productivity (rate of production) and how different types of primary producers and resources (diversity of production) affect trophic structure, community organization, and patterns of energy utilization by species. In our model, the diversity of consumers is controlled by two factors: (1) the interactive influences of productivity and population dynamics on the likelihood of a system forming and persisting (feasibility), and (2) the diversity of resources on which systems are based. Via modeling, we show that primary productivity, trophic structure, and trophic dynamics are interrelated in a nonlinear fashion, such that $F_{[sys]} = F_{[NPP]} \cap F_{[\lambda]}$. This relationship limits any attempt to make general statements about how species richness might be affected along a productivity gradient. This confirms the observation that over a wide gradient of productivity more systems (and species) are likely to be at intermediate levels of primary productivity. Next, we argue that communities are shaped by the diversity of production as well as the rate of production. Theory and empirical evidence indicated that communities are organized into assemblages of species that begin with different food types and are separated by different rates of utilization. At low levels of productivity, diversity is low owing to the dearth of available energy for higher trophic levels (low $F_{[NPP]}$). At a high level of productivity, diversity is low owing to the trade-offs between different life history strategies and the instabilities associated with oscillatory dynamics. The increase in diversity at the intermediate levels of productivity occurs through the potential for a higher number of stable assemblages.

Part III

Dynamic food web architectures

In the last part of this book, we discuss the variable and dynamic nature of the natural world and how this affects food web structure and dynamics. Variability in nature is not captured by the static representations of food webs that we have developed to this point. Indeed, many ecologists view more-or-less static representations of biological communities like those we have dealt with as being inappropriate. We also address the criticism that food web descriptions like ours are incomplete in that they do not fully account for all the species and links that are present. We will turn this argument around with the idea that variability in structure in space and time is a critical component to the maintenance of systems. Species and species interaction can be separated in space and time; hence, not all nodes and links are realized. Seemingly complex systems when viewed through the lens of static architectures are actually a compilation of less diverse and complex interactions of the whole; some are stable and some are not, but in aggregate they persist.

The chapters in Part III advance the idea that variation and dynamics in architecture are key features of food webs, and is important to understand their complexity and persistence. Stable food webs are not static entities, but are open and flexible systems that can change in species attributes, composition, and interactions. From a mathematical standpoint, this translates to a Jacobian matrix that changes in dimension and complexity, and whose elements change in magnitude. This view is closer to reality, as most food webs are in a continuous process of change in species composition and abundances. And within species there will be ongoing variation in population life-history parameters, growth rates, size distributions, and behavior.

What does this variability in food webs tell us? We reflect on our models and analyses that were based on the assumption of steady-state and asymptotic stability. Two questions come to mind. First, how has neglecting variability and dynamics in our steady-state models had an impact on the outcome of the analyses? Second, how can we incorporate aspects of variability? Moreover, the notions of variability and dynamic architecture demand a different approach to stability. The persistence of species, and the role of species in biological communities, should be examined in the context of dynamic food webs in a changing environment. In this way, we should look at stabilizing effects of the flexibility in food web structure, i.e. food webs as open and flexible structures that accommodate changes in species composition, attributes, and dynamics. These are features of ecosystems that are critical to our understanding of community resistance and resilience to environmental change and disturbance.

In Chapter 10 we discuss the role of variation and dynamics along various lines, challenging the steady-state assumption that is the basis for static populations and static architectures. We juxtapose the metaphor or the static arch and keystone species with an alternative metaphor based on the game Jenga, where species enter and leave the system and the keystone species shifts from one species to another depending on the status of the structure at any given point in time. The shifts in interactions invariably translate to changes within the configuration of the Jacobian matrix.

Keeping with our central theme, the energetic organization of the food web, defined in terms of connectance, energy flux, and the distribution of interaction strengths collectively define the dynamic properties of the system. We offer several biologically based mechanisms linked to energetics that can lead to changes within the Jacobian matrix. Immigration and emigration of species, movement of resources and species across landscapes, and seasonal cycles in resource availability and populations all operate to shift both the magnitudes of the elements (read interaction strengths) within the Jacobian matrix and the dimensions of the Jacobian matrix, depending on the spatial and temporal scale of observation. The traits of constituent species within the system and changes in environmental conditions are drivers of variability in the energetic organization and dynamics of the system. The genetic capital of species affords the system the degrees of freedom in structure and dynamics to change under constant environmental conditions, and in terms of their responses to stochastic events and directional change in conditions as witnessed and projected with climate change. Finally, we discuss how feedbacks within the system affect and hone its structure and dynamics.

In Chapter 11 we pick up on the theme of dynamic architectures and present three case studies to illustrate our points, each representing a different take. The case studies involve following changes in food web architecture along natural and human-induced successional gradients and gradients of productivity. It becomes clear that we use the same tools and theoretical foundations that we used under the steady-state assumptions and asymptotic stability in Parts I and II.

The first case study investigates the role of productivity and the diversity of productivity in the form of detritus inputs in shaping the structure of the food webs that develop within the sediments of Wind Cave, South Dakota. A steep gradient of productivity exists between the productive soils and sediments at the cave entrance and the energy-starved sediments of the remote reaches of the cave. The trophic structure along this gradient follows the step-like development of trophic levels with increased productivity envisioned by Elton (1927) and others (Oksanen et al., 1981, Moore et al., 1993) that we presented in Chapter 7. Manipulations of the rate and diversity of production of the incipient webs in remote regions support this too, but also they support the model of community development and diversity that we presented in Chapter 9.

The second example follows the primary successional development of soil food webs on the island of Schiermonnikoog, one of the Wadden barrier islands off the north coast of the Netherlands. The Wadden islands are drifting east, with their leading end subsumed by the sea and their trailing end emerging from the sea. The plant and soil communities are sparse at the younger trailing edge, increasing in diversity and complexity and moving inward with increased productivity and age. Analyses, akin to the ones presented in Chapter 6, of the Jacobian matrices of the food webs at points long this gradient of productivity reveal that the energetic organization plays an important role in their stability and development (Neutel et al., 2007). The interaction between the availability of energy ($F_{[NPP]}$) and the dynamic state of the system ($F_{[\lambda]}$) that we envisioned in shaping community complexity and diversity ($F_{[SYS]}$) in Chapter 9 unfolds along this successional gradient. Biomass and energy flux shift to higher trophic positions and the weights of trophic loops increase. The critical loop with a food web at one point along the productivity gradient foretells a threshold in the form of an entry point for additional species at successive points up the gradient.

The third case study explores the real and potential changes in arctic tundra food webs (vegetation and soils) at Toolik Lake, Alaska in response to warming and nutrient additions. The study highlights how changes in environmental conditions can alter the structure and dynamics of a food web. Warming and nutrient additions increase productivity and alter the plant communities, which in turn feed back to affect the belowground communities in the form of shifts between fast and slow energy channels. The reconfigurations in the energetic organization of the food webs under the different scenarios stress the importance of their compartmentalized structure and the linkages of compartments to their dynamics and stability, as discussed in Chapter 8.

In Chapter 12 we present alternative frameworks to studying food webs that do not rely on asymptotic stability. The concepts of *transient dynamics* and *permanence* present possibilities in which the food web is in a constant state of flux from one structural and dynamic state to another, few of which are asymptotically stable. Species in these systems have evolved coping mechanisms that allow for pauses in their activities to overcome resource limitations, respond to disturbances, or adapt to changes in underlying environmental conditions. These mechanisms, coupled with the state of flux in structural states, are what optimize diversity. In other words, uncertainty and variation in environmental conditions when operating on ecologically relevant temporal and spatial scales and within the physiological tolerances of species are what promote diversity and provide continuity in structure. This leads to the idea of how variability can be viewed in terms of *apparent complexity*. Temporal and spatial averaging of food web structure provides a composite view of its energetic organization that may overstate how interactive and complex the system is at any given point in space or time.

10

Species-based versus biomass-based food web descriptions

10.1 Introduction

The previous sections focused on static food web architectures. We defined the structure in terms of the biomass of species or groups of species. We allowed the densities of populations to fluctuate within or stray from a steady state, but fixed the connectedness structure of the web and the sign structure and placement of elements within the Jacobian matrix underlying it. In this chapter we explore the concept of dynamic food web architectures, in which connectedness structure, and hence the Jacobian matrix, are allowed to change in their species composition, dimensions, and in the number and configurations of their interactions. In doing so, we will compare species-based food web descriptions to biomass-based descriptions, and the implications of choosing one over another in theory and in practice.

Change in food web architecture over space and through time is arguably a routine occurrence. Species enter and leave systems, become inactive during times of stress or resource limitation, and exhibit seasonal and diurnal rhythms in activity. The food web approach we have taken to this point seems ill-equipped for studying changes in architecture given its reliance on stability or persistence of species within a static architecture. Entertaining dynamic architectures leads to some interesting questions and challenges our notions of stability. During the process of community assembly or succession, change in the species composition and the interactions in which species engage are part and parcel of the processes. Moreover, at the endpoints of assembly and succession, changes in species composition and food architecture are expected both in nature and theory. Herein lies the conundrum. We have defined stability in terms of a system's ability to recover to its original state following a minor disturbance, yet we when we consider succession and community assembly we allow for the species composition and connectedness structure of the system to change.

In this chapter we have three goals. First, we present a metaphor for the concept of the dynamic architecture and its implications for stability and persistence. Second, we explore cases in which the trophic structure and distribution of biomass within an ecosystem form the stable subunits—yet are not about species per se, but rather the energetic and functional attributes of species. Third, we will revisit and expand our treatment of stability, and explore how we can use it to study dynamic architectures.

10.2 Dynamic food webs—playing Jenga

From the dynamic architecture framework, the network of interactions that we call the "food web" may actually be a collection of networks that shift and adapt to changes in underlying conditions. Our ability to observe the changes is dependent on the temporal and spatial scales that we settle on in the description.

To illustrate the relationship between species, community structure and stability, Paine (1966) used the metaphor of a stone arch, with the stones representing species, the loading forces among stones representing interactions among species, and the "keystone" representing the species that has a dominant role in regulating structure and stability (Figure 10.1). Some time later in his review of the manipulative studies he conducted in the rocky-intertidal zone, wherein species were removed and the responses of the community were gauged, Paine (1980) concluded that the changes in the community are not related to abundance of the species but rather are more closely aligned with its placement in the community, defined by its interactions. Food web structure and stability are not summative in terms of the species and the interactions they engaged in, but rather, in some instances they affected the structure well beyond their direct interactions.

The dynamic nature of food web architecture is an important aspect of the keystone species concept that is often lost. If we take this metaphor a step further and broaden the temporal and spatial scales to allow for an open system whereby species enter and leave, we would see a collection of stone arches at various levels of construction over a landscape or through time. From here we can make several conjectures:

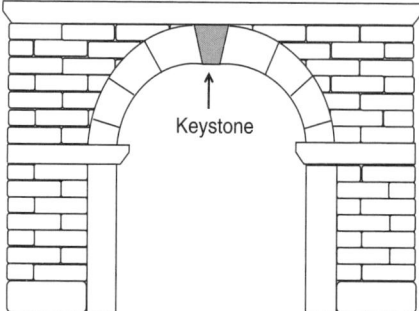

Figure 10.1 Like the keystone used in architecture, the concept of the keystone species was introduced as a metaphor to signify the importance of species that are low in number or biomass yet exert a high degree of control over the structure and stability of a food web (Paine, 1966, 1980).

1. not all the architectures in the collection are stable;
2. the vulnerability of the architectures to colonization or invasion might differ; each of the representations could have a different keystone;
3. with species entering and leaving the system, the architecture shifts and the relative importance of a species to the integrity of the structure will change as well.

In this case, the metaphor presented by de Ruiter et al. (2005) might better resemble the structures that are possible in a variation of the game Jenga (Figure 10.2). The simple rules of balance and energetics govern the stability of both the arch and Jenga structures, but unlike the static single arch or collection of arches, the Jenga structures are dynamic, constantly changing with additions and deletions of stones; the importance of a given stone's contribution to the structure's stability also changes.

The keystone and Jenga models both predict species effects on community dynamics that are disproportionate to abundance. The keystone concept assumes that certain species are inherently special in terms of their potential to influence community dynamics and stability. The Jenga model argues that this idea is context-specific, including species attributes, and abiotic and biotic features, including recent disturbances. Neither claims that all species are equivalent. In the keystone argument for a given structure, certain species are identified as load-bearers. In the

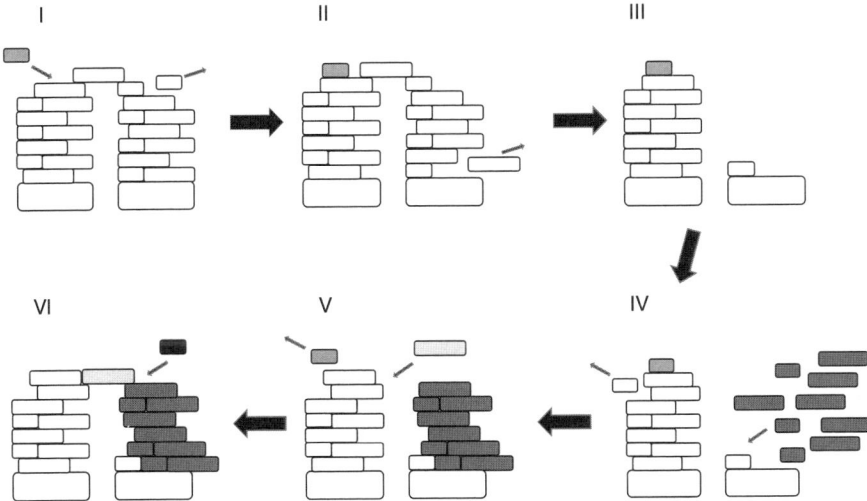

Figure 10.2 A variation of the game of Jenga might serve as a metaphor for a dynamic food web architecture. In a game of Jenga, players successively take away parts and place them on top until the structure becomes unstable and crashes. Each part can thus be a keystone. When parts are replaced at other positions, the stability of the Jenga structure can be maintained. If we extend this to a food web, the architecture of the web over space and time changes as species enter and leave the system, with each event having a different degree of impact on the system. The average structure exhibits maintains a familiar core but with variation.

Jenga analogy, some species are critical load-bearers, but as the structure changes the identities of these critical species can (and often do) change.

We also argue the following: (1) disproportionate species effects should not be linked to abundance or mass alone, but also to trophic context, (2) counterintuitive and disproportionate species effects emerge from pathways within networks of direct and indirect effects (Yodzis, 1988), and (3) disproportionate effects do not imply that biomass or energetics have no role in structuring communities or governing their dynamics. Energetic approaches have predicted the manner in which weak and strong trophic links govern community dynamics and stability (McCann et al., 1998; Neutel et al., 2002), and have highlighted the contextual nature of community responses to nutrient and detritus loading on trophic cascades (Polis and Strong, 1996).

The dynamic architecture of food webs can be viewed from one of two perspectives (Figure 10.3). One perspective is based on individual species in the traditional sense, where the nodes in the food web are represented by species and populated by individuals. Variation in the dynamics of the nodes and interactions among nodes in this species-based perspective is governed by the genetic variation among individuals within the populations of species. The other perspective is based on collections of biomass, where the nodes are the biomasses of the individuals from several species, which share common features. These species-based and biomass-based perspectives are not incompatible. Like the traditional species perspective, variation in the dynamics of nodes and interactions within the biomass-based energetics perspective is governed by the genetic variation among individuals within the populations of species, but includes the variation in species as well. In short, the energetic organization of a food web defined by the distribution and flow of biomass among species or groups of species is a central organizing feature of food webs.

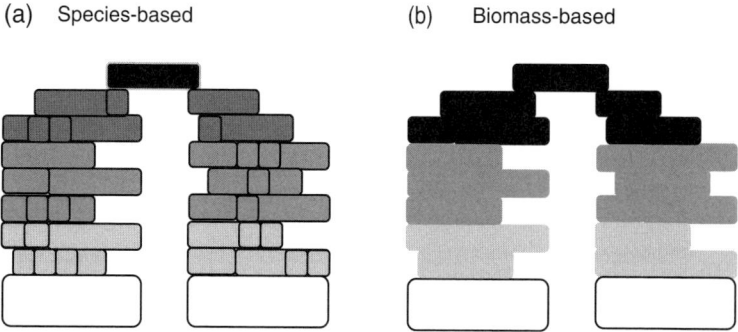

Figure 10.3 Dynamic food web architectures, like static representations, can be viewed from multiple perspectives. The species-based perspective might focus species and on numbers of individuals within a population. The biomass-based perspective might be based species or on groups of species and on the biomasses of the populations that constitute the group. The choices made will affect the structure and dimensions of the Jacobian matrix and hence our understanding of the system's dynamics and how we view stability.

10.3 Two case studies

We will use two case studies as organizing principles to discuss dynamic architectures, and approach this from the biomass-based energetics perspective. The first case involves the observations that led to the development of the theory of island biogeography and the empirical tests that followed. The second case involves the response of the well-studied Tuesday Lake food web to the removal of top predators. These studies provide a framework from which we can focus on changes in architecture that occur during the development phase that a system undergoes from one stage to the next and during the initial stages of colonization, and the changes in architecture that occur during a steady sate in development or when colonization and extinction are at equilibrium.

10.3.1 Island biogeography

The island biogeography literature offers some important insights into the concept of dynamic community architectures. In the theory of island biogeography, MacArthur and Wilson (1967) presented a conceptual and mathematical explanation for the pattern in number of species of different taxa on islands of different sizes and distances from a mainland based on colonization rates and extinction rates (Figure 10.4). For our discussion we deconstruct the idea into a community development phase that starts with an island devoid of species, and the steady state phase in which the island possesses its species richness.

During the community development phase in the run-up to a steady state in the number of species, the colonization rate (number of new arrivals per time) exceeds the extinction rate (number of extinctions per time). Every species is assumed to have the same potential to colonize the island. As illustrated by a colonization curve that presents the colonization rate versus the number of species present on the island, the colonization rate is highest early in the process of community development, but as species accumulate the likelihood of a new species arriving and successfully colonizing declines. Once established on the island every species has the same potential of going extinct. From the analogous extinction curve, the rate of extinction is low when few species are on the island and increases as more species arrive. A steady state or equilibrium in the number of species is reached when the colonization rate equals the extinction rate, or graphically at the point at which the curves intersect. The slopes of the colonization and extinction curves are dependent on the size of the islands and the distance of the islands from the mainland source of species. All things being equal, large islands have high colonization rates and lower extinction rates than smaller islands, and distant islands have lower colonization rates and equal extinction rates (Figure 10.4). These properties generate different equilibria in the species richness for islands of different sizes and distances from the mainland source of species.

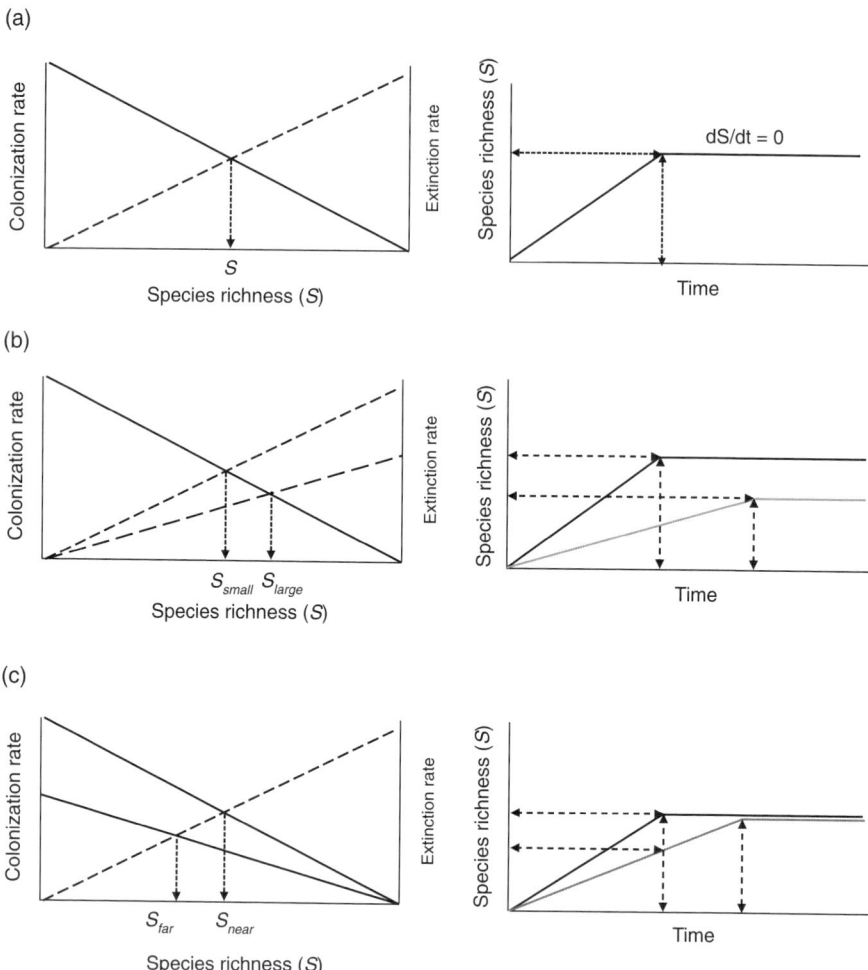

Figure 10.4 The theory of island biogeography is predicated on a dynamic food web architecture model both at the development phase and at the steady state (i.e., colonization rate = extinction rate). The figures on the left present the expected outcomes in species richness for the general model (a), the impact of island size (b), and the impact of island distance form the mainland (c). The accompanying figures on the right depict the dynamics of species richness through time. At steady state in species richness ($dS/dt = 0$ when colonization rate equals extinction rate), the model predicts that species turnover will occur.

We know that over time food webs change in terms of species composition, distribution of biomass, and trophic structure. The theory of island biogeography addresses the change in species composition in the development phase, as discussed above, and at the equilibrium phase. Once equilibrium in the number of species is reached, the system experiences a turnover in the species composition as each species is assumed to have an equal likelihood of extinction and a equal likelihood of colonization, or recolonization as it might be.

The phenomenon of species turnover described in the theory of island biogeography seems antithetical to the notion of a stable community in the mathematical sense embraced by our food web models. As species colonize the islands they engage in interactions with other species, and a food web emerges and evolves. Inherent in the process are changes in the composition of species, the number species, the interactions, and the connectance of the food web. Put another way, the structure of the food web changes during the development phase and when the community reaches a steady state in species richness. We will argue that changes in terms of community development do not negate the way we approach stability by means of Jacobian matrix representations of food webs (see section 10.4 Stability, disturbance, and transitions).

The experimental manipulations of small islands by Simberloff and Wilson (1969, 1970) provided results that supported the theory of island biogeography as proposed by MacArthur and Wilson (1967). Insect surveys were conducted on a series of mangrove islands off the coast of Florida. The islands were then fumigated with methyl bromide to remove the insects. The islands were resurveyed at several intervals to track the recolonization by the insects. Species richness increased on each of the

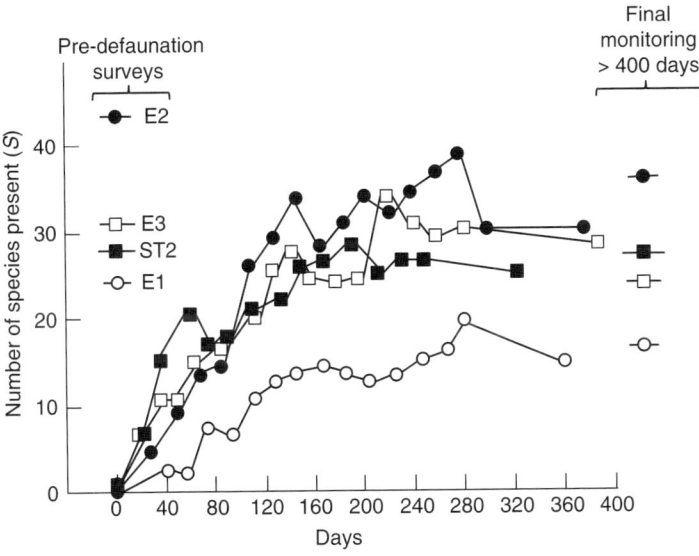

Figure 10.5 Colonization curves for arthropod species for four islands (E2, E3, ST2, and E1 - listed in increasing distance from the mainland) following defaunation (Day 0) with methyl bromide (Simberloff and Wilson 1969, 1970). The pattern of recolonization of the islands and eventual steady state in species richness followed the predictions from the theory of island biogeography as proposed by MacArthur and Wilson (1967).

fumigated mangrove islands in the manner predicted by the theory based on the size of the islands and their distance from the mainland (Figure 10.5).

A reanalysis of the studies by Simberloff and Wilson (1969) conducted by Heatwole and Levins (1972) provide the beginnings of how to reconcile the island biogeography theory and food web approaches by providing an added dimension to the process of colonization and extinction. Heatwole and Levins (1972) assigned the species that were present on the islands before fumigation and those that had recolonized the islands after 1 year into crude assemblages based on function (e.g., decomposers, predators, ants). Fewer species were present on the islands 1 year after fumigation than before. However, the allocation of species within each of the functional groups before the fumigation and 1 year following the fumigation were similar (Table 10.1).

Challenges to the approach that Heatwole and Levins (1972) took in assembling the datasets and the analysis notwithstanding, their studies reveal at least three important results. First, an underlying trophic structure emerged before the full complement of species repopulated the islands. Second, the trophic structure was based on functional diversity, which preceded species diversity. Third, the likelihood that a species successfully colonized the site and the likelihood of it going extinct depended in part on the functional status at the time of colonization and how the colonizing species utilized resources. Taken together, these results show a community matrix growing in dimension and changing in connectance as new species arrive in such a way that the functional diversity was restored before the taxonomic diversity.

Table 10.1. The trophic groups on the islands fumigated with methyl bromide by Simberloff and Wilson (1969) before fumigation and 1 year after fumigation (values in parentheses) as compiled by Heatwole and Levins (1972).

	Trophic Classes[†]							
Island	H	S	D	W	A	C	P	?	Total
E1	9 (7)	1 (0)	3 (2)	0 (0)	3 (0)	2 (1)	2 (1)	0 (0)	20 (11)
E2	11 (15)	2 (2)	2 (1)	2 (2)	7 (4)	9 (4)	3 (0)	0 (1)	36 (29)
E3	7 (10)	1 (2)	3 (3)	2 (0)	5 (6)	3 (4)	2 (2)	0 (0)	23 (26)
ST2	7 (6)	1 (1)	2 (1)	1 (0)	6 (5)	5 (4)	2 (1)	1 (0)	25 (18)
E7	9 (10)	1 (0)	2 (1)	1 (2)	5 (3)	4 (8)	1 (2)	0 (1)	23 (27)
E9	12 (7)	1 (0)	1 (1)	2 (2)	6 (5)	13 (10)	2 (3)	0 (1)	37 (29)
Total	55 (55)	7 (5)	13 (8)	8 (6)	32 (23)	36 (31)	12 (9)	1 (3)	164 (140)

[†] Trophic groups: H, herbivore; S, scavenger; D, detritus feeder; W, wood borer; A, ant; C, predator; P, parasite; ?, unknown.

10.3.2 Tuesday Lake

The collection of empirical studies of Tuesday Lake in Michigan, USA provides an interesting look at food web architecture, species composition, species abundance (numerical and biomass), and body size. The results of the studies complement the conclusions reached by Heatwole and Levins (1972), inasmuch as we see a distinction within food webs between an energetics-based trophic structure and a structure based on species interactions. We will approach this by presenting patterns in the energetics-based trophic structure following a discussion of a manipulative study that addressed the relationship between trophic structure and species composition.

Cohen et al. (2003) described the trophic interactions of the pelagic community of Tuesday Lake using the numerical abundances and average body sizes of individual species (Figure 10.6). The food web possesses a pyramid of numbers but not a pyramid of biomass, when spatial scaling is taken into consideration. The food web displays an inverse relationship between body mass and the numerical abundance of species. Species that are small tend to be found at lower trophic levels, whereas species that are large occupy upper trophic positions. Given that the size of an organism tells us much about its physiology, metabolism, and movement (Peters, 1983), the pattern in the distribution of species by size within the trophic structure produces an energetic organization with clear implications regarding the flow of energy through the food web, as large, long-lived predators consume and process prey at lower rates than smaller predators with shorter life spans.

Jonsson et al. (2005) compared the trophic structure of the Tuesday Lake food web before and after the removal of three planktivorous fish species, and addition of one piscivorous fish species. As a consequence, the lake's community structure changed remarkably in terms of linkages and species composition (Table 10.2). Yet, this manipulation had little effect on species richness (56 species in 1984; 57 in 1986), even though about 50% of the species were replaced by new incoming species within a period of less than 2 years (Jonsson et al., 2005). Despite a major change in species composition, the energetic setup defined by the system's trophic structure and distribution of biomass as described in Figure 10.6 remained roughly the same. Tuesday Lake possesses an underlying trophic structure that appears more resistant to change and invariant to the assemblages of actual species present in the system.

10.3.3 Summary of the case studies

The mangrove islands off the coast of Florida studied by Simberloff and Wilson (1969, 1972) and commented on by Heatwole and Levins (1972), and the series of studies at Tuesday Lake in Michigan (Cohen et al., 2003, Jonsson et al., 2005), reveal an important aspect of food web structure. In each case, the species composition and connectedness structure of the food web was altered, but the distribution of energy in the food web, in terms of body size, abundance, and biomass over trophic levels, remained roughly the same. Apparently, the new web structure allowed the

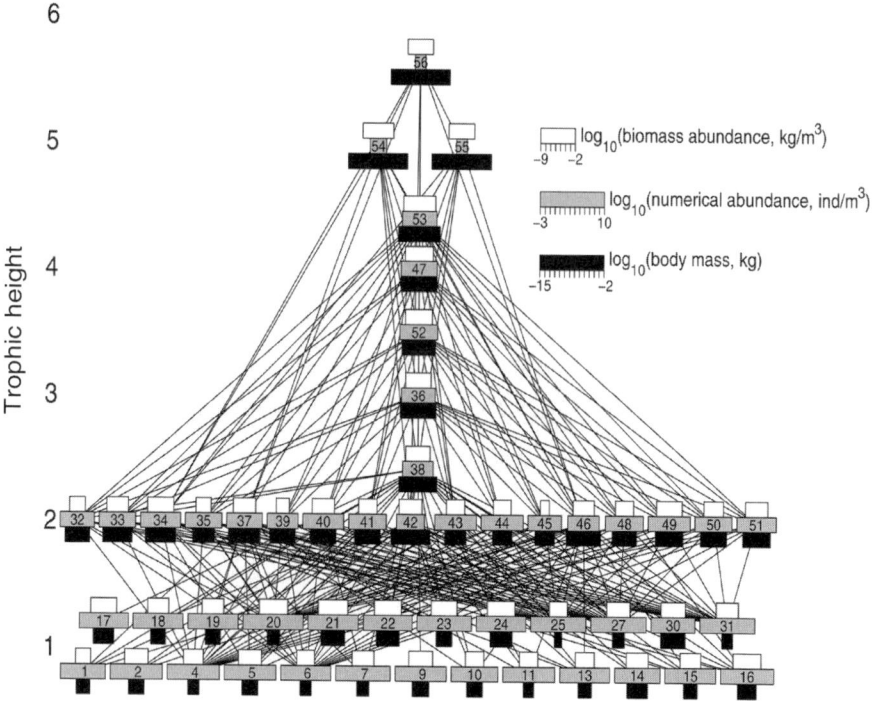

Figure 10.6 A hybrid connectedness energy flux food web of Tuesday Lake (1984), where fluxes are not quantified, but the groups of species are depicted using different metrics. The width of the horizontal bars shows the body mass ($\log_{10}$ kg), number ($\log_{10}$ individuals per m^3), and biomass ($\log_{10}$ kg/m^3), respectively, of each species. The vertical positions of the species show trophic height (Cohen et al., 2003). The changes in the architecture illustrate the points made in Figures 10.2 and 10.3. Despite a major change in species composition, following a manipulation, this energetic setup of the food web remained roughly the same (Johnsson et al., 2005).

community to conserve key ecological and functional features in the face of a major disturbance. Put another way, the functional attributes of species and the energetic organization of the community trumped species composition.

10.4 Stability, disturbance, and transition

How can we reconcile the dynamic nature of a food web's structure and the concept of asymptotic stability that is based on a static architecture? To answer this question we return to our definition of stability and expand it to allow for a dynamic architecture. Next, we discuss what aspects of the food web are changing and map these directly to the Jacobian matrix. In doing so, we can arrive at analytical approaches and tools to assess stability in a dynamic context.

Table 10.2. Summary statistics for the unlumped and trophic species webs of Tuesday Lake (from Jonsson et al., 2005).

Statistic	Unlumped Web 1984	Trophic Web 1984	Unlumped Web 1986	Trophic Web 1986
Diversity Metrics				
Species (S)	56 (50)[a]	27 (21)[a]	57 (51)[a]	26 (20)[a]
Phytoplankton species	31 (14)[b]		35 (18)[b]	
Zooplankton species	22 (6)[b]		21 (5)[b]	
Fish species	3 (3)[b]		1 (1)[b]	
Basal species (S_R)	25	9 (3.1)[c]	29	6 (3.6)[c]
Intermediate species	24	12 (14.7)[c]	20	12 (12.8)[c]
Top species	1	1 (3.1)[c]	2	2 (3.6)[c]
Food Chain Metrics				
Food chains	4836	214 (263)[c]	885	59 (115)[c]
FCL_{mean}	4.64	3.68 (5.08)[c]	4.21	347 (4.30)[c]
FCL_{max}	7	6 (10)[c,e]	6	5 (8)[c,e]
Links	269 (264)[f]	71 (67)[f]	241 (236)[f]	56 (52)[f]
Basal–intermediate links	166	31 (14.74)[c]	158	20 (12.75)[c]
Basal–top links	0	0 (3.12)[c]	7	2 (3.6)[c]
Intermediate–intermediate links	87	27 (34.39)[c]	68	27 (22.9)[c]
Intermediate–top links	11	9 (14.74)[c]	3	3 (12.75)[c]
Complexity Metrics				
Connectance (C) [a,f]	0.2155	0.3190	0.1851	0.2737
Consumers per resource species[a,f]	5.39	3.35	4.83	2.89
Resource per consumer species[a,f]	10.56	5.15	10.73	3.71
Consumers per phytoplankton species[a,f]	5.35		4.71	
Consumers per zooplankton species[a,f]	4.36		3.38	
Resource per zooplankton species[a,f]	10.68		11.10	
Resources per fish species	9.67		3	

[a]Isolated species excluded.
[b]Number of unique species in parenthesis (i.e., species that occurred in that year only).
[c]Number in parenthesis indicate cascade model predictions.
[d]Number of links.
[e]Longest food chain with an expected frequency greater than one.
[f]Cannibalistic links excluded.

10.4.1 Stability

In Chapter 2 we defined local or neighborhood stability in terms of the eigenvalues of the community matrix A for a system of equations that defined the dynamics of the deviation (x) from an equilibrium state of the form

$$\frac{dx}{dt} = Ax \tag{10.1}$$

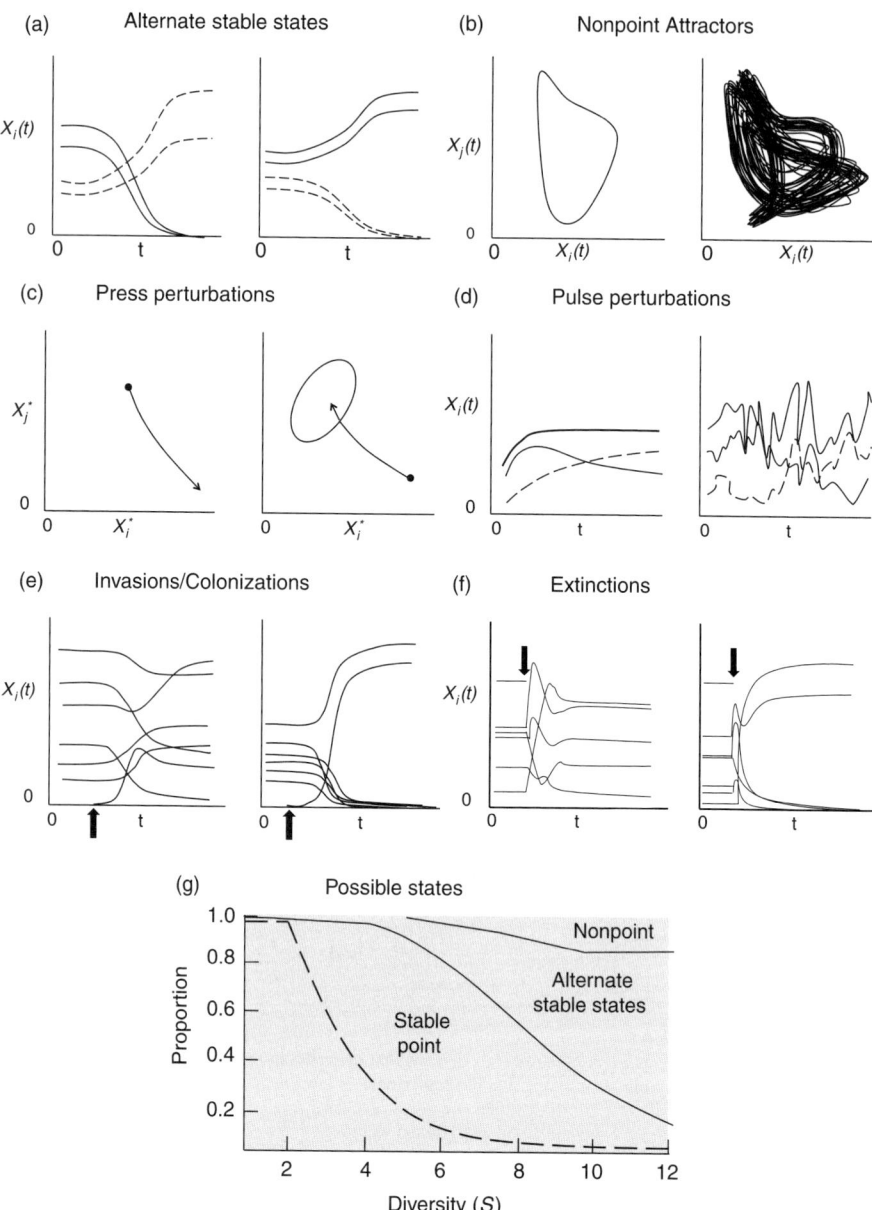

Figure 10.7 Different types of stability for systems that differ in the types of perturbations they experience (a–f), and a summary of dynamic states with increased diversity (g), as compiled by Ives and Carpenter (2007). (a) The initial densities of species determine which species persist, shown here as pairs of alternatively persisting or nonpersisting species. (b) A stable (left) and a chaotic (right) attractor. (c) For systems with a stable equilibrium, the

with elements, α_{ij}, of A defined as follows:

$$\alpha_{ij} = \left(\frac{\partial \frac{dX_i}{dt}}{\partial X_j}\right)^* \qquad (10.2)$$

Recall that if the real parts of the eigenvalues of matrix A are negative, the system is stable. If one or more of the eigenvalues possess positive real parts, then the system is not asymptotically stable. The important point is that the stability rested on the magnitudes and relationships among of the elements, α_{ij}, of matrix A defined by the connectedness architecture of the food web. The α_{ij} elements of matrix A are specific rates of interaction, defined in terms of life history, consumption, and physiological parameters, i.e., the specific rates of increase, specific death rates, consumption coefficients, and energetic efficiencies. The hypothetical disturbances that we induce to assess stability changed one or more of the population sizes of the species or collections of species, thereby creating one or more deviations from the asymptotic state, be it a stable equilibrium point or stable limit cycle. In mathematical terms, the disturbance was defined in terms of changes in the vector x, while the community matrix determines whether the changed state would return to its equilibrium. Hence, the disturbance changes neither the structure of the community matrix A, nor the magnitudes of the interaction rates α_{ij}.

A concept of a dynamic architecture would broaden our concept of stability in a way that accommodates changes in the configuration of the food web. Pimm (1991) provided a starting point by applying a concept of stability used in engineering to ecosystems that shifts the focus from species to both species and processes. Stability in this sense refers to system-level processes or characteristics such as species

---

disturbance can cause the equilibrium point to deviate to a point where species j ultimately goes extinct (left panel), or can force the system past a dynamic threshold wherein the equilibrium point bifurcates into a limit cycle or a stable nonpoint attractor (right panel). (d) Following a single pulse perturbations (left panel) or repeated (stochastic) pulse perturbations (right panel), the relative abundances of species (light and dashed lines) and absolute abundances (heavy line) can change. (e) Invasion (invasion marked by arrow) and successful colonization. A successful colonizing species can incorporate into the system but alter the relative abundances of resident species (left panel), or can precipitate the extinction of one or more resident species (right panel). (f) In this case, the extinction of a single species (marked by arrow) can lead to changes in the relative abundances of species (left panel) or the loss of one or more species (right panel) one or more additional species. (g) A summary of the prevalence of stable points, alternative stable states, and nonstationary attractors for randomly constructed simple systems (n =10,000 for each level of S). The dashed line gives the proportion of communities that were feasible (F_{NPP}), a requirement for each dynamic state. Consistent with our findings, feasibility declines and the likelihood of alternative state states and nonpoint attractors with increases with increased diversity.

diversity, the distribution and steady flows of energy and materials within the food web, nutrient retention, and species composition, rather than to the more rigid or absolute concept in which the connectedness structure does not change. With this concept, change is an inherent component of the food web architecture. Stability is discussed in terms of the resistance and resilience of the system to disturbance, to change and the variation in dynamics and dynamic states of populations and processes, and to changes in species composition. Different metrics are used, but the key is that the system retains its major features.

This more relaxed concept of stability is not incompatible with the more formal definition that we have used through the book. Ives and Carpenter (2007) reviewed the literature and summarized concepts of stability that were of this looser ilk and concluded that each depends on the inherent dynamics of the system and the types of perturbation that it experiences. As with the traditional definitions of stability we have used, the system of equations that are used to describe the interactions among species defines the inherent dynamics and the potential responses of the food web to disturbance (Figure 10.7). The structural or dynamic state of a system can exhibit simple or gradual change from state to another, an abrupt transition when a threshold is breached, or a fold bifurcation or hysteresis transition. The simple Lotka–Volterra food chain models maintain a stable equilibrium that increases with raised intrinsic rate of increase across the entire spectra, yet transitions from a system that returns to equilibrium in a monotonic manner to one that oscillates. The suite of more complex models of food chains presented by Rosenzweig (1971) transitioned abruptly from systems that possess stable equilibria to ones that oscillated with increased rates of increase. Simple single-species models using difference equations to multispecies food chains using differential equations that possessed complex functional responses transition from asymptotic states that include equilibria, fold bifurcations, limit cycles, and chaos (May, 1976, Hastings and Powell, 1991, Ludwig et al., 1997).

10.4.2 Agents of change

To this point we have provided several examples from modeling and empirical studies that have demonstrated that the changes in life history, behavioral and physiological parameters, altering the densities of populations, and removal or addition of a species from the food web can induce nonadditive effects, affecting asymptotic states (Paine, 1966, Rosenzweig, 1971, May, 1973, Pimm, 1979b, McCann et al., 1998). In each case some aspect of the system is altered, which ultimately affects the structure of the Jacobian matrix, and in turn its stability.

There are at least three ways that we can change the structure of the Jacobian matrix. We can change the magnitudes of the elements of the matrix in a way that preserves the dimension and connectedness structure of the matrix. We can preserve the dimension of the matrix and change the connectedness structure by adding or deleting species interactions, i.e., one or more nonzero elements of the matrix become zero, or vice versa. We can substitute one species for another species. We can change the dimensions of the matrix by adding or deleting species. In other words, the change agents include the parameters that represent the species attributes that make up the elements of matrix A, the elements of matrix A that represent the interactions among species, and the species that comprise matrix A. In this way, the changes are defined in terms of A and its elements α_{ij}, which defines its stability—whether or not a particular state will return to the equilibrium state or a state that can be stable.

A review of the more salient results presented in this book and elsewhere that have used one or more of the aforementioned approaches to study the dynamic stability of food webs have yielded common insights. First, as discussed, changes in the community matrix define a new state for the system. This point is not as trivial as it seems. Changes in state ultimately mean a change in the distribution of biomass among species, which in turn, affects changes in the processes associated with species interactions (mineralization and immobilization of organic matter), changes in the dynamic properties of the system, and changes in the system's tendency to change in dimension (add or delete species). Take, for example, the different two-species producer–consumer models we have used throughout the book. A common ploy to increase the productivity of the model system is to increase the intrinsic rate of increase of the producer species at the base of the food web. In studies that adopted this strategy, some aspect of the asymptotic state of the system was altered.

Second, species traits are change agents that affect interactions and material flow within food webs, and they are subject to evolutionary processes. This introduces a least two factors to consider about birth and death rates, consumption coefficients, and energetic efficiencies. First, for each species the parameters will have mean values and variation. The values are constrained in some fashion and operate over a plausible range. The ecological efficiencies lie between 0 and 1. The intrinsic rate of birth cannot be less than the intrinsic rate of death. Second, it is likely that the parameters did not evolve independent of one another. We do not wish to apply a mechanism here but simply recognize that the relationships exist. To quote Wade and Goodnight (1991) "... the ultimate fate of a gene is determined by its average effect with respect to fitness within large populations and that co-dependent gene complexes arise largely as a by-product of the effects of selection on single genes."

Third, the change agents are affected by changes in environmental conditions. May (1973) discussed this in the context of the uncertainty and stochasticity associated with randomly fluctuating environments. The tenor of the argument is centered on comparing the deterministic environments with constant population densities to an environment with variation about a mean that induces fluctuations in parameter values about a mean. We can carry this further to include regularities in environmental conditions that occur seasonally or over landscapes, or directional changes that occur in environmental conditions like those predicted under climate change scenarios. Changes in environmental conditions may affect only a subset of species or all species to different degrees, resulting in alterations in the nature interactions and the strength of interactions within the food web (Barton and Schmitz, 2009), the likes of which are captured in the Jacobian matrix.

Finally, change agents are affected by feedbacks that occur among the change agents themselves and the environment. Put another way, for a given species selective pressure arises from interactions with other organisms, and interactions with the physical environment, but also arise from the state of development that the systems is in at a particular point in time. Basic theories of community development are predicated on species within the system providing an entry point for the successful colonization and establishment of other species (Hairston et al., 1960, MacArthur and Wilson, 1967, Heatwole and Levins, 1972, Connell and Slatyer, 1977, Oksanen et al., 1981, Neutel et al., 2007). Invasion ecology is replete with a similar set of concepts that addresses why certain species prevail and others do not (Elton, 1958, Levin and D'Antonio, 1999, Davis et al., 2000, Fridley, 2011). How species have adapted to local conditions, the availability of resources, the nature and frequency of disturbance, and the supply and frequency of incoming species all contribute to the vulnerability of the system to change.

10.5 Summary and conclusions

In this chapter we entertained the idea that the performance and persistence of species, the role species play in biological communities, and stability should be approached from the perspective of a dynamic food web structure. We presented the concept of a dynamic architecture by juxtaposing the often-used metaphor of a stone arch and the keystone species (Paine, 1966) to represent a static architecture with a modification of the game of Jenga in which tiles are removed and added to represent the dynamic architecture (de Ruiter et al., 2005). The Jenga metaphor captures the observation that real food webs are dynamic in terms of species composition, in the nature and strength of their interactions among species, in the population abundances of species, and in the life histories, behaviors, and physiologies of species.

Two cases studies followed to illustrate that although changes in species and changes in the nature and strength of interactions may occur, the systems retained a core energetic organization defined by key functional groups of species and the distribution of biomass. This led to a discussion of a broader concept of stability that could accommodate these changes. The metrics we used to assess stability of the static architectures in Parts I and II could be applied to the dynamic architectures as well.

In some sense we have allowed for a dynamic architecture all along. Our earlier treatment of food webs fixed the connectedness structures of systems, comparing different underlying forms (e.g., primary-producer-based webs versus detritus-based webs), configurations (e.g., food chains of various lengths, compartmentalized versus random network), different parameterizations (e.g., endotherms versus ectotherms), or pushed their limits of stability by altering productivity or the directional flow of materials. Instability in this sense equated to uncertainty or change, an increased likelihood that one or more of the systems species would go extinct, or an invitation for new species to be added if the structure of the system or underlying conditions remained the same.

II

Dynamic architectures and stability of complex systems along productivity gradients

II.1 Introduction

Up to this point we have argued that productivity, in the form of primary productivity or detritus availability, shapes the energetic organization of ecosystems by affecting the distribution and flow of biomass among species or groups of species within food webs. The models presented in Part II show that the feasibility ($F_{[SYS]}$) of food chains depends on two aspects of productivity. On the one hand, there should be enough energy to support all trophic levels ($F_{[NPP]}$), but on the other hand, when productivity is increased to high levels, the dynamics of the populations in the food chain may begin to show large oscillations that destabilize the food chains ($F_{[\lambda]}$). Moreover, communities build on the diversity of production at the base of food webs starting in the same way as individual simple food chains. The complexity that ensues stems from cross-linkages among the food chains, many of which have a stabilizing effect (McCann et al., 1998), leading to a compartmentalized architecture. This two-fold effect of productivity, wherein $F_{[SYS]} = F_{[NPP]} \cap F_{[\lambda]}$, coupled with the stabilizing effect of cross-linkages, generates the familiar "hump-shaped" curve describing the relationship between productivity and diversity, but also maps the relationship between the vertical distribution of biomass with increased trophic position and food chain stability.

In Chapter 6 we studied how the energetic structure of real food webs generate patterns in the interaction strength that strongly enhance food web stability. By means of analyzing these patterns in terms of the weights of trophic loops, the stabilizing effect of the patterning was linked to the distribution of biomass, specifically trophic pyramids of biomass. We followed with a series of chapters that deconstructed the pattern, highlighting the relationships between the distribution of biomass and dynamics and the importance of a compartmentalized architecture, and the role of cross-linkages between compartments, to stability. The connectedness structures in these analyses were static, not dynamic. This brings up the questions of how we view stability and the role stability plays during the development and maintenance of simple and complex ecosystems. Is stability the central tendency that ecosystems gravitate toward; is it a measure of the propensity of a system to maintain its current configuration, or is it a measure of the vulnerability of the system to change?

If we take the key results from Part II, the stability and maintenance of ecosystems is founded on the way in which productivity influences population size, and subsequently the patterns in distributions and flow rates of matter and energy. At relatively low productivity levels, population sizes increase and are organized in the form of trophic pyramids with decreasing biomass over trophic levels. As the rate and diversity of production increases, communities become more diverse and complex, with longer food chains and a more web-like structure. At relatively high productivity levels, however, the pyramidal biomass structure may disappear and eventually turn into an inverse biomass pyramid. Inverse pyramids of biomass are accompanied with large destabilizing oscillations. If at such high productivity levels a new higher trophic level group can enter the food chain, then the models predict that the population size distributions will take a pyramidal shape again, preventing destabilization. In fact, this suggests that a food chain or collection of food chains maintain a relatively high level of stability as long as its length is "in balance" with the level of productivity.

In this chapter we will analyze how changes in environmental conditions may influence food web structure and stability of real food webs through detailed case studies of their natural histories and stability. The food webs dealt with are from the largely detritus-based cave system at Wind Cave, South Dakota, a primary succession gradient at the Dutch Wadden island of Schiermonnikoog, and an arctic tundra ecosystem at Toolik Lake, Alaska. Each system is perched on an existing productivity gradient formed by different means but offers a variant to a common theme of how changes in the availability of energy and diversity of organic forms can affect food web structure and stability. We will review a variety of experimental manipulative studies and modeling exercises to explore the concepts by Odum (1969), Connell and Slatyer (1977), and Oksanen et al. (1981) that ecosystem development correlates with higher energy inputs (productivity or detritus inputs), higher complexity and system stability, and how this relates to the theoretical considerations by May (1972, 1973) in that complexity may constrain stability. We will conclude by presenting a general framework for community development that incorporates these concepts.

11.2 Food web structure in a cave ecosystem

Wind Cave is located in the Black Hills of South Dakota. The cave is a phreatic-dissolution 3D-maze cave within the Pahasapa Limestone, capped by shale, slate, and carbonates, with over 221 km of passages under 3.5 km^2 of surface and a depth of 76 m (Figure 11.1). The cave has two natural entrances and is fairly well sealed by the overlaying Minnelusa Formation. The cave is largely dry except where it lies underneath modern drainages or where it hits the water table at its deepest point (Horrocks and Szukalski, 2002). The surface vegetation of the landscape overlying Wind Cave is mixed grass prairie (80%) with forests (20%) dominated by *Pinus ponderosa* Dougl. (Coppock et al., 1983, Whicker and Detling, 1988). The majority of the cave has had minimal to no human contact beyond the initial surveys conducted by cavers, with less than 2% of the surveyed cave open to the public for regularly scheduled tours.

The rate of organic input into the caves forms a gradient at the extreme low end of global productivity ($<0.1-10$ g C m^{-2} yr^{-1}). The organic matter (OM), P and NO$_3^-$N contents of the sediments follow similarly steep gradients (Table 11.1). The species diversity, length of food chains, and distribution of biomass with trophic level are related to the rate of detritus inputs (Table 11.2 and Figure 11.2). Regions of both caves at the high end of productivity, which receive detritus inputs from mammals, tourists, and surface drips, are able to support longer food chains than regions of the cave that receive reduced levels of these inputs. Moving down the input gradient

(a)

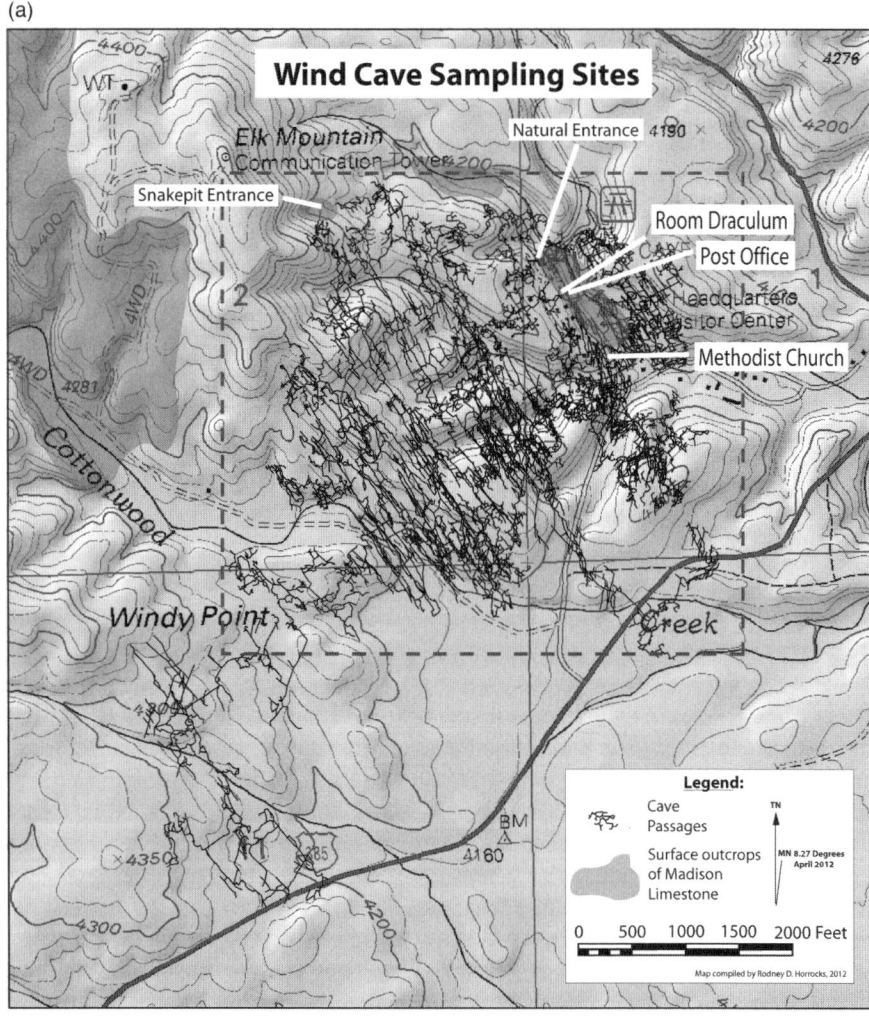

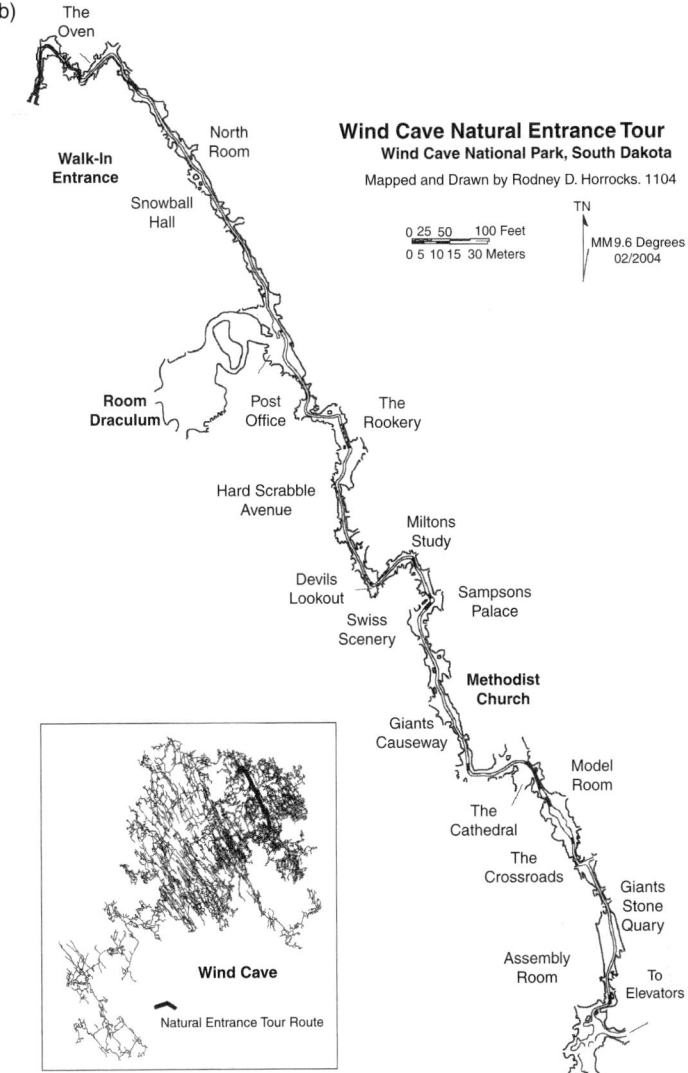

Figure II.1 (a) Topographic map with an overlay of the current known extent of Wind Cave (adapted from Horrocks and Szukalski, 2002). (b) Detailed map of the natural entrance tour route of Wind Cave. Wind Cave is located in Wind Cave National Park on the southeastern flank of the Black Hills of South Dakota, near the town of Hot Springs. The cave has over 221 km of mapped passageways situated within approximately 3.5 km^2 of land surface and 74 m vertical relief. Wind Cave is a phreatic dissolution type cave that formed within the Pahasapa Limestone and Minnesula formation. The Pahasapa Limestone was deposited about 330 million years before present (mybp) during the life and evaporation of a warm shallow sea, while the Minnelusa formation, consisting of sandstone, shales, and carbonates, was deposited 300 mybp during a subsequent advance and retreat of a later sea (Palmer and Palmer, 2000). The current caverns of Wind Cave formed during the uplift of the Black Hills between 65 and 40 mybp and continued to be modified during the erosion that has occurred over the past 40 million years. The region was not covered by glaciers during the glacial period, making the soils >1.8 million years old. The upper levels of the cave started draining 350,000 years ago and were completely air-filled by 155,000 years ago (Ford et al., 1993). Created by Rodney D. Horrocks.

Table 11.1. The chemical and physical characteristics (mean and S.E.) of sediments collected from within the sites characterized in Figure 11.2 and the manipulative study. The data were summarized from Jesser (1998) and Chelius et al. (2009).

Region	pH	Organic Matter (%)	NO_3-N (ppm)	P (ppm)
Natural entrance (soil outside)	6.3 (0.1)	11.20 (0.8)	33.9 (0.7)	23.2 (1.2)
Natural entrance (sediment inside)	7.7 (0.1)	0.30 (0.3)	105.0 (17.0)	13.2 (4.7)
Tour route	7.7 (0.2)	0.50 (0.1)	7.2 (1.7)	5.7 (0.6)
Room Draculum	7.5 (0.2)	0.10 (0.13)	3.17 (0.23)	7.9 (2.1)
Remote	7.6 (0.3)	0.02 (0.02)	3.2 (0.65)	11.2 (7.2)

Table 11.2. Summary statistics for the connectedness food webs along the productivity gradient in Wind Cave. Values of number of functional groups (S), the number of basal resources ($S_{Resources}$), connectance (C), linkage density (SC), mean food chain length (FCL_{Mean}), and maximum food chain length (FLC_{Mean}).

	S	$S_{Resources}$	C	SC	FLC_{Max}	FCL_{Mean}
Natural entrance	25	7	0.14	3.50	5	3.30
Tour route	12	3	0.197	2.36	3	2.92
Room Draculum	10	2	0.267	2.67	3	2.42
Remote areas	4	2	0.667	2.67	1	1

reveals a gradient in diversity and trophic structures. For example, the natural entrance at Wind Cave receives all the inputs listed above and possesses a complex food web with a maximum chain length of five links (Figure 11.2a). Entrances, defined here as the first 100 m into the cave, represent an ecotone between the outside world and the cave. Temperature and humidity at the entrance fluctuate with those outside. The areas are frequented by rodents (deermice, *Peromycus maniculatus*, and bushy-tailed woodrats, *Neotoma cinerus*) and amphibians (tiger salamanders) and support a rich assemblage of invertebrates. Further in, the climate stabilizes, there is no evidence of rodent activity, and the diversity declines (Figure 11.2b). The "Methodist Church" section is along the tour route; it receives the light, lint, and drip inputs listed above and possesses algae-, fungi-, and bacteria-based food chains, which are connected by a predatory mite (*Tydeus*). Lost are the cryptostigmatid mites (e.g., *Ceratozetes, Oppiella, Euphthiracarus*), many prostigmatid mites (e.g., *Speleorchestes*), and an astigmatid mite (*Tyrophagous*). Other well-traveled areas possess similarly structured communities connected by a predatory mite (*Cyta*). Still farther in the caves beyond the tour routes, the diversity decreases and the food chains shorten. A 50 m vertical climb brings one to "Room Draculum." This chamber, adjacent to the tour route, has reduced lint inputs but possesses numerous surface water drips. The food chains drop to three links or fewer, as no predators have been collected in Room Draculum (Figure 11.2c). At greater depths (e.g., the "Fairy

Palace" and "Chamber Pot Room") are regions that possess drip-dominated communities that are essentially lint-free. These remote regions support bacteria, fungi, and protozoa, but no nematodes or arthropods (Figure 11.2c). Here, the fungi-based food chains have fallen to two links, and the bacteria-based food chains remain at three links. In "dry" areas (moist sediments but no active drips) beyond the entrances and on tour routes, the sediments support bacteria, fungi, and protozoa, crickets (*Ceuthophilus* spp.), and isopods. In dry areas that are off the tour routes, the sediments contain only bacteria, fungi, and protozoa (Figure 11.2d).

The communities exhibit patterns in their energetic organization consistent with trophic cascades (Figure 11.3). Bacteria and fungi peak at an intermediate level of organic input, with fungi reaching their peak slightly ahead of bacteria. The apogees are coincident with the appearance of protozoa, nematodes, and arthropods, suggesting that these consumers are regulating the microorganisms. A high number of sites possess relatively high densities of bacteria and fungi and no protozoans, nematodes, or arthropods.

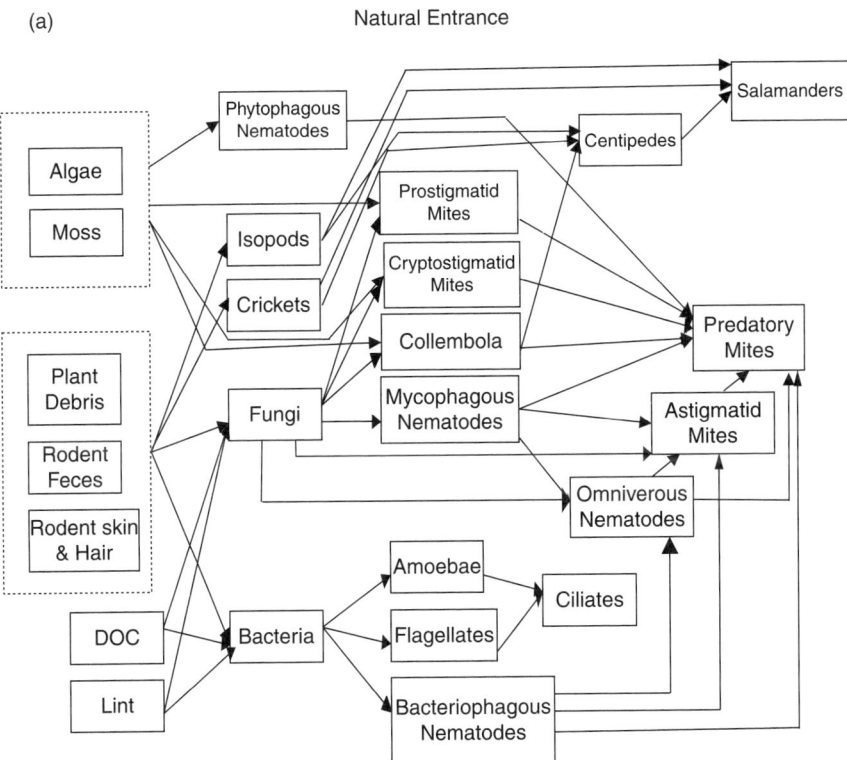

Figure 11.2 *Continues overleaf.*

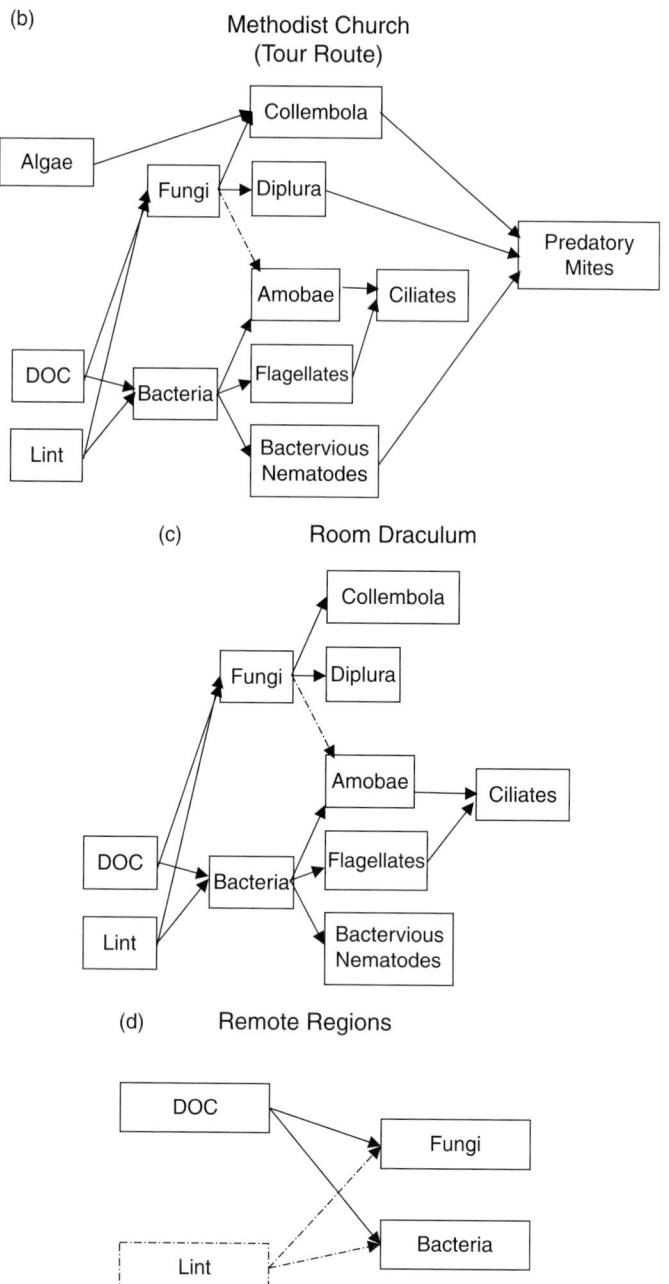

Figure II.2 A series of connectedness food web descriptions for the communities within sediments of Wind Cave by region. (a) Natural Entrance, (b) Methodist Church (Tour Route), (c) Room Draculum, and (d) remote areas.

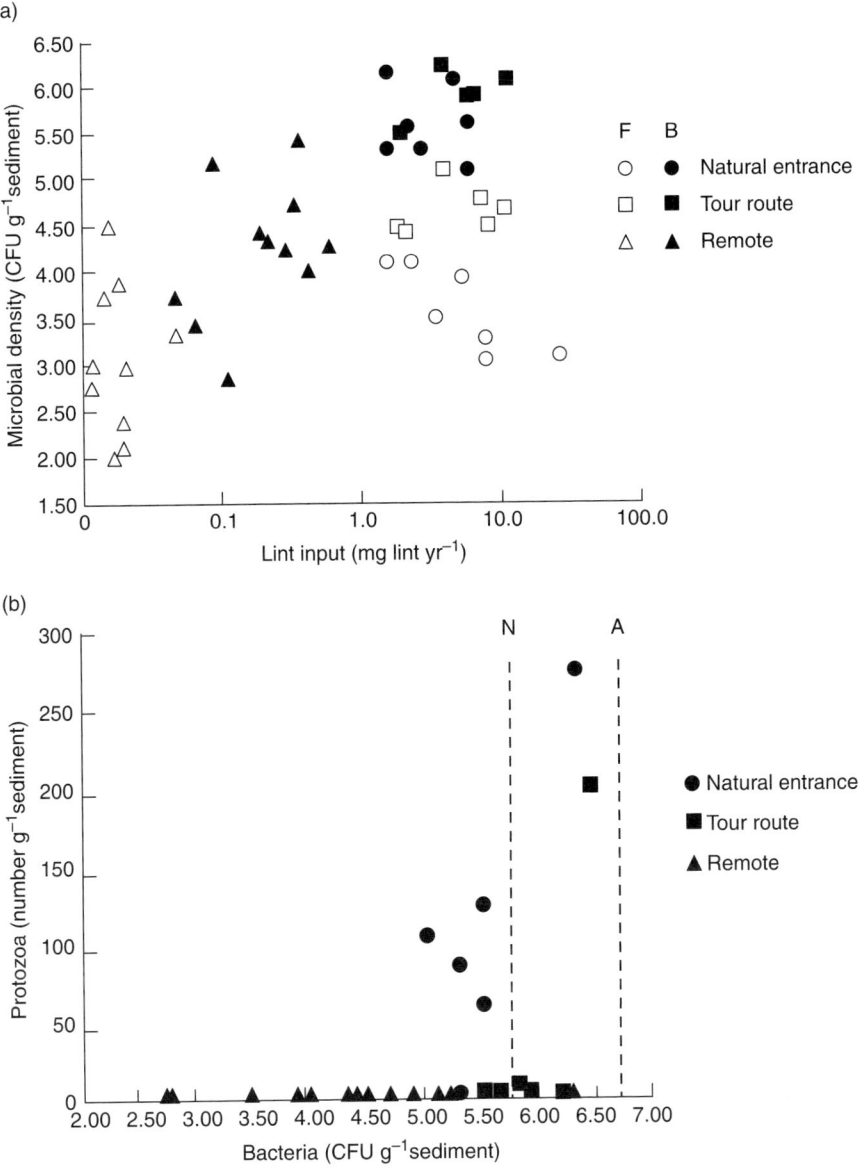

Figure II.3 (a) The relationships between bacteria density (●) and fungal density (○) expressed in terms of colony forming units (CFU) per gram dry sediment with the annual input rate of lint (human hair and clothing fibers) in Wind Cave, South Dakota. (b) The relationship of protozoa (numbers g^{-1} dry sediment) to bacteria (CFU g^{-1} dry sediment). Dashed lines indicate the levels of bacteria densities where nematodes (N) and arthropods (A) were first observed.

Interpreting the observations in Figures 11.2 and 11.3 strictly in terms of the rate of carbon input is problematic. Another explanation to explore is that the diversity of input may play a role. Regions that receive higher rates of inputs also receive a greater diversity of inputs. The loss of predatory arthropods, mycophagous arthropods, and microbivorous nematodes coincides with the loss the different types of inputs. Aspects of colonization could also explain the patterns seen in Figures 11.2 and 11.3. The dearth of biota in remote sections of the cave could be due to the distances that the biota would have to travel or to the fact that sediments in remote regions could be unsuitable for many biota. The caves are 30–40 million years old, and were not covered by glaciers during the Pleistocene, making the soils >1.8 million years old. In this time, the caves have experienced floods from surface runoff, and at lower depths, by the rise and fall of the aquifer. Owing to the cave's age and dynamic nature and the vagile nature of the fauna, there have been ample opportunities for colonization. Given the similarity in the chemistry of the diverse array of sediments near to and far from the entrance and the stable temperature and humidity within the cave, the unsuitability of these habitats to organisms that would occupy higher trophic levels seems equally unlikely.

Three mechanisms could explain the observed patterns in species diversity and trophic structure presented in Figures 11.2 and 11.3: 1) the rate of input, 2) diversity of inputs, and 3) colonization. A manipulative study by Chelius et al. (2009) involving introduced carbon- and nitrogen-rich substrates confirmed elements of each mechanism and their interactions (Figure 11.4). Pulses of organic substrates increased the total biomass within the systems and added a trophic level in a manner consistent with the models presented by Oksanen et al. (1981). Ambient levels of detritus inputs supported low population densities of microorganisms and few individuals of their consumers. The pulses initiated growth in the trophic structure in a step-like fashion, and the complexity of the system increased with an increase in the diversity of substrates (Figure 11.5). The results also revealed that microbial biomass and diversity is responsive to the levels of organic substrate input and diversity, consistent with the ideas discussed in Chapters 8 and 9 about different basal resources initiating compartments within the food web.

11.3 Food web structure and stability along the primary succession gradient at the Wadden island of Schiermonnikoog, The Netherlands

The island of Schiermonnikoog is located in the Waddensea north of the mainland of Netherlands (Figure 11.6). Like other Wadden islands, Schiermonnikoog is gradually moving toward the east at a rate of approximately 1 m yr^{-1}. As a consequence, the west side of the island is relatively "old" and the east side "new." This enables the study of different stages in ecosystem development with the age of the soils on the island. The productivity gradient, which includes above- and belowground organic matter production, starts at the outer east side, near the coastline, where there is only

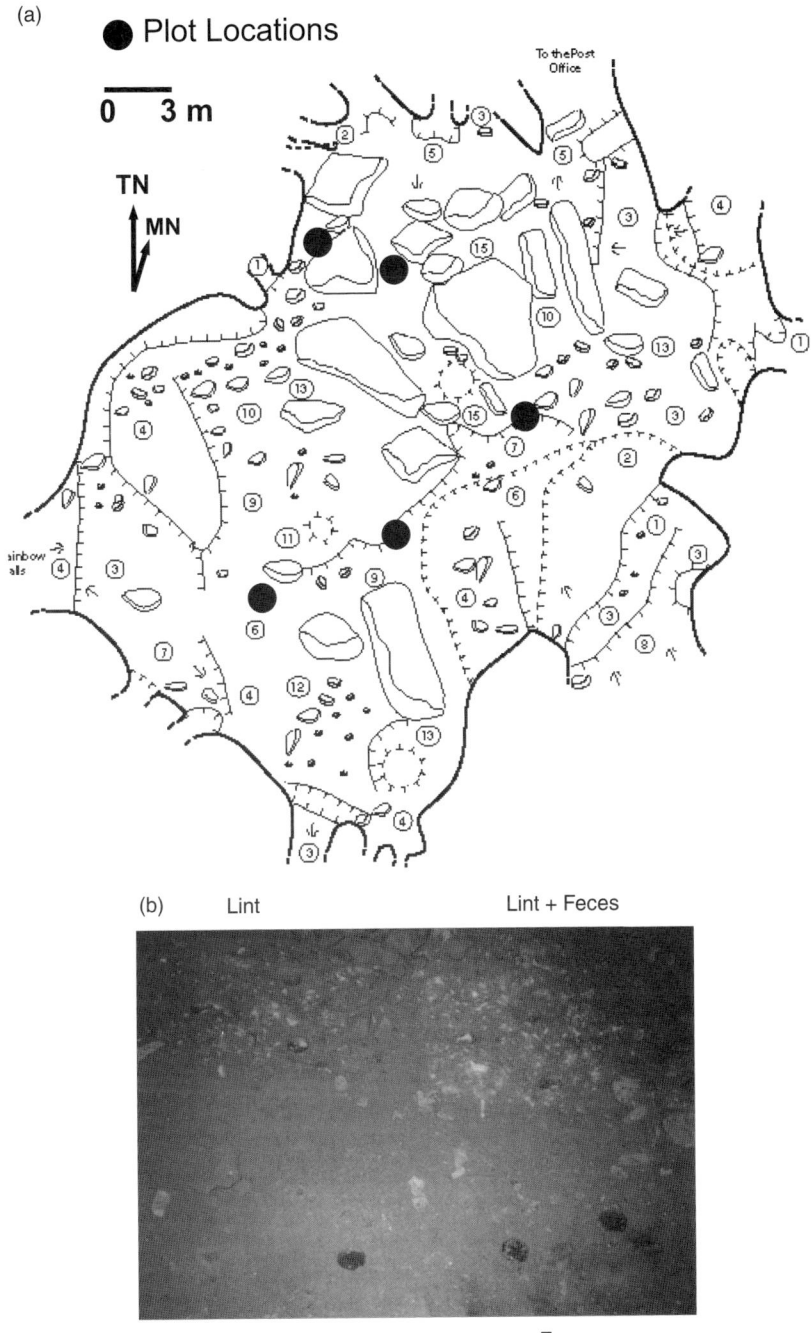

Figure II.4 (a) Map of plot locations within Room Draculum of Wind Cave. Drawn by Rodney D. Horrocks. (b) Photograph of one of the five plots (1.21 m^2) used in the manipulative study. The treatments within the subplots (0.25 m^2) are labeled on the photograph: Control (C), Lint (L), Feces (F), and Lint + Feces (LF). The pink flagging tapes mark the corners of the 0.25 m^2 subplots. The black discs visible in each subplot are small Petri dishes (60 mm) lined with a 9:5 mixture of charcoal and plaster of Paris (Snider et al., 1969), placed to estimate the number of small arthropods.

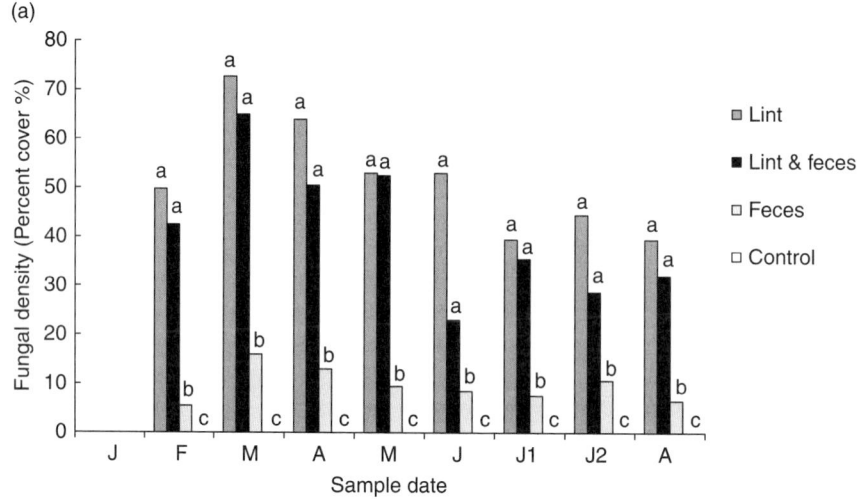

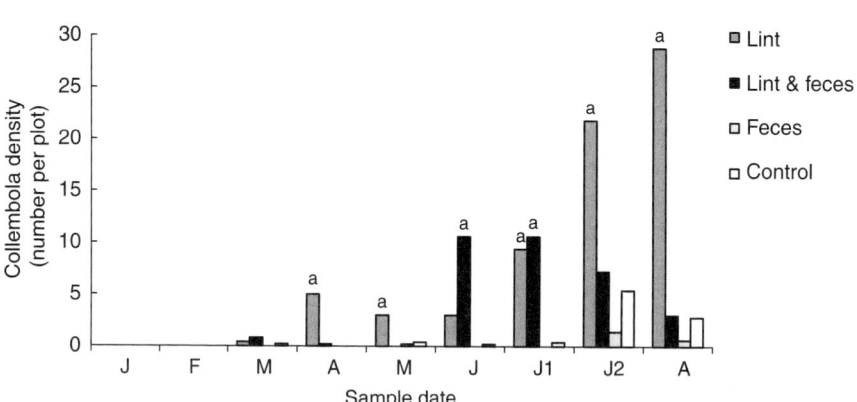

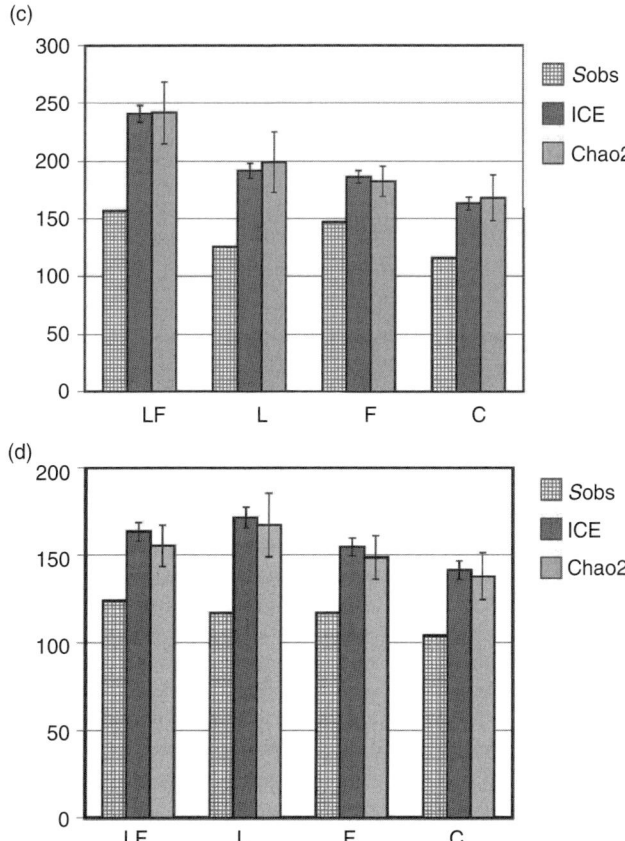

Figure II.5 Results of the manipulative study in Room Draculum of Wind Cave. (a) Fungal density (percent cover) for each treatment (Lint, Lint & feces, Feces, and Control) by sample date, (b) Collembola density for each treatment by sample date, (c, d) Species Richness with standard errors estimated from the Observed subsample species richness (*Sobs*) and operational taxonomic unit (OTU) richness estimated using Incidence-based Coverage Estimator (ICE) and Chao2 from data generated with restriction enzymes, Msp I and Rsa I respectively (from Chelius et al., 2009). ICE and Chao2, are non-parametric estimators of total OTU richness ICE is a model based on the assumption that the detection probabilities are heterogeneous among OTUs (Chazdon et al., 1998). Chao2 uses the number of unique and duplicate OTUs to estimate the number of missing species (Chao et al., 2005).

bare sandy soil ($0\,\mathrm{kg\,ha^{-1}\,yr^{-1}}$) to more than $100\,\mathrm{kg\,ha^{-1}\,yr^{-1}}$. The vegetation starts with mainly *Ammophila arenaria* and *Festuca rubra*. On the older soils (100 years old) the vegetation consists mainly of *Hippophaë rhamnoïdes*, *Sambucas nigra*, and *Urtica dioica* (van de Koppel, 1996, Neutel et al., 2007).

The soil food webs were established at different levels of productivity from sites that were estimated to be approximately 0, 10, 25, and 100 years of age. Food web structures were assessed with respect to food web complexity (number of trophic groups, the frequency of interactions indexed as connectance, and link density), maximum and mean food chain length (Table 11.3), and biomass (Table 11.4). For these food webs, stability was investigated by looking at the maximum loop weight (Equations 7.1 and 7.2) and by evaluating the eigenvalues of Jacobian community matrix representations of the food webs, following the same procedure as described in Chapter 6. Stability and maximum loop weights are established for matrices including the interaction strength values based on the observations ("real matrices") and for randomization of these matrices ("random matrices"). The stability measure used is the "critical" value of the parameter s reflecting the minimum degree for

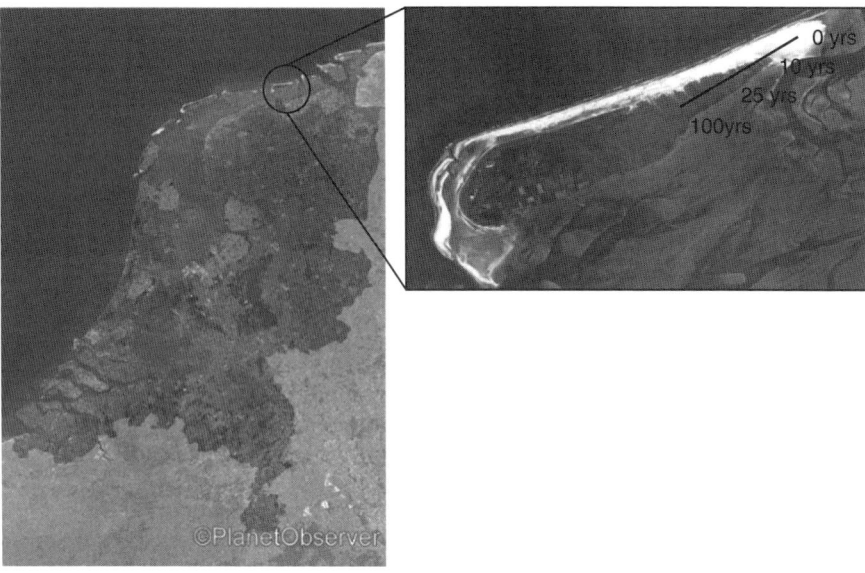

Figure 11.6 Aerial photographs of the Netherlands and Schiermonnikoog Island. The island is one of Wadden islands off the north coast of the Netherlands. The island is drifting in an eastern direction creating a younger eastern edge and older western edge creating a primary succession gradient from east to west. The soil ages are approximated along the transect line. Ages are 0, 10, 25, and 100 years of the stages 1–4, respectively. From PlanetObserver (www.planetobserver.com).

density-dependent death required for matrix stability. This value serves as a relative measure, which enables comparisons of stability between different food webs.

Along with productivity, the food webs become more complex, with increasing food chain length and complexity (Figure 11.7a–d). As for stability, the values for maximum loop weight and critical values of s of the food webs along this

Table II.3. Summary statistics for the connectedness food web of the soil communities along a primary succession gradient at the island of Schiermonnikoog.

	S	$S_{Resources}$	C	SC	FCL_{Max}	FCL_{Mean}
Stage 1: 0 years	8	1	0.286	2.286	3	2.20
Stage 2: 10 years	13	2	0.218	2.833	4	2.60
Stage 3: 25 years	14	2	0.253	3.538	5	3.33
Stage 4: 100 years	17	2	0.257	4.375	6	4.06

Ages are 0, 10, 25, and 100 years of stages 1–4, respectively. Values are based on four food web replications for each stage, based on three sampling dates.

Table II.4. Biomass (g ha^{-1} cm depth^{-1}) of the functional groups at four stages in the primary succession gradient at the island of Schiermonnikoog. Ages are 0, 10, 25, and 100 years of stages 1–4, respectively. The values are averages of four web replications, based on three sampling dates (modified from Neutel et al., 2007).

	Stage 1 0 years	Stage 2 10 years	Stage 3 25 years	Stage 4 100 years
Microorganisms				
Bacteria	1135.00	4850.00	19000.00	17750.00
Fungi	33.75	129.25	212.50	290.00
Protozoa				
Amoebae	0.0107	0.03425	0.2	0.35
Flagellates	0.2825	4.15	43.75	111.75
Nematodes				
Phytophagous	0.0045	3.225	0.8625	0.025
Bacteriovores	0.0595	9.45	39.25	71.75
Fungivores	0.0002	0.0045	0.0195	0.0056
Predators[1]	0.0003	0.25	0.0505	0.0518
Arthropods				
Bacteriophagous mites	0	0	0	0.53
Fungivorous Collembola	0	0.31	29	110
Cryptostigmatid mites	0.0001	0.0128	0.0228	0.1325
Non-cryptostigmatid mites	0.0002	0.057	0.075	0.1325
Nematode-feeding mites	0	0	0.0001	0.0114
Predaceous Collembola	0	0	0	0.0132
Predaceous mites	0	0.0012	0.0155	0.076

[1] Includes predators and omnivores.

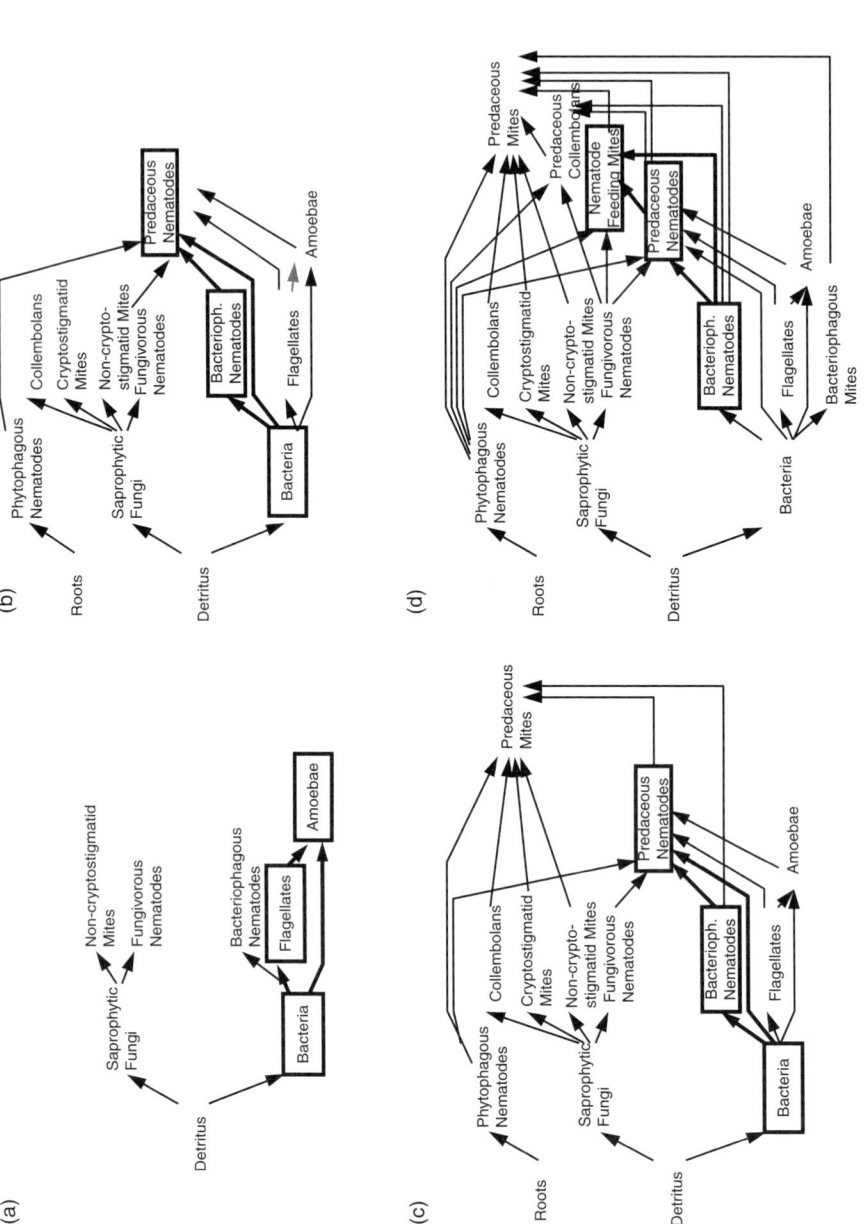

Figure 11.7 Food web structure along the primary succession gradient on the Wadden island of Schiermonnikoog (Neutel et al., 2007). Soil ages of the respective stages are 0, 10, 25, and 100 years. Each diagram is constructed as a representation of four food-web replications constructed on three sample dates: (**a**) stage 1, (**b**) stage 2, (**c**) stage 3, (**d**) stage 4. Thick arrows and boxed groups indicate the omnivorous loop with the maximum weight. Reprinted with permission from Macmillan Publishers Ltd: Nature.

productivity gradient reveal several patterns. First, over the entire productivity gradient the real matrix models of the food webs are much more stable than the random matrices. The critical s values of the real matrices are all below 0.5, whereas those of the randomized matrices range up to above 4 (Figure 11.8). This means that despite changes in the complexity of the food web (Figure 11.8), stability is maintained at a relatively high level. The same pattern emerges from the loop weight analysis. Also here, the real matrices have a much lower maximum loop weight (< 2.5 yr^{-1}) than the randomized matrices (up to 12 yr^{-1}). These results are very similar to the outcome of the comparison of stability and maximum loop weight between real and randomized matrix models of the Central Plains Experimental Range (CPER) site (see Chapter 6). Second, we see that during food web

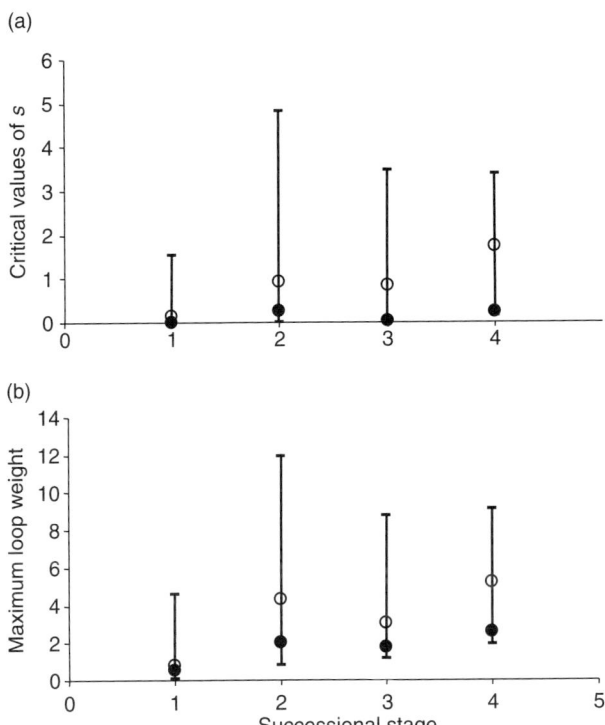

Figure 11.8 Maximum loop weight and stability of the real (closed circles) and randomized (open circles) food webs in the gradients of Schiermonnikoog. The stability measure s is determined as the value that leads to the minimum level of intraspecific interaction strength required for matrix stability according to $\alpha_{jj} = s\, d_j, j = 1\ldots n$., where d_j is a natural intrinsic death rate (Chapter 6). Loop weight is calculated by taking the geometric mean of the interaction strengths in the loop (Chapter 6). For the randomized matrices, the open circle with bars give means and standard deviation of 1000 replications of randomized webs.

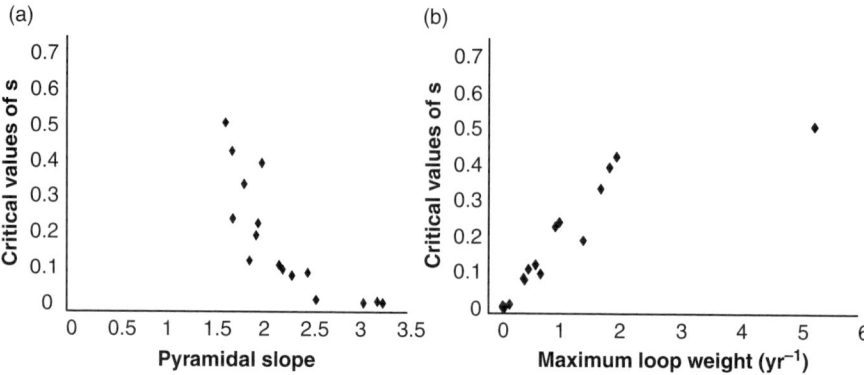

Figure 11.9 Pyramidal slope in the critical loop (i.e., the loop with the maximum loop weight), maximum loop weight and food web stability in the food webs along the productivity gradient at the Schiermonnikoog site. Each symbol represents a replicate. The pyramidal slope denotes the decrease over trophic levels of the species' biomass, expressed as a power of 10, i.e., pyramidal slopes of 1 and 2 mean a tenfold and a hundredfold decrease in biomass over trophic levels, respectively (adapted from Neutel, 2001).

development the critical loop (the loop with the maximum loop weight) tends to move up higher in the food chain (Figure 11.8). Introduction of higher-trophic-level organisms, such as the predatory nematodes in stage 2, are accompanied by changes of the critical loop, i.e. from *bacteria – flagellates – amoebae* in stage 1 to *bacteria – bacterivorous nematodes – predatory nematodes* in stage 2. The mechanism underlying the high stability during the course of development appears related to maintenance of pyramidal biomass structures (Figure 11.9). When productivity increases, biomass accumulates at the higher trophic levels. This leads to higher loop weights and greater instability. During this state of relative instability, productivity increases to become sufficiently high to support additional consumers. When these new groups colonize the systems, trophic pyramids are maintained and, along with this, low maximum loop weights and greater stability. This might be the case, for example, at stage 2. The introduction of the predatory nematodes may have prevented large population sizes of the amoebae, the highest trophic level in the critical loop in stage 1. This mechanism is supported by the correlation between the slope of the trophic pyramids in the critical loop (the loop with the maximum loop weight), the weight of these loops, and the stability of the food webs (Figures 11.8 and 11.9).

11.4 Food web structure in a changing Arctic

Toolik Lake, Alaska (Figure 11.10) is located north of the Arctic Circle and Brooks Range on the North Slope of Alaska. The soils of the Alaskan North Slope form a mosaic of tundra with different vegetation composition on surfaces of varying age, given the advance and retreat of several glaciers. The soils of the dominant moist

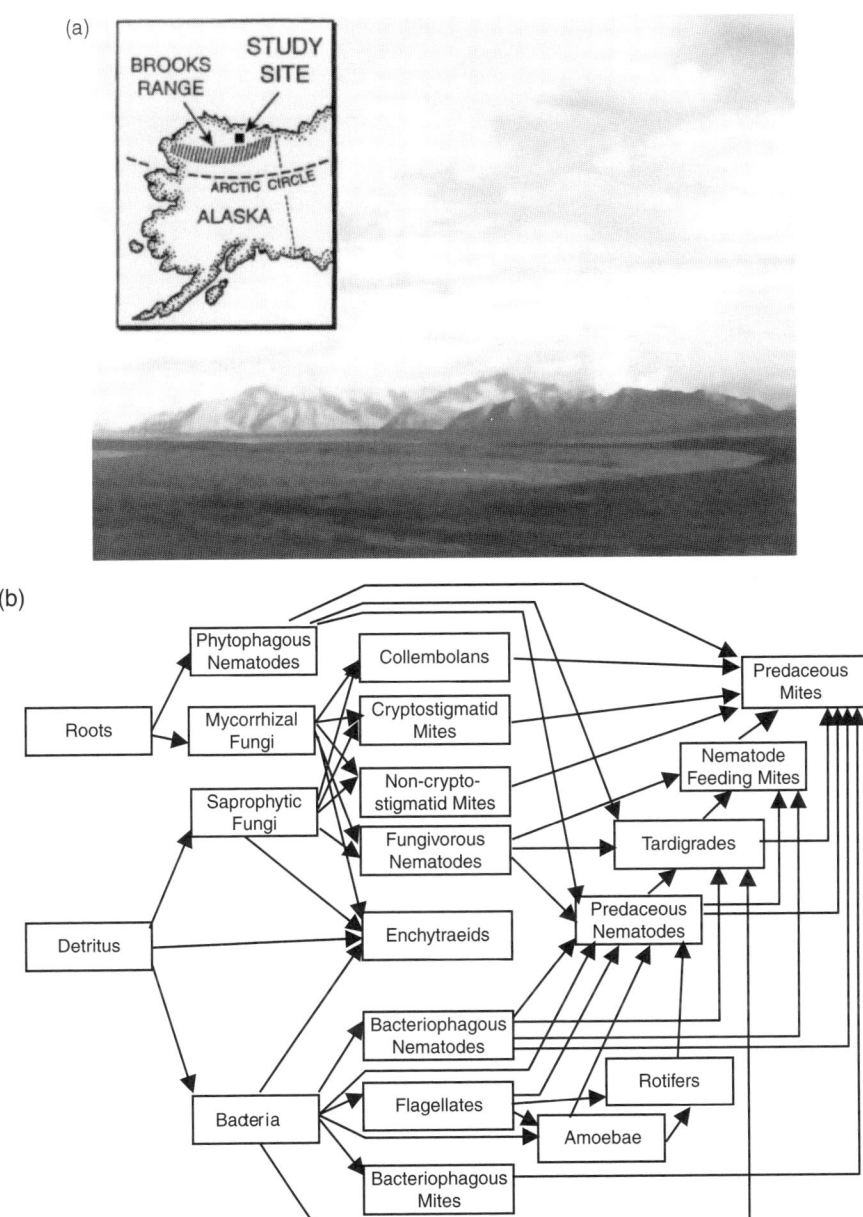

Figure 11.10 *Continues overleaf.*

(c)

Control   Fertilized (N+P)

Figure II.10 Summary of the moist acidic tussock (MAT) tundra study sites at Toolik Lake LTER, Alaska. (a) Toolik Lake is located in the northern foothills of the Brooks Range (68°38' N and 149°43' W, elevation 760 m) approximately 240 km north of the Arctic Circle. The mean annual air temperature at Toolik Lake is −7°, and 10°C during the growing season in June–August. Annual precipitation ranges from 200 to 400 mm, with 50% falling as snow. The site is typical of the northern foothills region near the Brooks Range. MAT is the most common vegetation type surrounding Toolik Lake. The MAT habitats possess poorly structured mineral soils covered by an organic mat 0–30 cm thick (Shaver and Chapin, 1991). The plant community is dominated by graminoids (*Eriophorum vaginatum* Linnaeus and *Carex bigelowii* Torr. ex Schwein), deciduous shrubs (*Betula nana* Linnaeus), evergreens, and mosses (Chapin et al., 1995). The grass species *E. vaginatum* forms dense, compact tufts called tussocks surrounded by moss-filled intertussock depressions. (b) The connectedness food web of MAT soil communities at Toolik Lake. The food web can be disaggregated into separate tussock and intertussock webs. (c) In 1989 manipulative studies were initiated to assess the impacts of nutrient additions (N, P, N and P), large herbivores, and warming. The photo on the left is an untreated control plot dominated by *E. vaginatum* and mosses, whereas the photo on the right is of a plot fertilized annually with N and P (10 g m^{-2} yr^{-1} as NH$_4$NO$_3$ and 5 g m^{-2} yr^{-1} as P$_2$O$_5$) dominated by *B. nana*, ten years after inital application. Photo © John C. Moore and Rod Simpson.

acidic tussock (MAT) tundra ecosystem are characterized by a thick organic horizon overlying a poorly developed mineral layer. The microtopographic features include tussocks formed by the growth of sedges and shrubs, and water-saturated intertussock depressions dominated by mosses. The organic layer that has formed is a result of the low pH of the plant debris (roots, stems, and leaves) and low decomposition rates due to cold temperatures, and anoxic soil conditions due to frequent water saturation, which act to inhibit the activity of soil microorganisms and invertebrates.

Over the past 30 years, the Arctic has experienced significantly increased regional warming (Chapin et al., 1995). Warming generally increases tundra soil nutrient availability by creating a more favorable environment for decomposition and nutrient mineralization. Coincident with warming has been a "greening" of the Arctic, as documented from remote satellite imagery and repeat aerial photography demonstrating an increase in abundance of several deciduous shrub species in multiple regions of the Arctic. Similar increases in birch and alder have been found in the pollen record during previous warm periods. With this increased woody biomass acting as long-term C storage, it is possible that these shrubs might offset the increased CO_2 production from greater rates of soil respiration, but this idea has recently been challenged by modeling and empirical studies (Shaver et al., 2006, Steiglitz et al., 2006) as changes in soil C cycling may go beyond a simple increase in decomposition rate (Weintraub and Schimel, 2005). However, with the shift to higher net primary productivity (NPP) and more abundant woody vegetation has been a loss of organic C in the soil (40–100 g C m^{-2} yr^{-1}) (Oechel et al., 2000, Mack et al., 2004). Given that Arctic soils contain 11–14% of the global soil C pool, understanding the mechanisms behind shifts of these magnitudes that have changed arctic soils from a net sink to a net source of atmospheric C is critical.

Research at Toolik Lake has aimed to determine how current and anticipated changes in climate affect abundances and dynamics of biota, both above and below ground, and their impacts on soil organic matter (SOM) in the dominant upland arctic tundra ecosystem. Several patterns have emerged from long-term nutrient additions and warming studies (Figure 11.11). Twenty years of nutrient addition has led to dramatic shifts in plant species composition (the species richness in MAT dropped from on average 10 sp/m^2 to 6 sp/m^2) with a loss of mosses, and increases in NPP and biomass (particularly among shrubs). The quality (higher C:N) of plant litter decreased, whereas its quantity increased, and plant roots increasingly proliferated from the organic horizon into the mineral layer. Coincident with these changes are losses of SOM in the organic and mineral horizons, and changes in the size distribution and quality of soil aggregates (Figure 11.11).

Much of the connectedness structure of the soil food web for the entire profile remained intact following 16 years of nutrient additions, with one notable exception. The organic horizon of the fertilized plots developed sizable populations of enchytraeids, small microviorous oligochaetes. Enchytraeids are present in the native control plots, but infrequent. Comparisons of the food webs within different soil horizons show a general decline in diversity and complexity within the mineral layers of the fertilized plots (Figure 11.12; Table 11.5). The energy flux descriptions of the soil food web based on the biomass estimates of the functional groups (Table 11.6) indicate that the community has undergone structural shifts as well (Figure 11.13). The base of the fungal energy was enhanced and activity within the bacterial energy channel declined within in the fertilized plots in the upper organic layer. There was reduced activity overall within the lower organic layer with fertilization, and enhanced activity within the bacterial energy channel within the mineral layer with fertilization.

The three aspects of the sequence of the changes in plants, soil biota, and soil aggregates are telling. First, nutrient additions to the N-limited organic horizon increased nutrient availability to plants, but also stimulated the activity of the existing soil community. The turnover rates of organic matter are dependent on the quality of organic matter and the structure of the decomposer community, with bacteria-based communities having higher turnover rates and subsequent carbon loss than fungi-based ones. Second, our observed changes in the belowground food

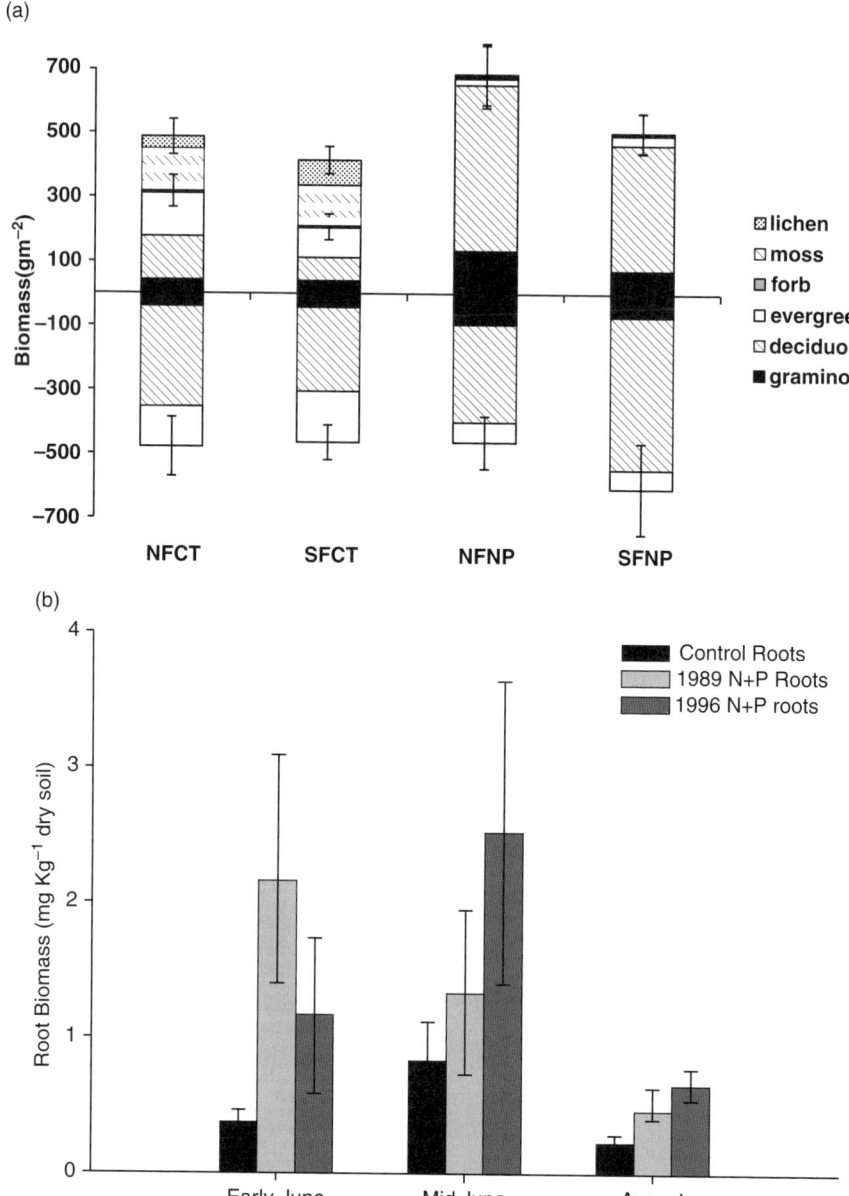

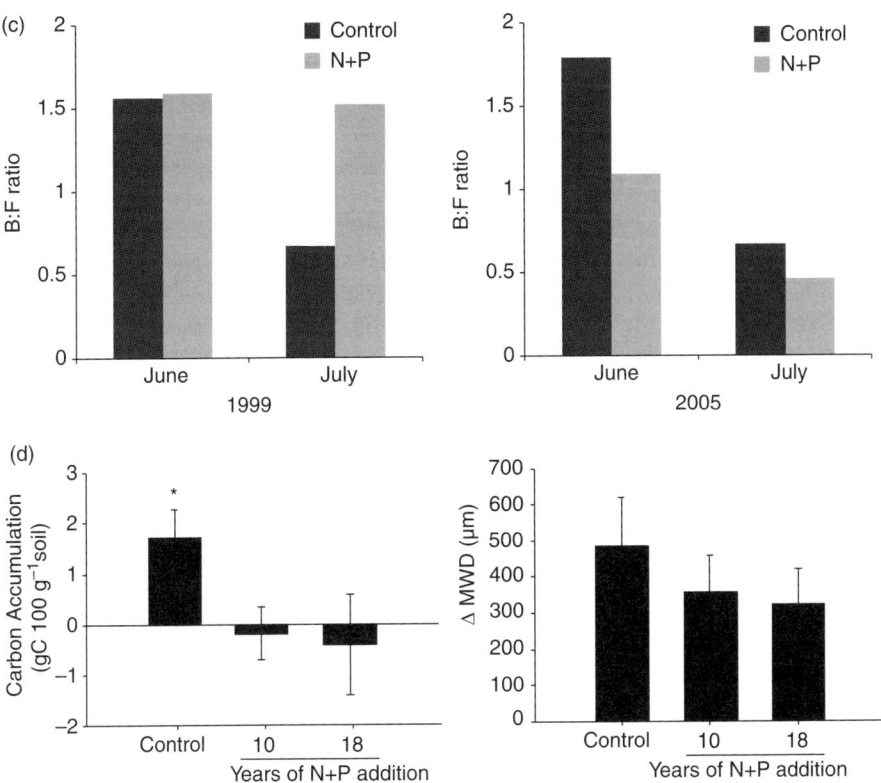

Figure 11.11 Responses of plants, microorganisms, and soils of moist acidic tussock (MAT) tundra Toolik Lake, Alaska to annual nutrient additions. (a) Aboveground plant biomass and species composition in 2006 (from Gough et al., 2012), (b) root biomass by depth, (c) the ratio of bacteria to fungi (B:F) in 1999 and in 2005, and (d) seasonal changes in soil carbon and aggregate size distribution expressed as mean weight diameter in 2006.

community with nutrient addition over time occur at a slower rate than vegetation shifts. This time lag may reflect the differences in the higher-quality materials of the pretreatment organic material from sedges to the newer lower-quality materials from shrubs. Third, the carbon loss over long-term nutrient addition may represent a transitory state, as there is evidence of a stabilization of carbon dynamics through a recovery of the fungi pathway in the soils. This later point is speculative, as while we have seen increases in fungal biomass, an increase in soil aggregation, which would be expected, has yet to be observed.

These results and those of others point to an ecosystem in transition from one state to another. For brevity we can separate the transition into three phases (Figure 11.14). The initial phase involves the current mix of plant growth forms commonly found in moist acidic tussock tundra in northern Alaska. Increased nutrient availability promote root growth through the upper and lower organic

(a) Toolik Lake Moist Acidic Tundra-Native Upper and Lower O Horizon

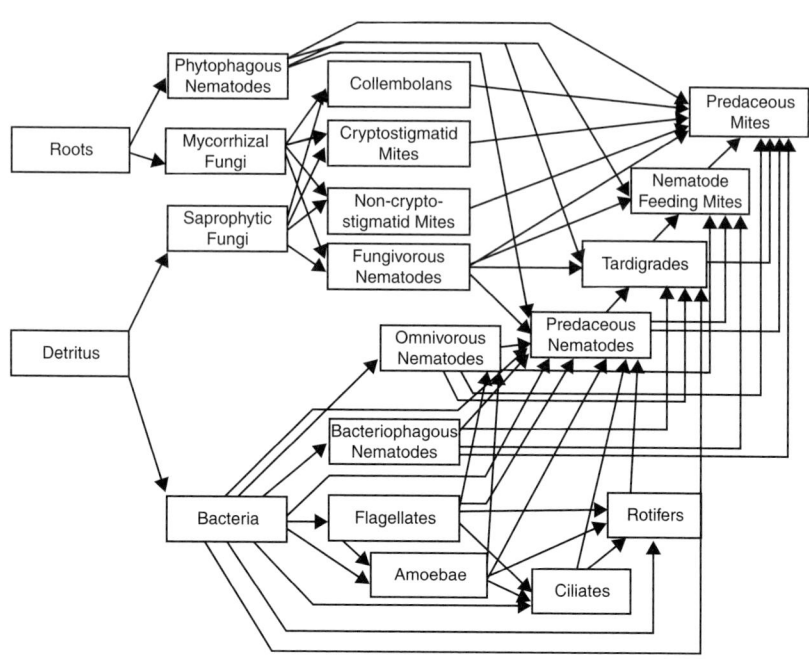

(b) Toolik Lake Moist Acidic Tundra-Native Mineral Horizon (0–5cm)

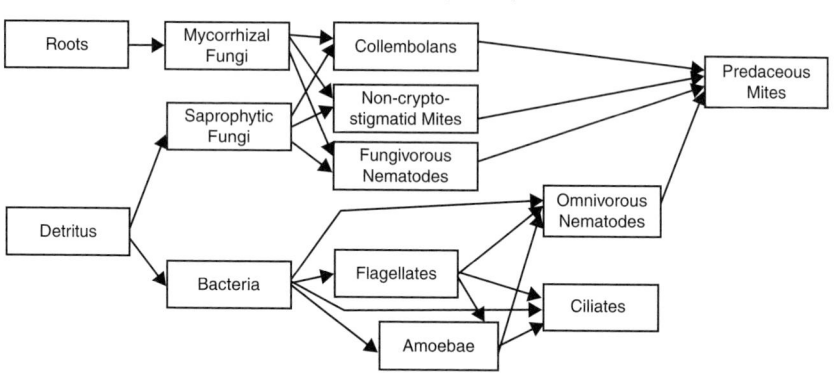

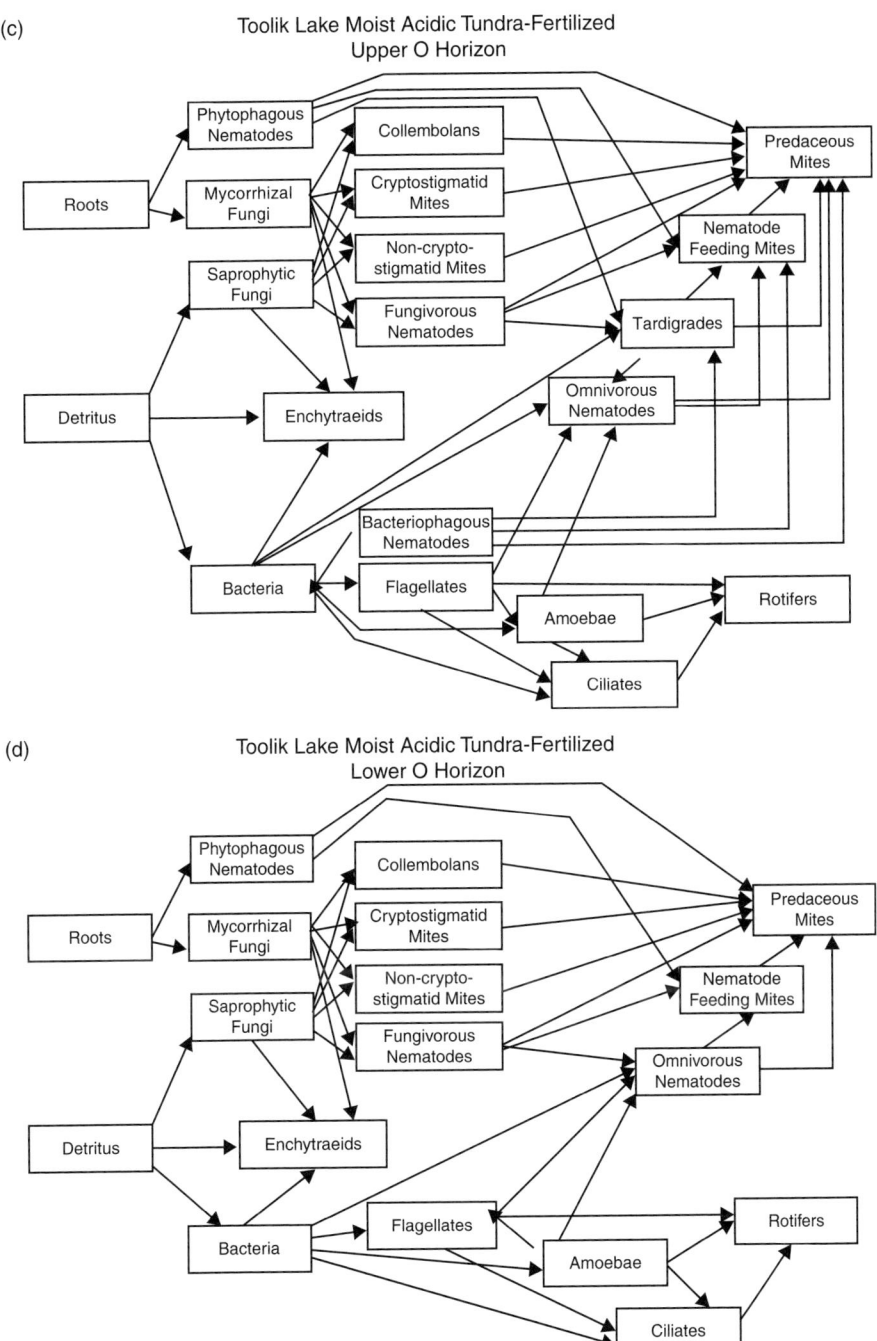

Figure 11.12 *Continues overleaf.*

(e)

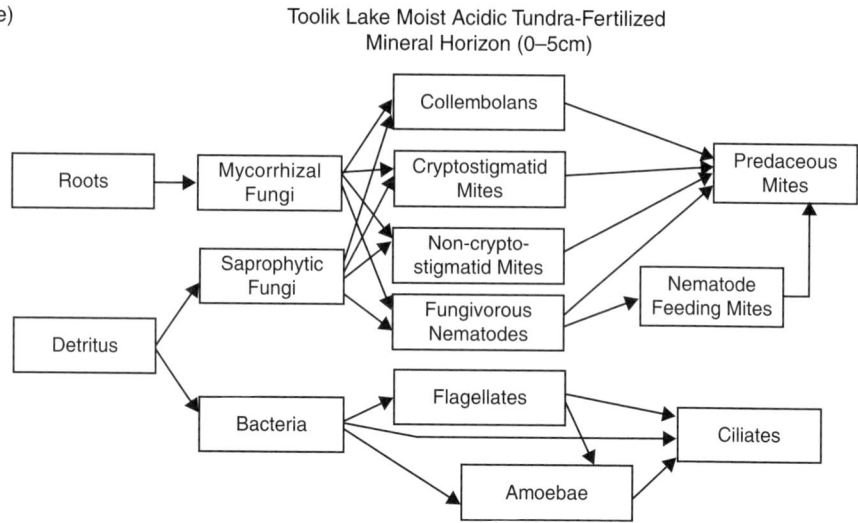

Figure II.12 Connectedness food web descriptions of soil communities by horizon from native and fertilized moist acidic tundra (MAT) at Toolik Lake, Alaska. (a) Upper (0–5 cm) and lower (5 cm–mineral horizon) organic (O) horizons – native tundra, (b) mineral horizon (0–5 cm) – native tundra, (c) upper (0–5 cm) organic horizon – fertilized tundra, (d) lower (5 cm–mineral horizon) organic horizon – fertilized tundra, (e) mineral horizon (0–5 cm) – fertilized tundra.

horizons, characterized by fresh plant materials and partially decomposed plant materials, respectively, and incursion into the mineral horizon, which was largely biologically dormant before these manipulations. After about 5 years of nutrient addition, the ecosystem enters a "transition" phase, during which the vegetation has shifted toward shrubs (with associated changes in litter quantity and quality), and NPP and N cycling have increased. The observed losses in soil C during the transition phase result from the metabolism of fresh organic C (largely the light/labile fraction) and existing soil organic material (light/labile and heavy/recalcitrant fractions) by the extant soil microorganisms and invertebrates, largely bacteria and their consumers. After approximately 15 years, the shift from sedges to shrubs is completed, and the lower quality and greater quantity of plant litter and roots take hold. During this "new stable state" phase the changes in litter and roots create a feedback that increasingly favors fungi and their consumers. The lower turnover rates, higher growth efficiencies, and more chemically complex structural morphologies and by-products of fungi and their consumers when compared with bacteria and their consumers translates into greater quantities of more stable SOM—but not as much SOM as had existed previously.

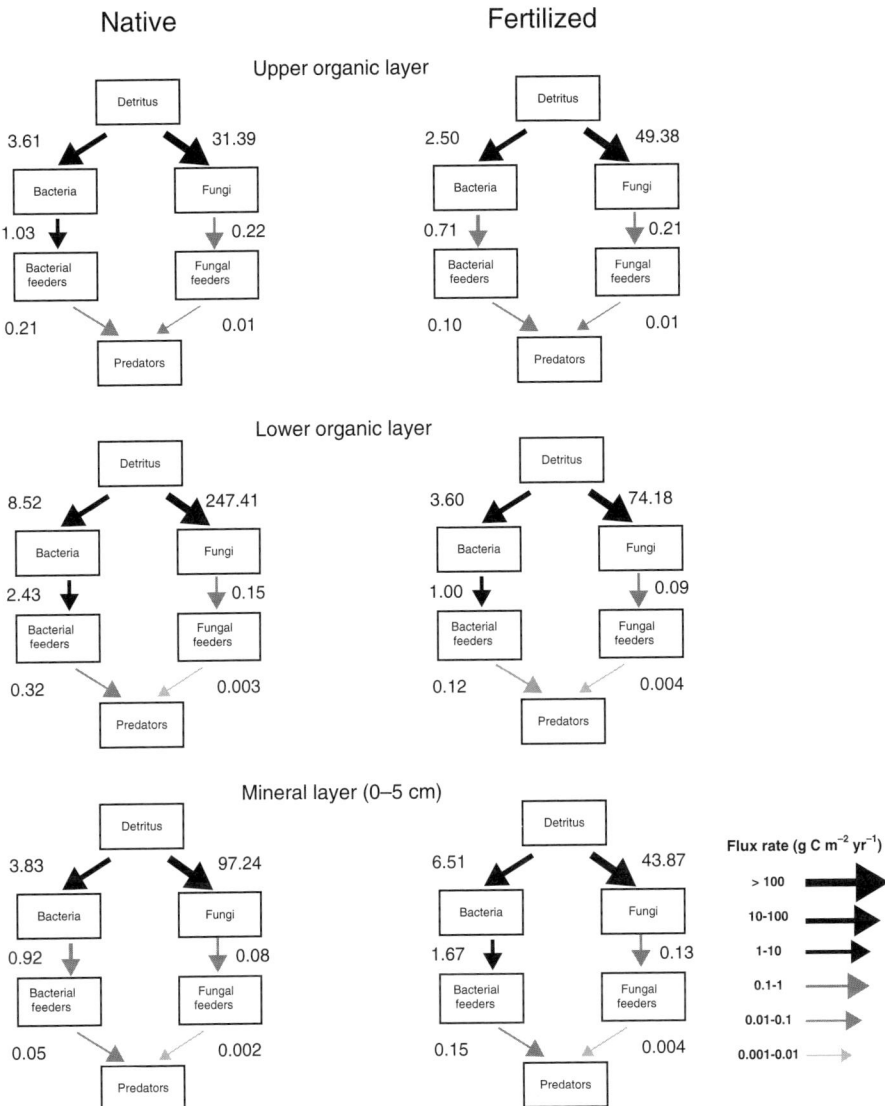

Figure II.13 Comparisons of abbreviated energy flux descriptions of soil food webs from native untreated control and fertilized (N and P) moist acidic tundra at Toolik Lake, Alaska. Arrows are scaled to the magnitudes of carbon fluxes (g C m^{-2} yr^{-1}) between the groups for each cm of depth.

Table II.5. Summary statistics for the connectedness soil food web descriptions from native and fertilized moist acidic tundra (MAT) at Toolik Lake. Values of number of functional groups (S), number of resources ($S_{Resources}$), connectance (C), link density (SC), and maximum (FCL_{Max}) and mean (FCL_{Mean}) food chain length.

	S	$S_{Resources}$	C	SC	FCL_{Max}	FCL_{Mean}
Native						
Upper O horizon	20	2	0.305	6.11	9	5.24
Lower O horizon	20	2	0.305	6.11	9	5.24
Mineral horizon (0–5 cm)	13	2	0.282	3.67	5	3.29
Fertilized						
Upper O horizon	20	2	0.253	5.05	7	4.97
Lower O horizon	18	2	0.255	4.59	6	3.48
Mineral horizon (0–5 cm)	14	2	0.253	3.54	4	3.14

Table II.6. Biomass estimates (g C m^{-2}) for the functional groups of the soil communities by soil horizons per cm depth in native and fertilized moist acidic tussock tundra (MAT), July 2008, Toolik Lake.

	Native MAT			Fertilized MAT		
Functional Groups	Upper O Horizon	Lower O Horizon	Mineral Layer 0–5cm	Upper O Horizon	Lower O Horizon	Mineral Layer 0–5cm
Microorganisms						
Bacteria	0.04725	0.10560	0.19529	0.03106	0.06956	0.23788
Fungi	7.66560	61.72812	24.23921	12.22101	18.47199	10.85987
Protozoa						
Amoebae	0.02706	0.11698	0.04352	0.03445	0.04837	0.07536
Flagellates	0.00110	0.00126	0.00109	0.00089	0.00088	0.00187
Ciliates	0.00352	0.01819	0.01715	0.00607	0.01274	0.03961
Rotifers	0.00030	0.00021	—	0.00004	0.00004	—
Nematodes						
Phytophagous	0.00008	0.00002	—	0.00001	0.00001	—
Bacteriovores	0.00006	0.00001	—	0.00001	—	—
Fungivores	0.00027	0.00013	0.00001	0.00004	0.00002	0.00001
Omnivorous	0.00155	0.00037	0.00010	0.00014	0.00011	—
Predators	0.00022	0.00016	—	—	0.00005	—
Enchytraeids	—	—	—	0.00023	0.00086	—
Tardigrades	0.00352	0.00550	—	0.00017	—	—
Arthropods						
Collembola	0.00186	0.00216	0.00037	0.00142	0.00045	0.00007
Cryptostigmatid mites	0.00143	0.00083	—	0.00056	0.00048	0.00004
Non-cryptostigmatid mites	0.00030	0.00012	0.00005	0.00012	0.00001	0.00001
Nematode-feeding mites	0.00083	0.00011	—	0.00048	0.00021	0.00012
Predaceous mites	0.00225	0.00061	0.00030	0.00100	0.00041	0.00023

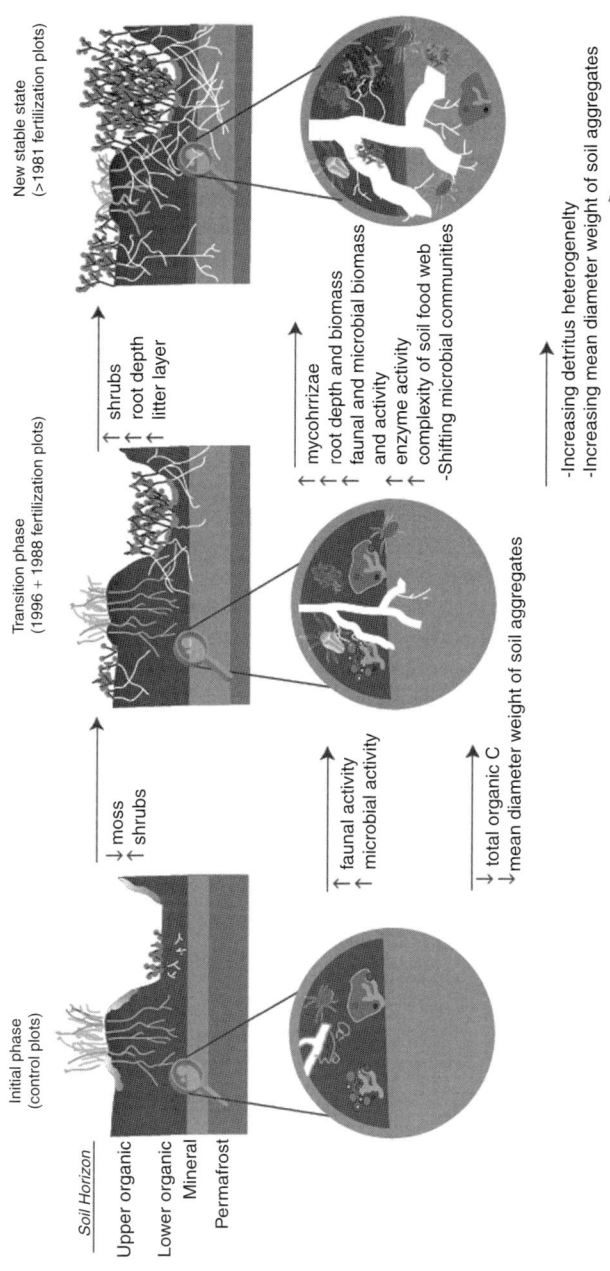

Figure II.14 Conceptual diagram with descriptions of the responses of the plant community, soil community, and soils in an Arctic tundra ecosystem in transition. The initial phase represents the untreated control plots, the transition phase is represented by the fertilized plots, while the new stable state is a hypothetical steady state (diagram by M. Wallenstein).

11.5 General framework

In this chapter we explored three variants of community development. In each case the underlying environmental conditions changed in ways that increased productivity, albeit by different mechanisms. With these changes the systems transformed by increasing the biomass of resident species, initially at lower trophic positions and then at higher trophic positions. What happened next depended upon the degree of complexity that the system had achieved or its stage of community development and the likelihood of additional species entering the system (Figure 11.15).

In the cases of Wind Cave and Schiermonnikoog, the food webs at the low end of the local productivity gradients reached a point at which sufficient biomass was available to support organisms at higher trophic positions. The transitions that occurred were initially gradual as biomass accumulated at the higher trophic positions, but then became abrupt or threshold-like with the addition of a new trophic level. These results could be explained on energetic grounds defined by $F_{[NPP]}$.

In the case of Schiermonnikoog, there is a strong case for dynamics stability as defined by $F_{[\lambda]}$. The observed movement of biomass from lower to higher trophic positions placed greater dependence on intraspecific density dependence or self-limitation to stabilize the system. In this case the diagonal terms within the Jacobian matrix increased. Concurrent with the addition of a species at an upper trophic position was an easing of conditions that if left unabated precede instability. The positions that were colonized were ones that were higher in the food chain and within the critical loop, i.e., the loop with the maximum loop weight. The introduction of organisms at these positions led to lower loop weights and more defined trophic pyramids. These results are consistent with the dynamic stability grounds defined by $F_{[\lambda]}$. Taken together, $F_{[SYS]} = F_{[NPP]} \cap F_{[\lambda]}$.

In the case of Toolik Lake, a complex and arguably climax community underwent a transition in state that was gradual at the initial phase, but over time was radical enough to resemble a fold bifurcation (Figure 11.15). As the plant species shifted from sedges

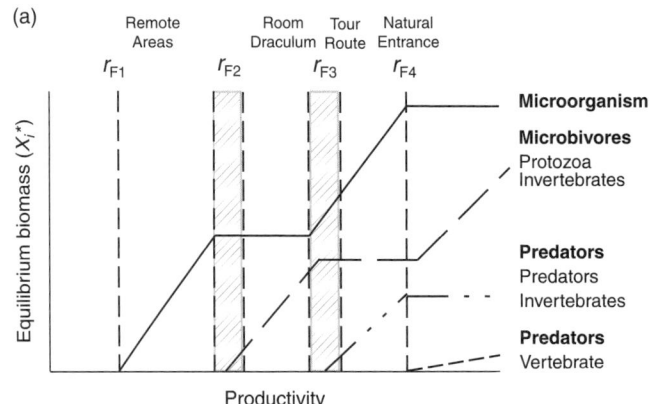

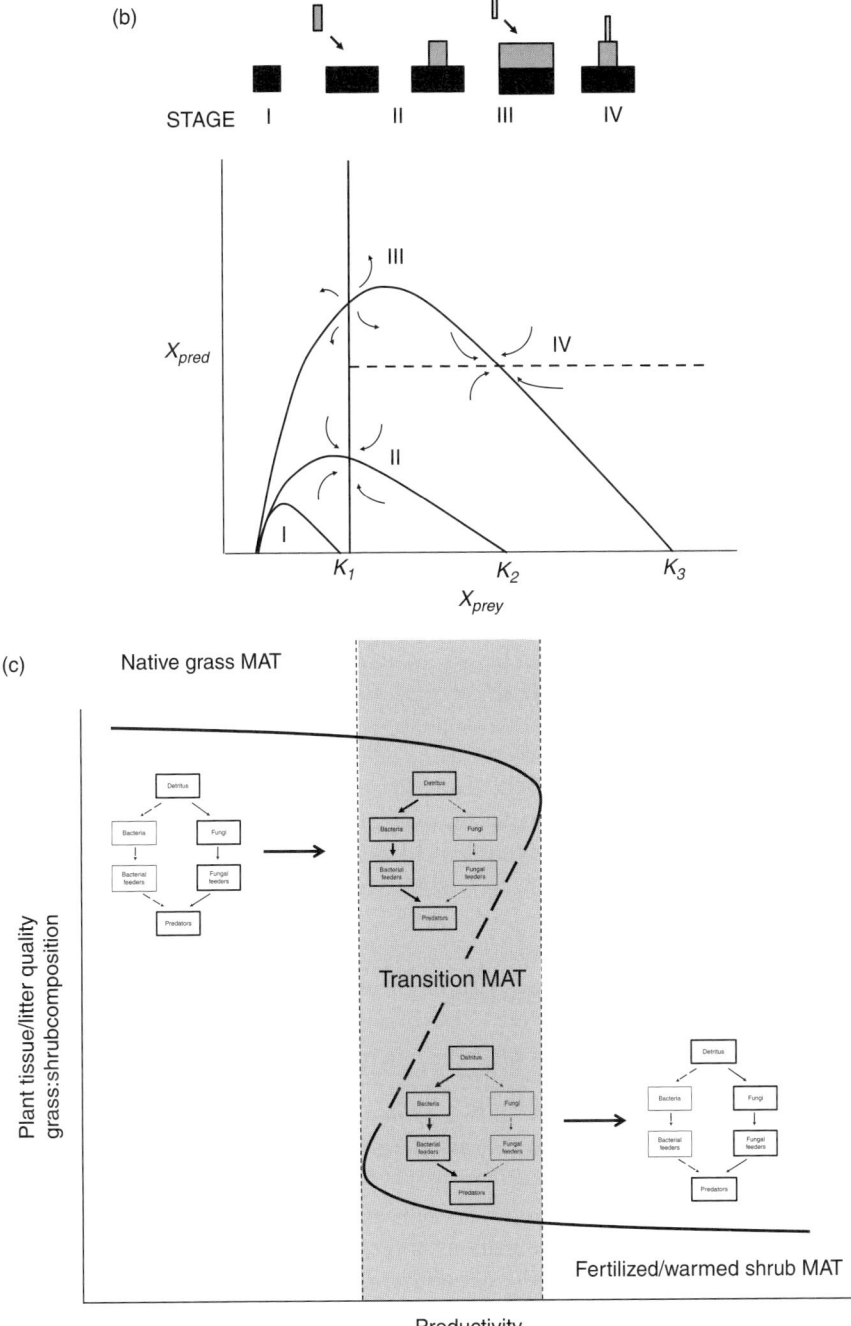

Figure 11.15 Summaries of the responses of the soil food webs to changes in productivity in the three case studies. (a) The influence of increased productivity on the trophic structure and distribution of biomass on the soil food webs of Wind Cave, (b) the influences of productivity on the diversity, complexity, biomass distribution and stability of the soil food webs at Schiermonnikoog, and (c) the observed transition of moist acidic tussock (MAT) tundra from a grass- to shrub-dominated system and restructuring of the soil food web represented as a fold-bifurcation.

to shrubs, the diversity and complexity of the belowground community declined, a new species entered the system, and flow of matter within the soil food web shifted between the bacterial and fungal pathways. The transition observed here did not involve the addition of a new trophic level, but rather a change in the dominance and quality of the basal plant resources and eventually the detritus they produce, and the dominance of the directional flow of matter through from one energy channel to another.

What does stability and instability mean in these cases? If we view systems from a static perspective, the onset of instability conjures uncertainty. Allen and Starr (1982) simplified the discussion by noting one of two fates: (1) a degeneration or collapse of the system, or (2) a regeneration or transition of the system to another state. The case studies presented here illuminate transitions of systems to another state that in many ways was recognizable from the earlier state in terms of underlying energetic properties if not species. The increase in productivity led to movement of biomass to higher trophic positions, additional trophic levels, or shifts in the directional flow of matter and energy within the food web affecting its complexity and dynamic state, and in a strict sense spelled instability. Coincident with these changes were changes in biogeochemical cycling rates and nutrient retention (sensu Odum, 1969). Despite this, stability in the looser sense (sensu Pimm, 1991) was maintained, by sustaining a balance between the energetically constrained number of trophic levels and productivity. Trophic pyramids were maintained, maximum loop weight was minimized, and resilience was maximized. Hence, the structural aspects of trophic pyramids embodied in the $F_{[NPP]}$ serves as a counterbalance to the dynamic constraints imposed by the increasing complexity of the food web embodied in $F_{[\lambda]}$. This interplay between productivity and food web stability governs and selects self-organizing pathways in community and ecosystem development, i.e., $F_{[SYS]} = F_{[NPP]} \cap F_{[\lambda]}$.

II.6 Summary and conclusions

O'Neill et al. (1986) argued that the formation of assemblages of species is a natural outcome of the evolution of complex systems. The parts of this argument have been around for some time, as many different forms of stable subsystems have been described (O'Neill et al., 1986). Wilson (1969) identified these assemblages of species as "assortative equilibria." Diamond (1975) referred to the assemblages as "permissible species combinations." In each case the assemblages were viewed as a small subset of the possible combinations of species that could occur given the larger number of species that they were drawn from. As an ecosystem develops, stable subunits should develop before further stages can develop (Simon, 1962). Although stability routinely has been used as a necessary condition of ecosystem structure and an end point in development, less is understood about the mechanisms behind community assembly that result in a stable ecosystem.

The process of community development goes far beyond the simple example of a predator with a narrow choice of prey being less likely to survive in a newly colonized habitat than one with a wider diet, or the species-centric niche-based examples common to the invasive species literature (Fridley, 2011). We propose that the energetic organization of the community plays an important role in governing the colonization and extinction process. As new species enter a community, the interactions in which they engage alter the pattern of material flow and the patterning of interaction strengths within the community. If the patterns of interaction strength were altered to an unstable configuration, one consequence would be the extinction of one or more species, and another would be the transition to an alternate state or configuration that retains many of the features of the original.

Given this, would a reticulate or compartmentalized community be more likely to develop? The answer clearly depends on how many basal resources can be established within the habitat. In the extreme case where the habitat possessed only a single basal resource, the diversity and structure of the system would be limited by the productivity of the basal resource and by dynamic constraints as successive trophic positions were added. If a second basal resource were to become established within the habitat, would the biota that subsequently colonize the habitat have a greater chance of utilizing the original resource or the new basal resource? With all things being equal, save the conditions allowing the new resource being able to establish itself, the arguments we presented in Part II would suggest that there are fewer restrictions to colonizing the new basal resource than the original resource. If this process were to continue, two things would occur: (1) the habitat would be comprised of many parallel, unconnected food chains, and (2) the species diversity of consumers within communities would be positively correlated with the number of basal resources within the community. The first condition occurs to a point, and as our case studies indicate, predators eventually enter the system and link the chains. The empirical evidence presented here and elsewhere (see Figure 8.7 in Chapter 8) supports the second condition, as the diversity of consumers is positively correlated to the number of basal resources.

A final thought focuses on the importance of the dynamics operating at different levels of organization and scale within food webs, e.g., species composition, population life-history parameters and abundances, and individual growth, size, and behavior. Dynamics acting at different levels of the biological hierarchy may affect food web structure and stability in different ways. Our field observations and theoretical models show that environmental heterogeneity creates stabilizing subsystems (compartments) of interacting species in food webs, especially at the lower trophic levels, representing dominant flows of energy that differ in transfer rates. Organisms at the higher trophic levels act as "integrators" linking the pathways in space and time, and stabilize dynamics of their resources via density-dependent adaptive foraging. These effects of heterogeneity and dynamics on stability explicitly relates to the idea that diversity may buffer against perturbations (MacArthur, 1955), and thereby override the constraints imposed by complexity on system stability (May, 1972).

12

Food web dynamics beyond asymptotic behavior

12.1 Introduction

The connectedness, energy flux, and functional descriptions of food webs are useful devices to study properties of food webs, but as we have noted in the foregoing chapters, they fail to capture the richness of change or degree of variation in structure that occur in space and time. In Chapters 10 and 11 we have discussed the dynamic character of food web structure and how it might change with seasons, colonization, and in response to disturbance. How can we reconcile the implications of this inherent variability with the concepts of steady state, equilibrium, and stability in the earlier chapters of this book? More to the point, what if the variation in space and time were an essential component of the stability or persistence of a system? Many ecologists have asked this question (de Ruiter et al., 2005, McCann et al., 2005, Hastings, 2004, 2010), and in doing so, have forced us to rethink the roles of the internal relationships that drive variation in food web structure and dynamics—as well as events external to the food web that also affect its structure and dynamics. Under a static architecture perspective, we look at the asymptotic behavior of the models. Stability is assessed near equilibrium in terms of how properties intrinsic to the interactions of the food web (endogenous factors) respond to external forces. Under a dynamic architecture perspective, we look at multiple facets of dynamic behavior. In this case stability is viewed in terms of how these external forces, in conjunction with the endogenous properties, are part of ecosystem functioning and dynamics in food web structure in ways that preserve species composition.

In this chapter we will explore how variability in food web structure in space and time can be critical to the stability, persistence, and function of ecosystems (Tilman, 1994, de Ruiter et al., 2005, Neutel et al., 2007). We will approach this in two ways. We will start with models based on static connectedness food web structures and discuss how these model results are sensitive to assumptions that the systems are in steady state and possess stable asymptotic solutions over space and time. We will apply this approach to the energy flux descriptions, which characterize energy flow and nutrient cycling from the connectedness food web descriptions, and to the functional descriptions, which assess dynamics and stability from the interaction strengths within the Jacobian matrix. Then we will discuss how the use of steady-state and asymptotic stability can be justified in the analysis of food web development along succession, as we described in Chapter 11. From here we will take the dynamics food web structure approach one step further by discussing the importance of transient behavior and spatial structure. We will then raise the possibility of persistence in an asymptotically ambiguous state, which may or may not possess a stable asymptotic solution but persists nonetheless.

12.2 Variability, equilibrium states, and asymptotic stability

Many of our analyses so far were based on the idea that the food webs have equilibrium states when stable, and that the webs would return to this state after a disturbance. In Part III of the book we have relaxed our insistence on this return-to-equilibrium behavior, as the mathematical derivation of the models allows only for predictions regarding relatively small deviations from the equilibrium state (i.e. local stability). Here, we go further and study how spatial and temporal variation in connectedness food web structure affects the energy flux, nutrient dynamics and stability. The food web models we have dealt with so far are neglecting variability at many levels of biological organization and at various temporal and spatial scales. Within functional groups we have lumped (sometimes many) species; these species differ from one another on all kinds of physiological and behavioral characteristics. On top of that, we model these characteristics as fixed properties. For example, we assume invariable death rate and energy conversion efficiencies, although we know that such properties may depend strongly on environmental factors. We neglected life-history-related variability of organisms, such as body size and related properties such as diet. Addressing life history and body size in our models may provide a different picture of the food webs. Along with age, body size and diet may change, and hence food web structure may change over time and may also become dependent on seasons. In fact, we disregarded this kind of variability by averaging our data, coming up with "averaged" trophic groups in terms of biomass and physiology.

Having said that, the question arises of how such variation may impact our conclusions reached on the basis of our models that assume steady-state and asymptotic stability. We will discuss this for our models that calculate energy flow and nutrient flows on the basis of observed population sizes, and for the models that analyze patterns of interaction strength and stability.

12.2.1 Variability, steady state, and the calculation of energy flow and nutrient cycling

The biomass data we used to model energy flow and nutrient mineralization were the outcome of extensive field sampling. For example, the data from the Lovinkhoeve Experimental Farm (Chapter 4) were annual average population sizes that we calculated on the basis of 3-week (microorganisms) and 6-week (fauna) samplings over a full 1-year period, and the data per sampling were averaged over large numbers of soil samples. In other words, we averaged over space and time, removing both spatial and temporal variability as an approximation of the asymptotic behavior of the system. The data collected for individual functional groups and process variables showed strong variations in space and time (Figure 12.1; Moore and de Ruiter, 1991, Zwart et al., 1994). This variability included variation with a random character, i.e. differences between individual samples over space and time, variation showing trends in space, such as along soil depth or distance from a root, and in time, such as along a season. By taking averages we summarized all this variation in single figures representing averages over a year and averages over surfaces and depths. By taking the annual averages we were allowed to assume steady state, as the data also showed that variation in annual biomass over the years was much smaller than variation in biomass between individual samplings within a year. When taking the steady-state approach, we could calculate feeding rates and mineralization rates by using the equation proposed by O'Neill (1969):

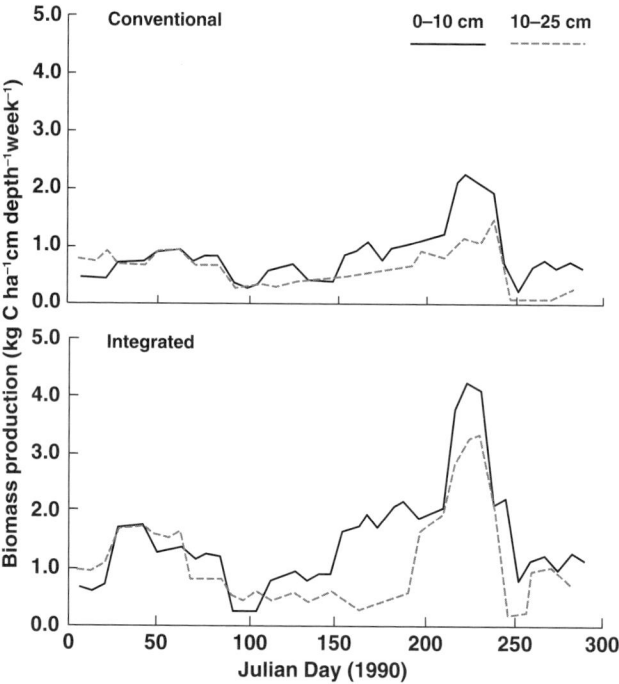

Figure 12.1 Temporal and spatial variations in the production rate of the total biomass of soil biota in two soil layers of the conventional (CONV) and integrated (INT) plot of the Lovinkhoeve Experimental Farm, Marknesse, The Netherlands. Redrawn from Zwart et al. (1994), with permission from Elsevier.

$$F_j = \frac{d_j B_j + M_j}{a_j p_j} \quad (12.1)$$

where F_j is the feeding rate (e.g., in terms of kg C ha^{-1} yr^{-1}), B_j is the biomass of the consumer (kg C ha^{-1}), M_j is the mortality of the consumer caused by predation (kg C ha^{-1} yr^{-1}), d_j is the specific death rate of the consumer (yr^{-1}), a_j is the assimilation efficiency (fraction), and p_j is the production efficiency ($B_j.B_i^{-1}$). Hence the data for the biomass of the consumer B_j reflect the steady-state biomass on the time scale of years.

The interesting thing here is that O'Neill (1969) also showed that this steady-state equation can be slightly modified in such a way that it captures the observed variation in population sizes, enabling the calculation of the dynamics in energy flow and nutrient cycling:

$$F_j = \frac{d_j B_j + M_j + \Delta B_j}{a_j p_j} \quad (12.2)$$

where ΔB_j refers to the change in biomass of the consumer over a particular time interval, e.g., the period between two consecutive sampling dates. In fact, this equation calculates the biomass production necessary to compensate for death and either increases this with production necessary for an observed increase in biomass, or decreases it to account for an observed decrease in biomass.

An example of the application of Equation 12.2 on the soil food webs from the Lovinkhoeve Experimental Farm is given in Figure 12.2 (de Ruiter et al., 1994). With Equation 12.2, feeding rates can be calculated for all periods between consecutive sampling dates using the same-steady state and thermodynamics-based assumptions. From these calculated feeding rates, the model can provide estimates of the dynamics in N-mineralization as well. The example shows that the model outcome also captures the observed dynamics in N-mineralization rather well, as periods of increase and decrease between observed and modeled mineralization rates are strongly correlated. Starting from steady-state assumption does not necessarily hamper the modeling of dynamics and change.

12.2.2 Variability, trophic interaction strengths, and asymptotic stability

The steady-state assumption was not only used to calculate energy flow and mineralization rates, but this assumption was also the basis for calculating interaction strengths and food web stability (see Chapter 6). The use of the steady-state assumption, and hence the denial of variation in these calculations, may have interfered with our calculations and model results in two ways.

The first is a point of methodological, and possible model, artifact. One might argue that by using a steady-state assumption for calculating interaction strengths (see Chapter 6), the high probability of stable Jacobian matrices that we observed arose, at least in part, from the steady-state assumption. If that is the case, the stability of the Jacobian matrices does not reflect the stability of the food webs, but

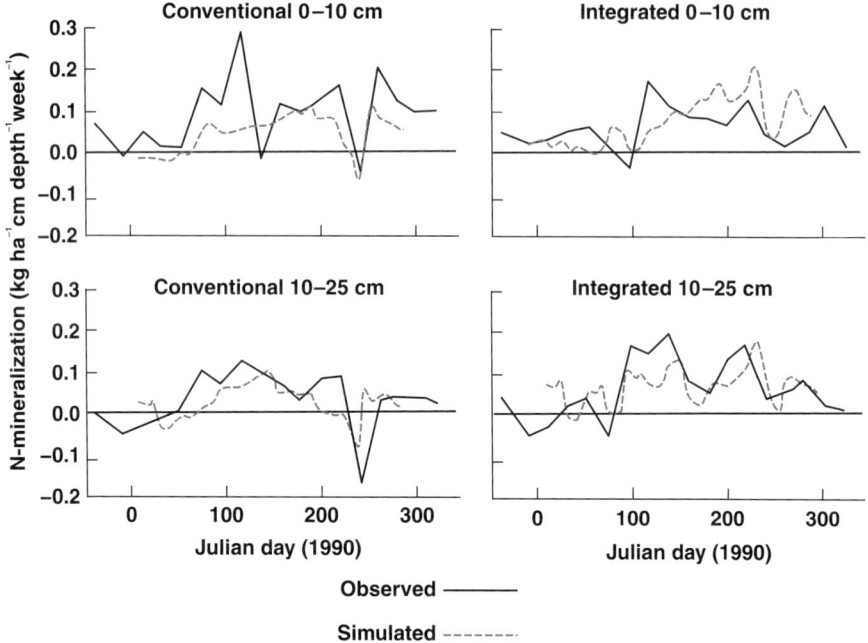

Figure 12.2 Observed and simulated nitrogen mineralization rates under integrated and conventional farming in the depth layers 0–10 cm and 10–25 cm at the Lovinkhoeve, Experimental Farm, Markenesse, The Netherlands. The vertical bar denotes the standard error of differences of means (SED). (Redrawn from de Ruiter et al., 1994.)

merely the mathematical way we constructed the matrices. As we have seen in Chapter 6, this is probably not the case. As a control analyses, we constructed matrices with randomized input parameters, but used the same steady-state assumption to calculate the interaction strengths and stability. This results in matrices that are likely to be unstable. Hence, the high probability of stability of the matrices using the observed data can be ascribed to these data, rather than to the steady-state assumption underlying the calculations.

The second point is how we justify the steady-state assumption when we analyze the stability of changing food web structures as we did when looking at food web development along the primary succession gradient at the Island of Schiermonnikoog in Chapter 11. In this analysis, the calculation of the feeding rates, interaction strengths, stability, and loop weights is based on the steady-state assumption, whereas the data describe the development of food web structure in terms of increases in the number and biomasses of the trophic groups. Hence, the data show change rather than steady state. It would have been possible to use Equation 12.2 as the starting point of calculating interaction strength, but in fact we used Equation 12.1. The reason for this is that we found that the increase in biomass over succession stages (ΔB_j) was negligible compared with natural death ($d_j B_j$) and death due to predation (M_j). For example, most trophic groups have a specific death rate (d_j) of more than 1 per year, whereas biomass increase along the succession gradient is mostly less than 0.1 per year. So the stability analysis of the food web along the succession gradient has been preceded by defining equilibrium states for each succession stage. Then we analyzed the stability of these equilibrium states by means of the Jacobian eigenvalues and the weights of the trophic loops. If we project the analysis of the consecutive equilibria on the gradient as a whole, the analysis shows "moving" equilibria along stable pathways of development. Hence, again we could make use of steady-state assumption and the asymptotical stability approach to analyze change and dynamics in food web structure.

12.3 Transient dynamics

In the previous sections, we analyzed the stability of food web models in terms of whether a feasible equilibrium exists and whether the system will return to this equilibrium after a disturbance. Asymptotic stability analysis of this type focuses on the long-term dynamic behavior of a model food web near its equilibrium or steady state. Focusing on the asymptotic stability in this way may have also constrained our study of food web structure and dynamics and thus been misleading, as they tend to ignore critical aspects of transient dynamics following a disturbance. All things being equal (i.e., no changes in parameters or fundamentals of the external drivers), we know that natural and anthropogenic perturbations may cause food webs to deviate from their asymptotic equilibrium state. In response to the perturbation, systems may exhibit short-term transient dynamic behavior that differs from its asymptotic behavior. When transient dynamics occur on ecologically relevant time scales, they can be as important or more important to understanding the dynamics and persistence of ecosystems than asymptotic behavior (Neubert and Caswell, 1997, Chen and Cohen, 2001, Hastings, 2001, Verdy and Caswell, 2008, Hastings, 2010).

Transient dynamics are likely to occur in a broad variety of ecological systems. Hasting (2004) emphasized the importance of processes and interactions operating on different spatial and temporal scales and stochasticity to understanding transient dynamics. There are distinct classes of ecological systems that lead to transient dynamics, many of which we have discussed throughout the book. These include systems that possess chaotic saddles, spatial heterogeneity, varying timescales, coupled oscillators, or high degrees of stochasticity. Spatial and temporal heterogeneity and the degree of asynchrony in processes and interactions are emphasized. In natural ecosystems, the species dynamics in the different compartments can be held out of phase by means of natural seasonal processes, while in agricultural systems many kinds of management practices can act as local disturbances, destroying and creating habitats.

We have discussed transient dynamics at length throughout the book, although not by name. We alluded to the importance of transient dynamics of asymptotically stable systems in terms of resilience and the nature of the transient dynamics. Our analyses of transient behavior to this point have focused on the resilience of the system or the time a system took to return to equilibrium and the frequency of oscillations that occur during the return to equilibrium using information gleaned from the real and imaginary parts of the dominant eigenvalue, respectively. Following a disturbance in a mathematically stable system, oscillations in the population densities can bring the population close to zero on the return to equilibrium, thereby increasing its likelihood of extinction. If return times are long relative to the frequency of disturbances, the system may be kept far from equilibrium (DeAngelis and Waterhouse, 1987). In this case the perturbations may build upon one another and amplify, leading to transitions to alternative states, or, worse yet, extinctions (Hastings, 2004, Carpenter et al., 2008, Carpenter and Brock, 2010, Hastings, 2010, Hastings and Wysham, 2010).

A framework to study transient dynamics is beginning to emerge that captures the aforementioned tendencies of perturbations and their consequences on systems. The framework moves beyond resilience and the tendency to oscillate to include the magnitudes, shape, and trajectories of perturbations. Additional metrics to characterize transient behavior include reactivity, maximum amplification, and the time at which amplification occurs (Neubert and Caswell, 1997, Chen and Cohen, 2001), as summarized in Box 12.1.

Box 12.1 Transient dynamics

Transient dynamics for our discussion represent short-term dynamic behavior that energetically feasible asymptotically stable systems exhibit in response to disturbance. A system of n interacting species is energetically feasible if all n species possess positive equilibria, and is locally asymptotically stable if the real parts of the eigenvalues of the Jacobian matrix, A, of the system are negative. As illustrated in Figure B12.1A, students of transient dynamics have defined the dimensions of the transient behavior in terms of the amplification and decay of a perturbation, which includes the magnitude and extent of the deviation and its temporal components (Neubert and Caswell, 1997, Chen and Cohen, 2001, Verdy and Caswell, 2008). The following metrics of transient dynamic behavior have emerged for feasible asymptotically stable systems like the ones used throughout the book:

Amplification envelope ($\rho(t)$): The maximum possible growth of a deviation from equilibrium following a disturbance at time $t \geq 0$. Recall from Chapter 2 that dynamics of the deviation is governed by $dx/dt = Ax$ where x represents a vector of the deviations, x, of individual populations from equilibrium and A represents the Jacobian matrix. The amplification envelope, $\rho(t)$, is calculated as follows:

$$\rho(t) = |||e^{At}||| \qquad (12.3)$$

where $|||e^{At}|||$ is the Euclidean norm of the Jacobian matrix, A.

Reactivity: The maximal instantaneous rate that a deviation, x, from equilibrium can be amplified. Reactivity is calculated as follows:

$$\text{reactivity} = \max\{\lambda_i((A + [A]^T)/2); i = 1, ..., n\} \qquad (12.4)$$

where λ_i are the real eigenvalues of matrix A. An equilibrium is reactive if and only if reactivity as calculated from Equation 12.4 is positive. A feasible system is reactive if and only if it is locally asymptotically stable and its positive equilibrium is reactive.

Maximum amplification (ρ_{max}): The largest growth of the deviation from equilibrium following a disturbance.

$$\rho_{max} = \max_{t \geq 0} \rho(t) \qquad (12.5)$$

Time at which maximum amplification occurs (T_{max}):

$$T_{max} = min\{t \in (0, \infty) \mid \rho(t) = \rho_{max}\} \qquad (12.6)$$

Resilience: the rate that a feasible locally asymptotically stable system returns to equilibrium following a disturbance, as measured by the absolute value of the largest (most negative) eigenvalue of the Jacobian matrix, A.

Return time: the duration of the transient behavior measured as the inverse of resilience. The time required for a disturbance, i.e., deviation from equilibrium, to decay to within e^{-1} of the equilibrium.

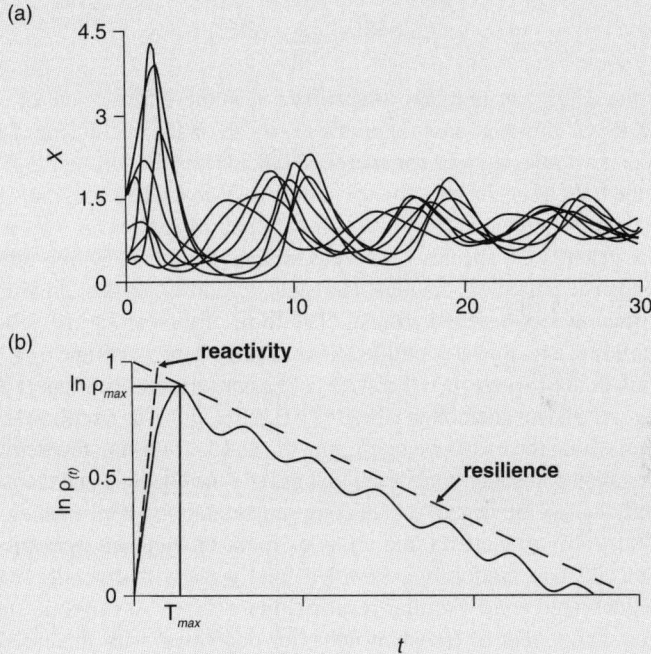

Figure B12.1A (a) The transient dynamics of a system (n = 8 species) that has been disturbed and that eventually returns to equilibrium. (b) A summary of the components of the transient behavior of a reactive system in terms of the amplification envelope $\rho(t)$ through time. The deviations from equilibrium amplify to a maximum ρ_{max} at time T_{max} with a rate defined by its reactivity during the short-term transient phase and eventually decays to the long-term asymptotic phase at a time defined by its return time at a rate defined by its resilience (adapted from Chen and Cohen, 2001).

Transient dynamics are sensitive to model structure and parameter selection (Chen and Cohen, 2001, Moore et al., 2004, Verdy and Caswell, 2008). Our own assessment of simple food chains using Lotka–Volterra models demonstrated clear differences between the transient behavior of detritus-based and primary-producer based models that varied in length, and in the energetic efficiencies of the consumers. The concept of system feasibility (F_{SYS}) included a component of dynamics (F_λ) that was based in large part on the nature of the transient dynamics using the general arguments presented above in relation to the populations being modeled. Studies that have used the framework provided in Box 12.1 take these analyses a step further.

Chen and Cohen (2001) provided a comprehensive analysis of the influence of different types of trophic interactions on the asymptotic stability and transient dynamics food webs of different dimensions and complexity using the Lotka–Volterra cascade model (LVCM) developed by Cohen et al. (1990). The model for the i species has the familiar form:

$$\frac{dX_i}{dt} = r_i X_i + \sum_{j=1}^{n} c_{ij} X_i X_j \qquad (12.7)$$

where X_i is the density or biomass of species i, r_i is the intrinsic rate of increase or decrease ($-d_i$ in our formulations) for species i, and c_{ij} is the coefficient of interaction between species i and j, adjusted for energetic efficiencies for consumers. The trophic structure of the LVCM is governed by the simple rules that all species i have a positive equilibrium (i.e., the system is F_{NPP}) and for each pair of species $i, j = 1, \ldots, n$, where $i < j$, species i cannot eat species j, whereas species j can eat i with a probability of c/n ($0 < c < n$). The paired off-diagonal elements, α_{ij} and α_{ji}, of the community matrix were established in a two-staged manner. The form of the paired interactions (links) were set as follows: a recipient-controlled link (r link) where $\alpha_{ij} < 0$ and $\alpha_{ji} = 0$, a donor-controlled link (s link) where $\alpha_{ij} = 0$ and $\alpha_{ji} > 0$, a consumer-victim link (t link) where $\alpha_{ij} < 0$ and $\alpha_{ji} > 0$, and no interaction where $\alpha_{ij} = 0$ and $\alpha_{ji} = 0$. The likelihood of each link type occurring within the matrix being r/n, s/n, t/n, and 1−(r+s+t)/n, respectively, where r, s, and t are nonnegative constants such that r+s+t $\leq$ n. The diagonal elements, α_{ii}, of the community matrix are negative, reflecting self-limitation for all species.

Asymptotic stability analyses and an assessment of transient dynamics with the metrics in Box 12.1 of randomly assembled food webs with diversity (n) using the above rules generated some familiar results (Figure 12.3). Asymptotic stability and the resilience component of transient behavior decreased with increased diversity and connectance. A more thorough analysis of transient behavior exposed how the response of the transient dynamics to increased diversity and connectance depended on the type of links within the food web. For example, food webs dominated by consumer-victim links were more resilient and less reactive, and possessed lower mean maximum amplification and shorter times to maximum amplification, than webs with more donor-controlled links and recipient-controlled links. Likewise, as connectance increased, the responses of each of the transient dynamic metrics was dependent on the types of links that dominated the food webs. These results echo our earlier comparisons of feasibility and dynamic behavior of primary-producer-based and detritus-based food chains.

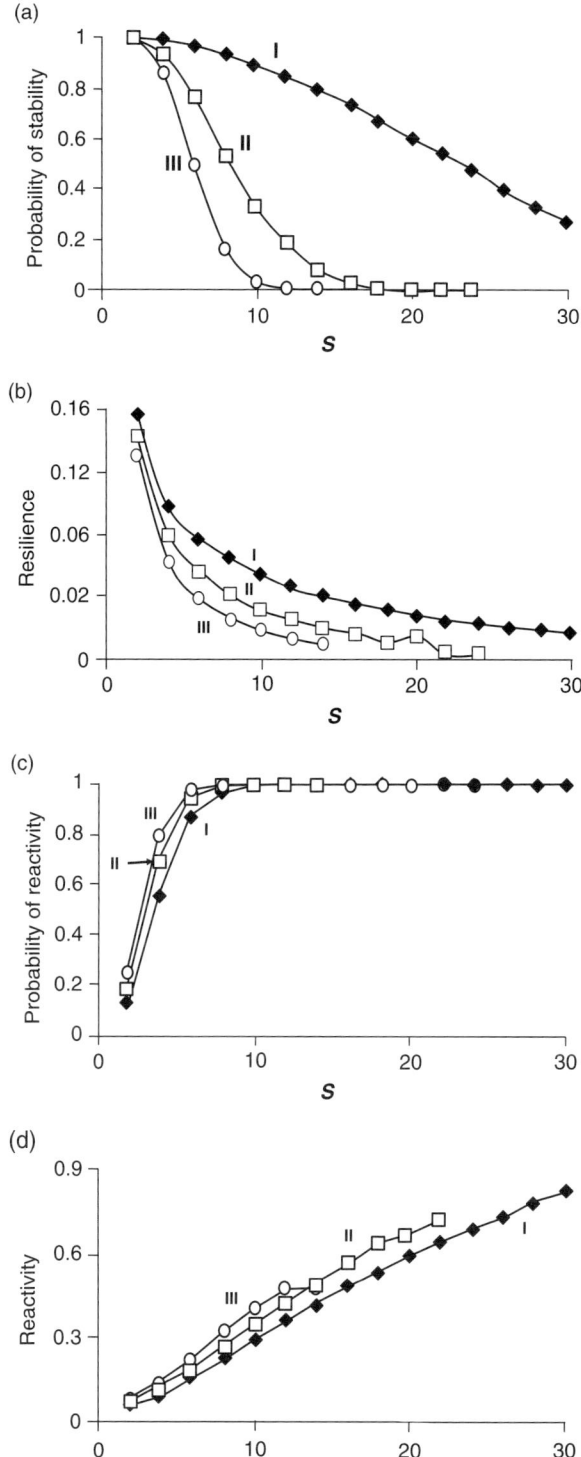

Figure 12.3 *Continues overleaf.*

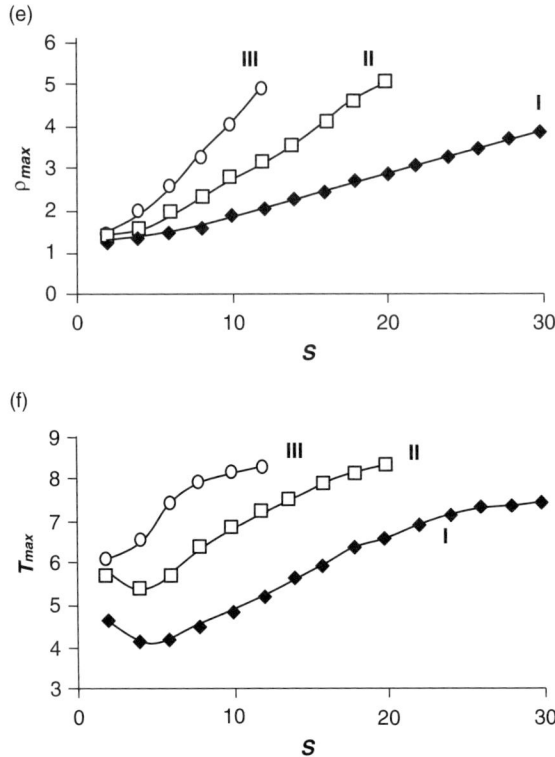

Figure 12.3 Asymptotic stability (a) and attributes of transient dynamics (b,c – see Box 12.1 for definitions) as a function of the number of species (S) for three variations of the Lotka–Volterra Cascade Model presented by Chen and Cohen (2001). I: $t/n = 0.4$ and $r/n = s/n = 0$; II, $t/n = r/n = s/n = 0.2$; III, $t/n = 0$, $r/n = s/n = 0.4$. See text and Chen and Cohen (2001) for details on the model. The number of species (S) should not be confused with the model parameters.

Verdy and Caswell (2008) demonstrated the importance of enrichment, predation rates, the transfer efficiency between trophic levels, and the distribution of biomass among trophic levels on transient dynamics using two stable parameterizations (Table 12.1) of the Rosenzweig–MacArthur predator–prey model (Rosenzweig and MacArthur, 1963). The model takes the following form:

$$\frac{dP}{dt} = rP\left(1 - \frac{P}{K}\right) - g\frac{P}{P + P_h}Z \qquad (12.8)$$

$$\frac{dZ}{dt} = eg\frac{P}{P + P_h}Z - dZ \qquad (12.9)$$

where P is the biomass and K is the carrying capacity of the prey, Z is the biomass of the predator, r is the specific rate of increase of the prey, e is the energetic efficiency

Table 12.1. Parameters used by Verdy and Casswell (2008) for the Rosenzweig MacArthur predator–prey models (Rosenzweig and MacArthur, 1963), presented in equations 12.8 and 12.9.

	Parameters	Parameter Values		Units
		Equilibrium A	Equilibrium B	
e	Energetic efficiency	0.15	0.8	—
g	Predation rate	2.3	0.8	day^{-1}
d	Death rate	0.1	0.1	day^{-1}
P_h	Half-saturation prey density	1.0	1.0	µmol m^{-3}
r	Prey-specific growth rate	1.0	1.0	day^{-1}
K	Carrying capacity	1.25	1.25	µmol m^{-3}
P^*	Prey equilibrium biomass	0.41	0.18	µmol m^{-3}
Z^*	Predator equilibrium biomass	0.41	1.26	µmol m^{-3}

of the predator, and d is the predator's specific death rate. The predator exhibits a Holling type II functional response wherein g is the maximum predation rate, and P_h is the half-saturation density of the prey.

Their study revealed a suite of energy-based adaptive strategies that form predator-driven and prey-driven mechanisms, which regulate transient behavior that hinges on the distribution of biomass at the different trophic levels, and the efficiency of energy flow between the trophic levels. Reactivity increased when the maximum predation rate and specific death rate of the predator, or carrying capacity of the prey, increased or when the energetic efficiency of the predator, half-saturation prey density, and the intrinsic rate of increase of the prey decreased.

To further investigate the mechanisms behind these phenomena, Verdy and Caswell (2008) reformulated the model (Equations 12.7 and 12.8) in terms of the dimensionless parameters e (energetic efficiency), g/r (scaled predation rate), d/r (scaled mortality rate), and K/P_h (prey enrichment ratio). They studied reactivity in terms of energetic efficiencies, e, and the scaled predation rate, g/r, in the energetically feasible (F_{NPP}) near equilibrium for the two stable parameterizations presented in Table 12.1.

Two regions of high reactivity emerge (Figure 12.4): one occupied by the stable equilibrium for parameterization A (Region A), where the trophic structure is a pyramid of biomass in which the predator possesses a low predation rate and high energetic efficiency; and the other is occupied by the stable equilibrium for parameterization B (Region B), where the structure is an inverted pyramid of biomass in which the predator has a high predation rate and low energetic efficiency. Increases in enrichment and the predation rate tend to increase reactivity in Region A and B, while the effects of energetic efficiency and specific death rate of the predator on reactivity within the regions were context dependent (Figure 12.5). The transient dynamics in Region A are predator-driven by the low predation rates of the predator, so small deviations from an already low predator population size lead to rapid increases in prey (reactivity and amplification), until which time the predators increase and the

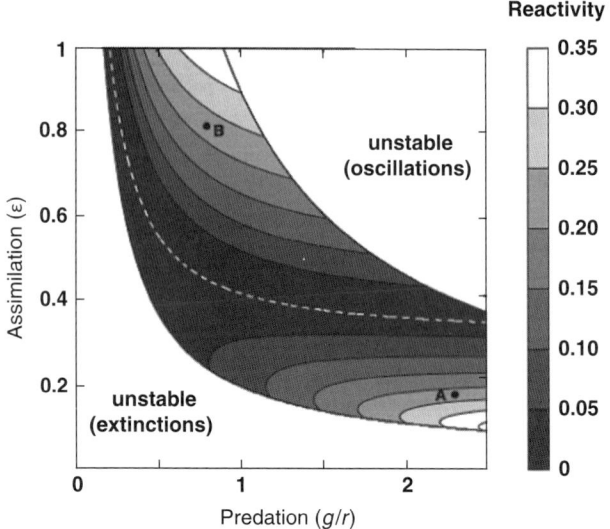

Figure I2.4 Reactivity, v_0 of the Rosenzweig–MacArthur (1963) predator–prey model presented in Equations 12.4 and 12.5 in the parameter space for assimilation efficiency (ϵ) and predation (g/r). The shades of gray give the value of v_0 in regions where the coexistence equilibrium for the predator and prey is stable. The white dashed line indicates a ratio of $\bar{Z}/\bar{P} = 3$ (above the curve, ratio is higher). The reactivity values for the stable equilibria A and B (see Table 12.2) are identical, but as explained in the text the mechanisms that are responsible for transient growth are different. (Redrawn from Verdy and Caswell, 2008.)

perturbation decays. The transient dynamics in Region B are prey-driven, so increases in the biomass of the prey are readily exploited by the predators.

12.4 Spatial systems

The spatial structure of a food web links the spatial distributions and movements of energy, resources, and organisms that operate across different spatial scales (Loreau et al., 2003). Holt (2002) identified the following four ways in which spatial structure can influence food web structure and dynamics: (1) The spatial context of community assembly, (2) spatial strategies and utilization by individual species, (3) dynamical implications of space and spatial utilization, and (4) spatial fluxes of energy, nutrients, and species. Most food web descriptions and models do not explicitly account for the variation in spatial structure, but rather provide a spatial average of the structure within a habitat. In doing so, the food web architecture may appear static and the dynamics within the architecture are described in terms of asymptotic behavior. However, species share membership in both a food web and a

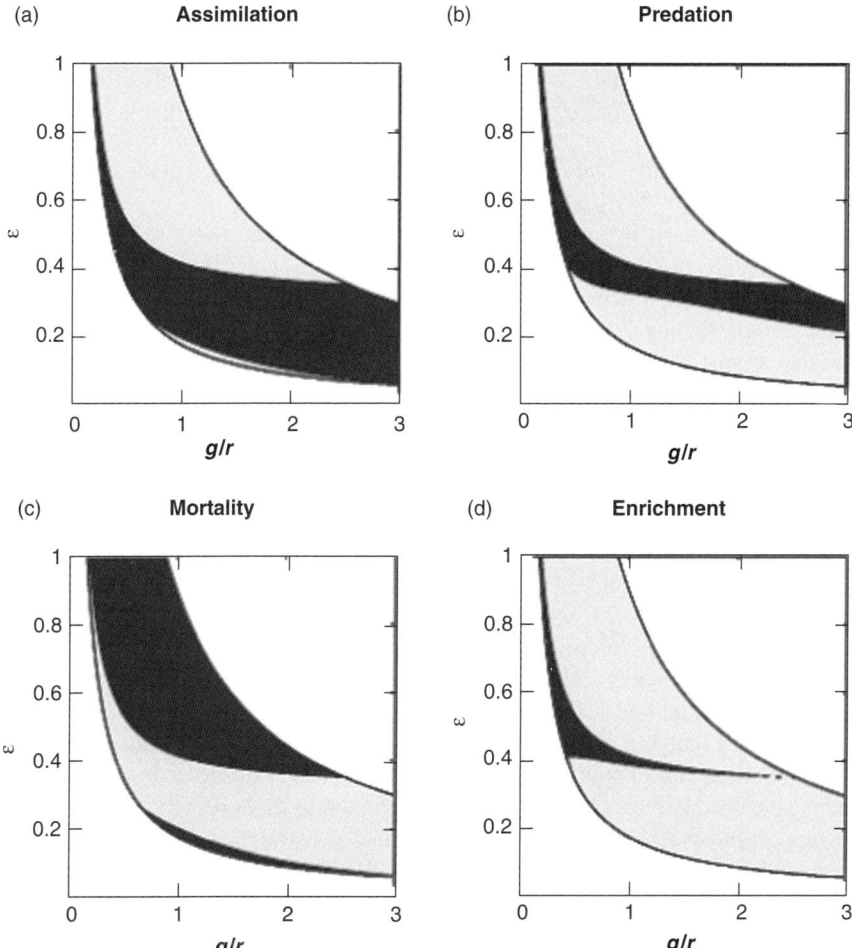

Figure 12.5 Sensitivity of reactivity of the Rosenzweig–MacArthur predator–prey model (Equations 12.4 and 12.5) to energetic efficiency (a), predation (b), mortality (c), and enrichment (d) rate in the region of parameter space that is asymptotically stable. The dark shaded regions indicate negative sensitivity, and the gray shaded regions indicate positive sensitivity. (Redrawn from Verdy and Caswell, 2008.)

spatial network. Consideration of the spatial structure adds an additional dimension to the study of food webs, and a dynamic component to the architecture.

This is not to say that one or more of the elements of spatial structure listed above have not been considered. Our detritus-based models included allochthonous sources of detritus that entered from outside the system; our discussions of compartments included a spatial component in the form of linked habitats and subhabitats. In addition, our discussion of food chain length, diversity, and system feasibility hinged on the diversity of resources and the movement patterns of consumers and higher predators, and our discussions on community development invoked concepts from island biogeography wherein the stage of the energetic organization and structure of the community affected further community development. In short, many have convincingly argued that the underlying spatial structure of a community can affect its structure, stability, and resilience (Cohen and Newman, 1991, Huxel and McCann, 1998, Holyoak, 2000, McCann et al., 2005).

We will focus on two elements of spatial structure that affect stability: (1) the distribution and movement of resources and consumers, and (2) the difference in the spatial extent of resources and consumers. We start with a general result that links our previous findings to spatial structure and then move to analyses that look at variants of spatial structure. We then deconstruct the results and propose strategies and mechanisms.

12.4.1 Spatial networks, structure, and resilience

Simple, spatially explicit, individually based simulation models of two- and three-species food chains suggest that the spatial structure of a food web across a landscape can affect its feasibility and resilience (Achter and Webb, 2006a, 2006b, Webb, 2007). Intermediate levels of spatial modularity (or aggregation) of the basal resource may enhance the feasibility and resilience of populations in familiar ways and for much the same reasons (Figure 12.6). The relationships presented in Figure 12.6(a–c) summarize our earlier discussions about simple food chains that did not explicitly include spatial structure, but rather focused on how increasing productivity affected both the likelihood of there being sufficient energy to support the system ($F_{[NPP]}$) and the dynamic constraints it imposed ($F_{[\lambda]}$). The relationships presented in Figure 12.6(d–f) include systems with explicit spatial structure. Given that productivity is defined as a rate of input per unit area, decreasing spatial modularity (or the degree of aggregation) over a landscape increases the average rate of productivity, with a few caveats. Decreasing modularity could also affect the degree to which the habitat was compartmentalized, a feature known to affect the dynamic stability of the system.

Figure 12.6 illustrates, with two different independent approaches, that the optimum in system feasibility occurs at intermediate levels of productivity (Moore et al., 1993) and intermediate levels of the spatial modularity of productivity. One conclusion from the results in Figure 12.6 is that system-level feasibility ($F_{[SYS]}$) is dependent on productivity being sufficiently high enough to maintain all species at positive densities ($F_{[NPP]}$). The spatial arrangement of productivity (modularity or degree of aggregation) over a landscape, being at a scale able to maintain all species ($F_{[NPP]}$), restrains species interactions (and hence dynamics) and curbs the propagation of disturbances and therefore the resilience of the system ($F_{[\lambda]}$). In other words, both the rate of production and the spatial modularity of productivity at the system

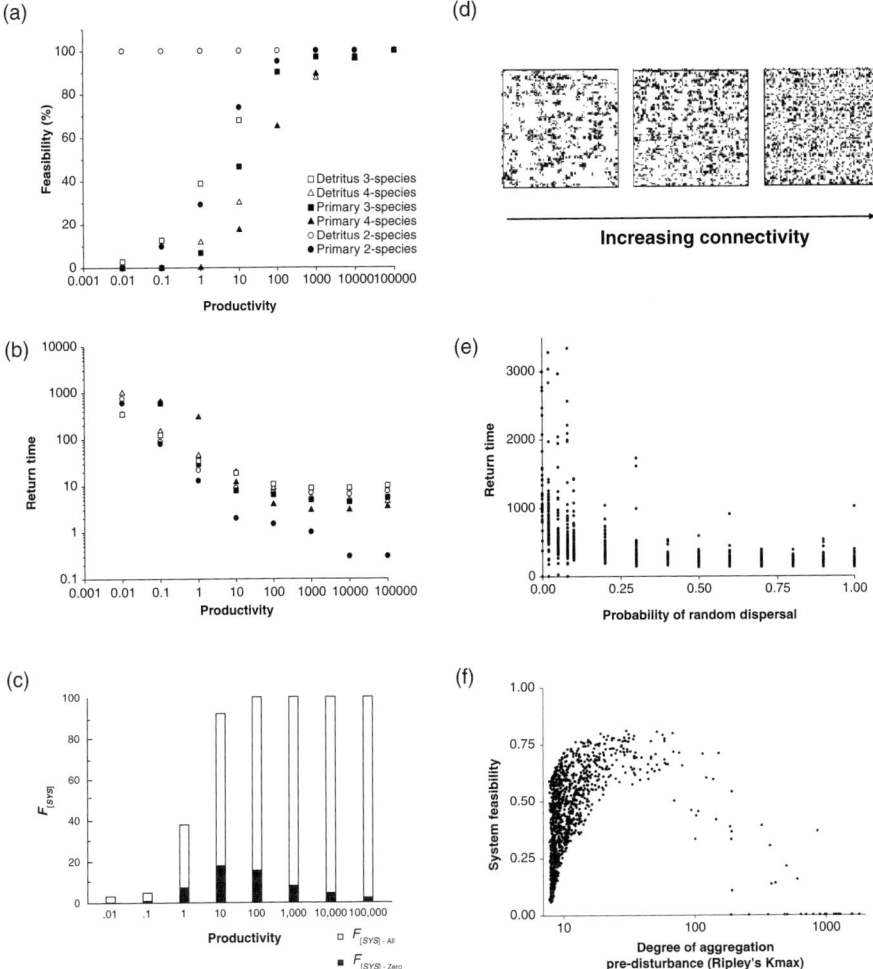

Figure 12.6 Model results illustrating how productivity and modularity impact resilience. The symbols represent actual data; the lines are for illustration. (a–c) Spatially implicit models. (a) Percentage of food webs that are feasible increases with productivity (all steady-state densities > 0). (b) Return time of the same food webs decreases with productivity. Arrow indicates onset of oscillations (eigenvalues with complex parts). (c) Feasibility (%) in terms of productivity and dynamics (zero to low oscillations) tradeoff to constrain system feasibility (area below dark line), which is optimal at intermediate levels of productivity. (d–f) Productivity does not vary. (d) Spatial patterns of interest: random, intermediate modularity, modular. (e) Spatial patterns in d were generated using an individual based simulation where random dispersal was varied. Return time decreased with increasing random dispersal. Arrow indicates value of random dispersal that corresponds to optimum in f. (f) System feasibility or the % of the population remaining post-disturbance exhibits an optimum at intermediate levels of modularity aggregation). The optimum results from a tradeoff between low resilience with high levels of global connectivity (solid line) and low resilience with high levels of local connectivity (dash-dot line) that occurs in highly modular spatial distributions. Little tradeoff occurs between return time (b,e) and system feasibility (c,f).

and landscape levels are products of species interactions, but interact to constrain the space of resilient food webs.

Both components of productivity are in part the products of ecosystem and spatial feedbacks that originate from the input of resources, and the energetic, life history, and movement attributes of individual species and the interactions among species. We have dealt with the ecosystem feedbacks throughout the book. The spatial feedbacks originate from the degree of modularity of the patches, which can influence movement patterns of predators and the overall connectivity of the system, and the spread of disturbances and their impacts on system stability. We will discuss each below.

One way to view a spatial network over a landscape is as a set of resource or habitat patches that represent nodes of interaction. In this regard, spatial networks are similar to food webs in their structure, and the properties that govern their dynamics (Keitt and Johnson, 1995, Keitt, 1997, Urban and Keitt, 2001). In the preceding analysis, the degree of modularity and connectivity in the spatial network affected the feasibility and resilience of the network. In the spatial context, modularity equates to the food web concept of a compartment. Spatial compartments or modules appear as patches within a landscape and are formed by habitats that are relatively easy to migrate among (Urban and Keitt, 2001).

The spatial distribution of prey or hosts affects both movement and feeding preferences of predators, pathogens, and parasites (Hassell and May, 1974, Peters, 1983, Comins and Hassell, 1996, Aguiar and Sala, 1997, Bonsall and Hassell, 1998, 2000, Boots et al., 2004, Aguiar et al., 2003, McCann et al., 2005). At the landscape level, the average movement distances of consumers match the pattern of resource or prey availability in several differing species (Thrall and Burdon, 2003, Fortuna et al., 2006, Rooney et al., 2008). McCann et al. (2005) provide an explanation that links these observations to dynamic stability.

McCann et al. (2005) use a spatially implicit model to show that predators that operate on larger spatial scales than their prey stabilize food webs. In this case the resources and consumers are restricted to localized patches, and predators can migrate from patch to patch. In the absence of predators, the consumers in the localized patch could potentially overexploit resources. The spatial arrangement of the patches in relation to the movement capabilities and behavior of the predator is key to the stability and persistence of the food web. If the ecosystem is too expansive, the spatial modularity increases to a point that the distances between patches are beyond the capabilities of the mobile predators. In this case the patches become the system, which is quickly overexploited by the predator. If the ecosystem is too compressed, the spatial modularity decreases, and the patches are highly coupled, leading again to overexploitation and instability. Just as with the results presented in Figure 12.6, an intermediate level of spatial modularity in patches, or productivity from the predator's perspective, leads to greater stability and long-term feasibility. Recent empirical arguments regarding apex predators support much of this proposition (Estes et al., 2011).

The spatial distribution of prey or hosts can be impacted by spatially spreading disturbance, or what Acher and Webb (2006a, 2006b) referred to as spatially distributed mortality factors. The degree of modularity affects the resilience of the food web by containing the spread of a disturbance within modules (Pimm, 1979a, 1979b, Rozdilsky et al., 2004, Krause et al., 2003, Teng and McCann, 2004, Rooney et al., 2006). Bolker (2003) used a somewhat similar approach, but with disturbance that mimicked habitat destruction. He found that spatial aggregation also affected resilience by enhancing recovery through spatial refuges instead of through limiting the spread of a disturbance.

12.4.2 Dynamical implications of space

From the discussion above it is clear that there are endogenous and exogenous factors that affect food webs and their spatial structure. Species interactions do not occur in all places at all times, but are localized. Additionally, disturbances or events (e.g., pulses of precipitation or nutrient additions) may destroy or create habitats within a spatial network. These factors interact to create asynchronies in the timing and placement of interactions occurring across both homogenous and heterogeneous landscapes to varying degrees that are governed by the suitability of the habitats within the spatial structure, and the movement capabilities and patterns of resources and species.

Such spatial variability within a system may result in a persistent transient dynamics state that can be stabilizing. The stabilizing mechanisms arise from the spatial heterogeneity in encounters that provide temporary refuges for resources and prey species, allowing for populations to escape predation. In the cases where the dynamics are reactive, predators shift their attention to regions of increased prey, attenuating unstable runaway production. The spatial average of this type of scenario may mimic asymptotic behavior, whereas the spatial structure displays transient behavior.

Holt (2002) provides two ways in which the spatial structure of the landscape can promote stability of otherwise unstable configurations. In the first example, persistence is contingent not on the inherent stability of the trophic structure and interactions, but rather on the size and structure of the spatial landscape. If the spatial landscape is large enough and subdivided into enough habitats relative to the movement patterns and resource needs of the species, local extinctions are off-set by dispersal and colonizations of other habitats. This scenario is dependent on the size of the spatial network, resulting in a persistent transient state when viewed from a metapopulation perspective, i.e., spatial structure promotes stability (Holyoak, 2000).

In the second example, Holt (2002) presents a three-species food chain in which the spatial structure of trophic interactions can generate a stable state over a landscape that might otherwise be unstable through the actions of the top predator. In this scenario the top predator experiences self-limitation through territoriality or other means, whereas its prey does not. The asymptotic behavior of the two-species resource–prey food chain in the absence of the predator is prone to cyclic dynamics whereas that of the three-species resource–prey–predator food chain is in stable equilibrium. Depending on the amplitude of the oscillation and the frequency of disturbances the systems experiences, the two-species resource–prey system might

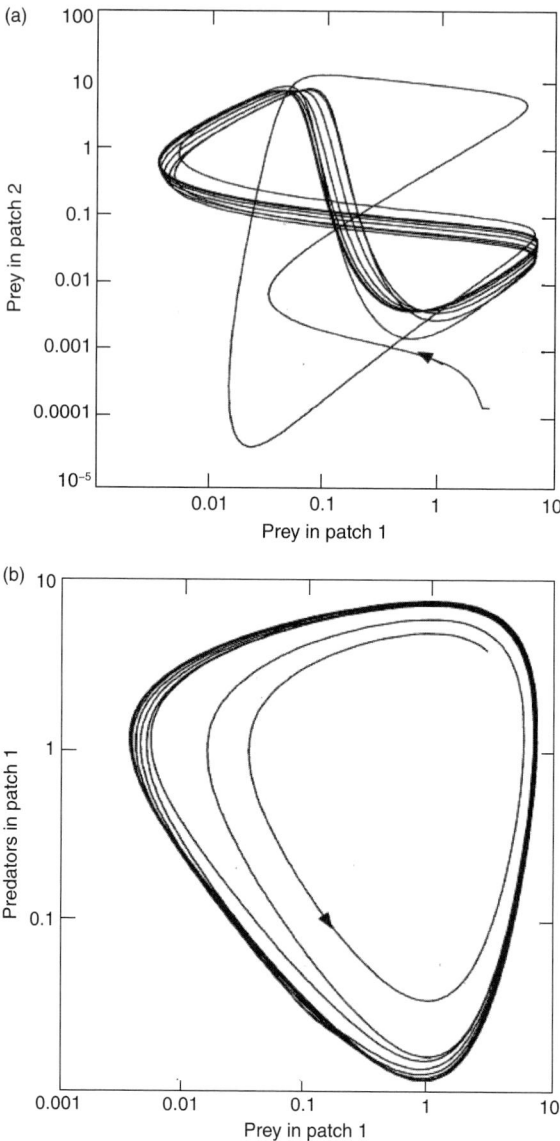

Figure I2.7 The phase-space of a two-patch predator–prey system (Hastings, 2001). (a) The prey numbers in the two patches, and (b) the predator and prey numbers in one of the patches. The model system used a familiar Lotka–Volterra form with a Holling type II functional response. The equations for both the predator and prey included additional terms to account for the random movement of each between the two patches.

experience higher rates of local extinction than the three-species resource–prey–predator system.

Hasting (2001a) stressed that the stability conferred by spatial structure was due in large part to the degree of spatial and temporal asynchrony in the spatial structure, illustrated in the analysis of a predator–prey system in which the predators and prey are distributed over two patches (Figure 12.7). When the species are spatially distributed over these patches, the system can persist over ecologically relevant timescales as long as the population dynamics at the different patches are "out of phase." When they become more "in phase," the conditions for persistence approach the conditions that are required for asymptotic stability. The interesting point here is that apart from the dynamics of the interactions themselves, there can be environmental factors that ensure that the dynamics will stay out of phase, e.g., processes or events that destroy or and create patches. As in the examples presented above, the disturbances should act on a spatial scale so that not all patches are affected, but only one, or a few of them are impacted at any given time. Otherwise the disturbance will not drive the patches out of phase. Hence, (random) disturbances on a relatively small spatial scale may promote persistence over long periods of time, even when the system itself is not stable.

12.5 Asymptotically ambiguous states

There are ways to look at dynamics independent of whether the system is asymptotically stable or not (Hastings, 2004). Populations may rise and fall with the regularity of seasonal forcing or in response to spatial patterning of resources over a landscape, never reaching a hypothetical asymptotic state, thus existing in a perpetual transient state. We characterize these situations as asymptotically ambiguous states. The ambiguity can arise from underlying mathematics in cases where there are no simple solutions. Numerical analyses that explore different regions of parameter space can be a powerful tool. The ambiguity can also arise from our inability to separate the empirical from the theoretical. In these cases the empirical observations are not well matched to the timescales that the system is operating under and that theory would require for validation. We can present two related cases in which these ambiguous states might happen: *permanence* and *apparent complexity*.

12.5.1 Permanence

"Permanence" is a stability criterion that has been proposed as an alternative for asymptotical stability. The key criterion of permanence is that all species should increase in abundance when they become rare. In this way permanence is different from asymptotical stability in two ways. The first is that the set of permanent systems includes systems that are persistent but not necessarily in an asymptotic way. In permanent systems, the population abundances may show dynamics after a disturbance that tend to go away from the equilibrium instead of approaching the

304 • Energetic Food Webs

equilibrium, as long as none of the species becomes extinct (Figure 12.8). One can visualize a state-phase diagram with a threshold for each population close to zero. When a species becomes rare, i.e., below that threshold close to zero, than permanence requires that this species will increase in abundance. The second way in which permanence differs from asymptotic stability is that permanent systems are persistent after any disturbance, also those bringing the system far away from the equilibrium state. Hence, permanence predicts global, but not necessarily asymptotic, stability.

Several analyses of predator–prey systems show that the conditions for a system to be asymptotically stable are a subset of the conditions for a system to be permanent. These analyses are restricted to rather simple two- or three-species food chains like those presented in Part II of this book and not like the more complex food webs presented throughout the book (Law and Morton, 1996). There are also counterexamples showing systems that are asymptotically stable but not permanent (Hofbauer and Sigmund, 1989). When the simple predator–prey systems are indicative for complex food webs, then it is likely that food web stability as defined in this book not only includes asymptotically stable systems locally, but also globally. This does not necessarily mean that they are asymptotically stable systems. On the other

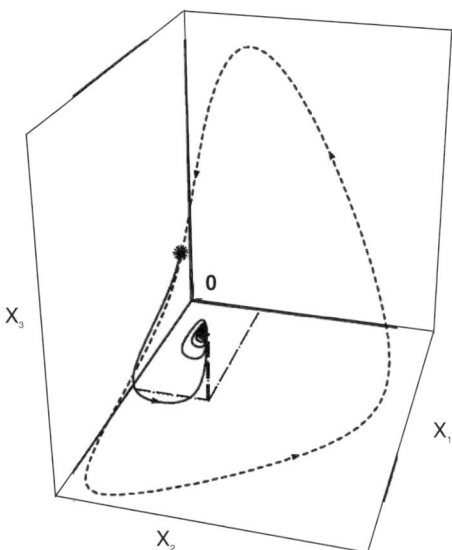

Figure 12.8 The phase-space for a three-species primary-producer-based Lotka–Volterra food chain (Equation 12.3) with a cyclic attractor (from Law and Morton, 1996). The model has the familiar form used throughout the book with some restrictions on the parameter values. The attractor is shown with the discontinuous line. The time average of the orbit is represented by the continuous line, starting at the starred point at time 0. The time average converges to the densities at the fixed point: I = (0.51, 2.08, 0.27).

hand, systems might have been classified as not stable, although they might be permanent. A typical class of ecological processes that are suitable to analyze with the permanence criterion is that of invasions and community assembly (Law and Morton, 1996). The relevant dynamics are not close to equilibrium and coexistence does not require asymptotic behavior tending toward the equilibrium, but merely that for a species to invade successfully it is necessary to increase in number when rare.

Permanence is a criterion that for some systems is mathematical tractable (Law and Morton, 1996). In summary, when we have a system description in terms of Lotka–Volterra differential equations, similar to those used in this book, a sufficient condition for permanence includes two criteria. The first is that populations will not grow infinitely. It is easy to see that Lotka–Volterra equations including self-limitation at the level of the basal resources will meet this criterion, as the higher trophic level consumers will be limited by resource availability. The second criterion is that we have a function:

$$P(x) = \prod_{i \in S_n} x_i^{h_i} (h_i > 0) \tag{12.10}$$

which is an average Lyapunov function. Here x is the vector of all densities of the species x_i in the set S. For a complete explanation see Law and Morton (1996).

12.5.2 Apparent complexity

"Apparent complexity" describes how a food web at a given spatial or temporal scale rarely fully realizes the extent of the interactions that the connectedness, energy flux, and functional descriptions depict. We tend to think of food webs as possessing a single connectedness structure with energy flux and interaction strength represented by average values that were functions of the population densities of species and their physiological and life-history attributes. In the preceding sections of this chapter we explored the consequences of variations in food web structure and explored their impacts on food web functioning and stability. Throughout the book we broached this topic when discussing compartments within food webs and weak links between compartments forged by consumers, and in our discussion of spatial and temporal compartments. In each case, as we had before when studying static architectures, we stressed that all populations possessed feasible population densities ($F_{[NPP]}$) at steady state ($X_i^* > 0$ for all i species). An important assumption in the models was that the population densities were greater than zero, and that the species that were present were all active. Here we will discuss the concept of apparent complexity and investigate situations in which species go from active to inactive states depending on conditions or when species interactions are spatially or temporally separated.

The dynamics of many systems are characterized by pulses of intense biological activity punctuated by periods of lesser activity. Systems that possess these pulse dynamics are often hierarchically organized with interacting subunits of species whose activities are constrained by a single factor (e.g., water availability) or multiple factors (e.g., water, temperature, and nutrients). When conditions are optimal, all the interactions within the connectedness structure are in place and the food web is fully operational. In this case, the rules governing food web dynamics and stability that we have discussed throughout the book are in play. When conditions are less than optimal, a subset of the interactions within the web fails to fully materialize. The rules are still operating but only on the subset that is active.

There are several examples and variants of the pulse-dynamic paradigm that encompass the notion of apparent complexity. Collins et al. (2008) presented a case study to illustrate the importance of spatial structure and pulse dynamics within the context of the pulse-reserve concept. The grasslands in the semiarid and arid water-limited US southwest possess a complex food web structure that is regulated to a large extent by the frequency and magnitudes of precipitation events and the degree to which different components of the system respond to these events (Noy-meir, 1973, Belnap et al., 2004). Rainstorms stimulate plant growth and biological activity in the soils, which generate stocks of biomass, propagules, and organic matter that lay in wait during intervening dry periods and prime the systems when the next rainfall event occurs. The system persists due to a mixture of living biomass that can endure varying degrees of environmental conditions, in this case moisture and dryness, and inactive propagules that can survive the dry spells, and a dependency on detritus in the form of organic material.

12.6 Reconciling asymptotic stability, spatial structure, and transient dynamics

Integrating the spatial and temporal heterogeneity, complexity, and ambiguities discussed previously in a way that reconciles our use of asymptotic stability in light of the very real perspective offered by transient dynamics may not be as difficult as it might seem. A first step is to identify the factors that contribute to the dynamics and stability of the system, in terms of the state variables themselves (endogenous factors), and in terms of external drivers of the systems (exogenous factors). The endogenous and exogenous factors operate at both the species level and at the community level. At the species level, the endogenous and exogenous factors operate as evolutionary agents of change for the species themselves. At the community level the endogenous and exogenous factors are what shape the hierarchical organizational structure of the community.

Hastings (2010) emphasized that the ecological dynamics that we observe are governed by the dynamics produced by the endogenous factors and the dynamics of the system in response to exogenous forcing. Endogenous factors are those inherent to the species and interactions in which they engage. We have explored a range of dynamic states that include, but are not limited to, stable equilibria with and without oscillations following a disturbance, limit cycles, and chaos. Hastings (2010) would add the initial transient dynamics of the ecosystem that occur in movement toward these states, the transient dynamic responses of the ecosystem to disturbances, and the potential long-term meandering transient dynamics from one state to another.

Exogenous factors include traditional ecosystem drivers that are external to the species interactions, but shape them nonetheless. These include daily, seasonal, and annual fluctuations in photoperiod and climate (temperature and precipitation), decadal oscillations in climate and ocean temperature, millennial changes in climate (e.g., ice ages and warming periods), landscape and landform development and resulting heterogeneity in geologic and ecological time frames, and anthropogenic influences on climate, landscapes, and species distributions.

Here and elsewhere it has been emphasized that species have evolved—as May (1973) put it best—devious strategies in response to, and that cope with, these endogenous and exogenous factors. These strategies have a direct links to the life histories, physiologies, and morphologies of (read "energetic properties") of species. In terms of life histories, the phenology of a species defines its temporal extent, including its responses to seasonal and interannual variation in climate. Within this includes the life-history attributes of dormancy, resting states, and other forms of cryptobiotic states, as well as the reproductive capacity embodied in the intrinsic rate of increase—all strategies to cope with changing environmental conditions. Couple the life-history attributes with morphological attributes, and spatial considerations come into play. For example, body size is strongly correlated with energetic efficiencies and energy requirements, and movement capacity. These attributes define the home ranges representing the spatial extent of a species, but they also define their abilities to migrate and colonize new habitats and to link habitats within a spatial network.

At the community scale, the concepts behind transient dynamics fit well within a dynamic hierarchical systems framework. The compartmentalization and the degree of asynchrony of interactions, activity, and distribution in space and time of food web interactions may lead to asymptotically stable states, while the narrow slices of spatial and temporal scales within the food web may indeed exhibit transient dynamics. All interactions within the food web are not engaged simultaneously at any given point in space and time. In other words, the spatial and temporal averaging that are used to study a static food web architecture hides the internal dynamics that occur within a dynamic food web architecture. Transient dynamics may not be the end-all or even the norm, but they represent the ways in which subsystems interact with one another within the broader context of asymptotic stability.

12.7 Summary and conclusions

In this chapter we discussed alternatives to using asymptotic stability as the principle guide to studying food web dynamics. Food web architecture, density of species, and flow of matter among species change on a regular basis. The variation in structure and dynamics that these changes create can be viewed in several ways. Rather than gauge the stability and persistence of the system in terms of its capacity to return to its original state, we asked whether systems migrate from one state to another, and whether the variation in space and time were incompatible with or an essential component to the stability or persistence of a system.

In answering these questions, the examples and arguments we presented about transient dynamics, spatial structure, permanence, and apparent complexity paralleled those that we have made earlier about the presence and importance of compartments to food web structure and stability. This is reassuring inasmuch as the tools, if not the perspective or endgame, that asymptotic dynamic stability provided are still relevant. We are still left with dynamic hierarchical systems defined by a single Jacobian matrix whose elements are compartmentalized through interactions, spatially compartmentalized over a landscape, or seasonally compartmentalized as collections of linked subsystems. The energetic properties of species, their life-history attributes, and their behavior define the boundaries of subsystems and the linkages between them.

Seasonality, pulse-dynamics driven by seasonal processes, and general environmental variability create situations in which species and species interactions in communities will be temporally and spatially separated, resulting in a structure in which not everything everywhere is doing the same things all the time. This view of food web structure provides a picture analogous to city lights at night showing highly variable and dynamic spots of activity and movement. Looking in such a way at the variation and dynamics within food webs will surely alter our fundamental notions and understanding of the functioning and stability of complex ecological communities.

Finally, permanence is a stability criterion that has attracted much attention, especially because it clearly refers to our intuitive notion of stability, but there are also other stability criteria (see Chapter 2); for a comprehensive overview the reader is referred to McCann (2000). Regardless of the criterion that is chosen, the study of food webs hinges on the energetic properties of organisms embodied in their morphologies, physiologies, and life histories, as these properties define the ecological temporal and spatial scales under which the system is operating. When these factors are taken into consideration, we see that coexistence and persistence within these ecologically bounded scales does not depend on or require that our models produce a stable equilibrium or stable asymptotic solutions or states. The devious strategies of organisms spoken of by May (1973) provide ample means for species to navigate through the energetically and dynamically feasible and nonfeasible spatial and temporal landscapes they confront. In many cases species have developed the adaptive capacity to seek out feasible landscapes or wait for conditions to change in their favor. In doing so, the community at the spatial and temporal scale of the species and processes may or may not be stable in the strict asymptotic sense, yet they persist. Alternatively, transient dynamics may represent a component of the internal dynamics that subsystems may operate under at narrower spatial and temporal scales than the system as a whole.

In a way we had deviated a little from this concept in our analysis of food web development during succession and in our discussion of changes in the dominance of one energy channel over another. Here we took this approach a step further. We strayed from the idea of "return to the original equilibrium state," and redefined stability as the ability to maintain food web structure in terms of species (or trophic group) composition after a disturbance, but allowing for other states than the state before the disturbance. In fact, we looked at persistence of food webs by analyzing the nature of dynamics of the population abundances following a disturbance. This is a relatively realistic way of looking at stability and persistence, as there are many examples of systems, including predator–prey systems, that show dynamics over relatively long and ecologically relevant timescales, without any species becoming extinct; at the same time, they would be classified as being stable (McKane and Drossel, 2005). This way of approaching stability was attractive for several reasons. First, it links the analysis to an intuitively realistic notion of stability, as it classifies systems as stable as long as all species and/or groups will persist after a disturbance. Also, it links the concept of persistence to timescales that are similar to the ones used in most empirical studies, which is in contrast to a criterion such as asymptotic stability, which predicts the outcome of the system after a period far outside the experimental time frame.

What we find by exploring these alternatives is that ultimately organisms require a minimum amount of energy to survive and reproduce and have evolved numerous strategies to obtain and maintain this energy. Evolving the capacity to acquire energy for growth and reproduction alone is not sufficient, as organisms live within an energetic network that we define as a food web, which in turn is embedded in both spatial and temporal networks. The organization of the food web within these networks is constrained by the underlying amount of energy available and the dynamic constraints that the organization structure imposes and the ability of the structure to respond to change. The devious strategies that organisms have evolved include the means to cope with these dynamic constraints as well. That is to say, $F_{[SYS]} = F_{[NPP]} \cap F_{[\lambda]}$.

References

Abrams, P. A. & Ginzburg, L. R. (2000) The nature of predation: prey dependent, ratio dependent or neither? *Trends in Evolution and Ecology,* 15, 337–41.
Achter, J. D. & Webb, C. T. (2006a) Mixed dispersal strategies and response to disturbance. *Evolutionary Ecology Research,* 8, 1377–92.
Achter, J. D. & Webb, C. T. (2006b) Pair statistics clarify percolation properties of spatially explicit simulations. *Theoretical Population Biology,* 69, 155–64.
Adler, P. B., Seabloom, E. W., Borer, E. T., Hillebrand, H., Hautier, Y., Hector, A., O'halloran, L. R., Harpole, W. S., Anderson, T. M., Baker, J. D., Biederman, L. A., Brown, C. S., Buckley, Y. M., Calabrese, L. B., Chu, C.-J., Cleland, E. E., Collins, S. L., Cottingham, K. L., Crawley, M. J., Davies, K. F., Decrappeo, N. M., Fay, P. A., Firn, J., Fater, P., Gasarch, E. I., Gruner, D. S., Hagenah, N., Hillerislambers, J., Humphries, H., Jin, V. L., Kay, A. D., Kirkman, K. P., Klein, J. A., Knops, J., Pierre, K. J. L., Lambrinos, J. G., Li, W., Macdougall, A. S., Mcculley, R. L., Melbourne, B. A., Mitchell, C. E., Moore, J. L., Morgan, J. W., Mortenson, B., Orrock, J. L., Prober, S. M., Pyke, D. A., Risch, A. C., Schuetz, M., Stevens, C. J., Sullivan, L. L., Wang, G., Wragg, P. D., Wright, J. P., & Yang, L. H. (2011) Get over the hump: productivity does not predict fine-scale species richness. *Science,* 330, 1750–3.
Aguiar, M. N. R. & Sala, O. E. (1997) Seed distribution constrains the dynamics of the Patagonian steppe. *Ecology,* 78, 93–100.
Aguiar, M. R., Camara, G., & Souza, R. C. M. (2003) Modeling spatial relations by generalized proximity matrices. *GeoInfo 2003—V Brazilian Symposium on Geioinformatics.* Campos do Jordao.
Ahl, V. & Allen, T. F. H. (1996) *Hierarchy Theory: A Vision, Vocabulary, and Epistemology,* New York, Columbia University Press.
Allen, T. F. H. & Starr, T. B. (1982) *Hierarchy: Perspectives for Ecological Complexity,* Chicago, University of Chicago Press.
Allen-Morley, C. R. & Coleman, D. C. (1989) Resilience of soil biota in various food webs to freezing perturbations. *Ecology,* 70, 1127–41.
Andrén, O., Lindberg, T., Boström, U., Clarholm, M., Hansson, A.-C., Johansson, G., Lagerlöf, J., Paustian, K., Persson, J., Petterson, R., Schnürer, J., Sohlenius, B., & Wivstad, M. I. O. A. (1990) Organic carbon and nitrogen flows. In T. Lindberg, K. Paustian, & T. Rosswall eds, Ecology of Arable Land-Organisms: Carbon and Nitrogen Cycling, *Ecological Bulletin,* 40, 85–125.
Arditi, R. & Berryman, A. A. (1991) The biological control paradox. *Trends in Ecology & Evolution,* 6, 32.

Arditi, R. & Ginzburg, L. R. (1989) Coupling in predator prey dynamics: ratio-dependence. *Journal of Theoretical Biology,* 139, 311–26.
Austin, M. P. & Cook, B. G. (1974) Ecosystem stability: a result from an abstract simulation. *Journal of Theoretical Biology,* 45, 435–58.
Barton, B. T. & Schmitz, O. J. (2009) Experimental warming transforms multiple predator effects in a grassland food web. *Ecology Letters,* 12, 1317–25.
Beddington, J. (1975) Mutual interference between parasites or predators and its effect on searching efficiency. *Journal of Animal Ecology,* 51, 597–624.
Beddington, J. R., Free, C. A., & Lawton, J. H. (1978) Modeling biological control: on the characteristics of successful natural enemies. *Nature,* 273, 513–19.
Belnap, J., Phillips, S. L., & Miller, M. E. (2004) Response of desert biological soil crusts to alterations in precipitation frequency. *Oecologia,* 141, 306–16.
Bender, E. A., Case, T. J., & Gilpin, M. E. (1984) Perturbation experiments in community ecology: theory and practice. *Ecology,* 65, 1–13.
Benke, A. C., Wallace, J. B., Harrison, J. W., & Koebel, J. W. (2001) Food web quantification using secondary production analysis: predaceous invertebrates of the snag habitat in a subtropical river. *Freshwater Biology,* 46, 329–46.
Berg, B. & Matzner, E. (1997) Effect of N deposition on decomposition of plant litter and soil organic matter in forest systems. *Environmental Reviews,* 5, 1–25.
Berlow, E. L. (1999) Strong effects of weak interactions in ecological communities. *Nature,* 398, 330–4.
Berlow, E. L., Neutel, A. M., Cohel, J. E., de Ruiter, P. C., Ebenman, B., Emmerson, M., Fox, J. W., Jansen, V. A. A., Jones, J. I., Kokkoris, G. D., Logofet, D. O., McKane, A. J., Montoya, J. M., & Petchey, O. (2004) Interaction strengths in food webs: issues and opportunities. *Journal of Animal Ecology,* 73, 585–98.
Bersier, L.-F., Banasek-Richter, C., & Cattin, M. F. (2002) Quantitative descriptors of food-web matrices. *Ecology,* 83, 2394–407.
Bird, R. D. (1930) Biotic communities of the Aspen Parkland of central Canada. *Ecology,* 11, 356–442.
Bloem, J., Lebbink, G., Zwart, K. B., Bouwman, L. A., Burgers, S. L. G. E., De Vos, J. A., & de Ruiter, P. C. (1994) Dynamics of microorganisms, microbivores and nitrogen mineralisation in winter wheat fields under conventional and integrated management. *Agriculture, Ecosystems, and Environment,* 51, 129–43.
Bolker, B. M. (2003) Combining endogenous and exogenous spatial variability in analytical population models. *Theoretical Population Biology,* 64, 255–70.
Bonsall, M. B. & Hassell, M. P. (1998) Population dynamics of apparent competition in a host-parasitoid assemblage. *Journal of Animal Ecology,* 67, 918–29.
Bonsall, M. B. & Hassell, M. P. (2000) The effects of metapopulation structure on indirect interactions in host-parasitoid assemblages. *Proceedings of the Royal Society of London B,* 267, 2207–12.
Boots, M., Hudson, P. J., & Sasaki, A. (2004) Large Shifts in Pathogen Virulence Relate to Host Population Structure. *Science,* 303, 842–4.
Borer, K. T. (1971) Control of food intake in Octopus briaeus Robson. *Journal of Comparative Physiology and Psychology,* 75, 171–85.
Borrvall, C., Ebenman, B., & Jonsson, T. (2000) Biodiversity lessens the risk of cascading extinctions in food webs. *Ecology Letters,* 3, 131–6.

Brewer, R. H. (1994) *The Science of Ecology*, 2nd Ed, Ft. Worth, TX, Saunders College Publishing.
Briand, F. (1983) Environmental control of food web structure. *Ecology*, 64, 253–63.
Briand, F. & Cohen, J. E. (1987) Environmental correlates of food-chain length. *Science*, 238, 956–60.
Brooks, J. L. & Deevey, E. S. (1963) New England. In Frey, D. G. ed. *Limnology in North America*. Madison, Wisconsin, University of Wisconsin.
Brussaard, L., Van Veen, J. A., Kooistra, M. J., & Lebbink, G. (1988) The Dutch programme on soil ecology of arable farming systems. I. Objectives, approach and preliminary results. *Ecological Bulletin*, 39, 35–40.
Camberdella, C. A. & Elliott, E. T. (1994) Carbon and nitrogen dynamics of soil organic matter fractions from cultivated grassland soils. *Soil Science Society of America Journal*, 58, 123–30.
Carpenter, S. R. & Brock, W. A. (2010) Early warnings of regime shifts in spatial dynamics using the discrete Fourier transform. *Ecosphere*, 1, art10.
Carpenter, S. R., Brock, W. A., Cole, J. J., Kitchell, J. F., & Pace, M. L. (2008) Leading indicators of trophic cascades. *Ecology Letters*, 11, 128–38.
Carpenter, S. R., Kitchell, J. F., & Hodgson, J. R. (1985) Cascading trophic interactions and lake productivity. *BioScience*, 35, 634–9.
Carpenter, S. R., Kitchell, J. F., Hodgson, J. R., Cochran, P. A., Elser, J. J., Elser, M. M., Lodge, D. M., Kretchmer, D., He, X., & Ende, C. N. V. (1987) Regulation of lake primary productivity by food web structure. *Ecology*, 68, 1863–76.
Chao A., Chazdon, R. L., Colwell, R. K., & Shen T. J. (2005) A new statistical approach for assessing similarity of species composition with incidence and abundance data. Ecology Letters, 8, 148–59.
Chapin III, F. S., Shaver, G. R., Giblin, A. E., Nadelhoffer, K. J., & Laundre, J. A. (1995) Responses of Arctic tundra to experimental and observed changes in climate. *Ecology*, 76, 694–711.
Chazdon, R., Colwell, R. K., Denslow, J. S., & Guariguata, M. R. (1998) Statistical methods for estimating species richness of woody regeneration in primary and secondary rain forests of NE Costa Rica. In F. Dallmeier & J. Comiskey eds, *Forest Biodiversity Research, Monitoring and Modeling: Conceptual Background and Old World Case Studies*. Paris, France, Parthenon Publishing.
Chelius, M. K., Beresford, G., Horton, H., Quirk, M., Selby, G., Simpson, R. T., Horrocks, R., & Moore, J. C. (2009) Impacts of alterations of organic inputs on the bacterial community within the sediments of Wind Cave, South Dakota, USA. *International Journal of Speleology*, 38, 1–10.
Chen, X. I. N. & Cohen, J. E. (2001) Global stability, local stability and permanence in model food webs. *Journal of Theoretical Biology*, 212, 223–35.
Clark, F. E. (1977) Internal cycling of ^{15}Nitrogen in shortgrass prairie. *Ecology*, 58, 1322–33.
Clarke, T. A., Fleehsig, A. O., & Grigg, R. W. (1967) Ecological studies during Project Sealab II. *Science*, 157, 1381–9.
Clements, F. E. (1916) *Plant succession: an analysis of the development of vegetation*, Washington, DC, Carnegie Institution of Washington.
Clements, F. E. (1920) *Plant indicators: the relation of plant communities to process and practice*, Washington, DC, Carnegie Institution of Washington.
Clements, F. E. (1936) Nature and structure of the climax. *Journal of Ecology*, 24, 252–84.
Cohen, J. E. (1978) *Food webs in niche space*, Princeton, NJ, Princeton University Press.

Cohen, J. E. (1990) A stochastic theory of community food webs. VI. Heterogeneous alternatives to the cascade model. *Theoretical Population Biology,* 37, 55–90.
Cohen, J. E., Jonsson, T., & Carpenter, S. R. (2003) Ecological community description using food web, species abundance, and body-size. *Proceedings of the National Academy of Sciences,* 100, 1781–6.
Cohen, J. E., Luczak, T., Newman, C. M., & Zhou, Z. M. (1990) Stochastic structure and nonlinear dynamics of food webs: qualitative stability in a Lotka–Volterra cascade model. *Proceedings of the Royal Society London B,* 240, 607–27.
Cohen, J. E. & Newman, C. M. (1985) A Stochastic Theory of Community Food Webs: I. Models and Aggregated Data. *Proceedings of the Royal Society of London B,* 224, 421–48.
Cohen, J. E. & Newman, C. M. (1991) Community area and food-chain length: theoretical predictions. *American Naturalist,* 138, 1542–54.
Coleman, D. C. (1996) Energetics of detritivory and microbivory in soil in theory and practice. In G. A. Polis & K. O. Winemiller eds, *Food Webs: Integration of Patterns and Dynamics,* New York, Chapman Hall.
Coleman, D. C., Reid, C. P. P., & Cole, C. V. (1983) Biological strategies of nutrient cycling in soil systems. In A. Macfayden & E. D. Ford eds, *Advances in Ecological Research,* New York, Academic Press.
Collins, S. L., Sinsabaugh, R. L., Crenshaw, C., Green, L., A., P.-A., Stursova, M., & Zeglin, L. H. (2008) Pulse dynamics and microbial processes in aridland ecosystems. *Journal of Ecology,* 96, 413–20.
Colton, H. S. (1916) On some varieties of *Thais lapillus* in the Mount Desert region, a study of individual ecology. *Proceedings of the Academy of Natural Sciences of Philadelphia,* 68, 440–54.
Colton, H. S. (1922) Variation in the dog whelk, Thais (*Purpurk auct*) lapillus. *Ecology,* 3, 146–57.
Comfort, A. (1979) *The Biology of Senescence,* Edinburgh, Churchill Livingstone.
Comins, H. N. & Hassell, M. P. (1996) Persistence of multispecies host–parasitoid interactions in spatially distributed models with local dispersal. *Journal of Theoretical Biology,* 183, 19–28.
Connell, J. H. (1961a) Effects of competition, predation by *Thais lapillus*, and other factors on natural populations of the barnacle *Balanus balanoides*. *Ecological Monographs,* 31, 61–104.
Connell, J. H. (1961b) The influence of inerspecific competition and other factors on the distribution of the barnacle *Chthamalus stellatus*. *Ecology,* 41, 710–23.
Connell, J. H. & Slatyer, R. O. (1977) Mechanisms of succession in natural communities and their role in community stability and organization. *American Naturalist,* 111, 1119–44.
Copeland, B. J., Tenore, K. R., & Horton, D. B. (1974) Oligohaline regime. In H. T. Odum, B. J. Copeland, & E. A. McMahan, eds. *Coastal Ecological Systems of the United States.* Washington, DC, The Conservation Foundation.
Coppock, D. L., Detling, J. K., Ellis, J. E., & Dyer, M. I. (1983) Plant-herbivore interactions in a North American mixed-grass prairie. *Oecologia,* 56, 1–9.
Coughenour, M. B., Ellis, J. E., Swift, D. M., Coppock, D. L., Galvin, K., McCabe, J. T., & Hart, T. C. (1985) Energy extraction and use in a nomadic pastoral ecosystem. *Science,* 230, 619–25.
Culver, D. C. (1982) *Cave Life,* Cambridge, Massachusetts, Harvard University Press.

Cushing, J. M., Costantino, R. F., Dennis, B., Descharnais, R. A., & Henson, S. M. (2002) *Chaos in Ecology: Experimental Non-linear Dynamics,* London, Academic Press.

Davis, M. A., Grime, J. P., & Thompson, K. (2000) Fluctuating resources in plant communities: a general theory of invasibility. *Journal of Ecology,* 88, 528–34.

Day, J. H. (1967) The biology of Knysna Estuary, South Africa. In G. H. Lauff, ed. *Estuaries—Publication number 83.* Washington, DC, American Association for the Advancement of Science.

de Ruiter, P. C., Bloem, J., Bouwman, L. A., Didden, W. A. M., Hoenderboom, G. H. J., Lebbink, G., Marinissen, J. C. Y., De Vos, J. A., Vreeken-Buijs, M. J., Zwart, K. B., & Brussaard, L. (1994) Simulation of dynamics in nitrogen mineralisation in the belowground food webs of two arable farming systems. *Agriculture, Ecosystems & Environment,* 51, 199–208.

de Ruiter, P. C., Moore, J. C., Bloem, J., Zwart, K. B., Bouwman, L. A., Hassink, J., De Vos, J. A., Marinissen, J. C. Y., Didden, W. A. M., Lebbink, G., & Brussaard, L. (1993a) Simulation of nitrogen dynamics in the belowground food webs of two winter-wheat fields. *Journal of Applied Ecology,* 30, 95–106.

de Ruiter, P. C., Neutel, A. M., & Moore, J. C. (1995) Energetics, patterns of interaction strengths, and stability in real ecosystems. *Science,* 269, 1257–60.

de Ruiter, P. C., Van Veen, J. A., Moore, J. C., Brussaard, L., & Hunt, H. W. (1993b) Calculation of nitrogen mineralization in soil food webs. *Plant and Soil,* 157, 263–73.

de Ruiter, P. C., Wolters, V., & Moore, J. C., eds. (2005) *Dynamic Food Webs,* Amsterdam, Academic Press.

DeAngelis, D. L. (1975) Stability and connectance in food web models. *Ecology,* 56, 238–43.

DeAngelis, D. L. (1980) Energy flow, nutrient cycling, and ecosystem resilience. *Ecology,* 61, 764–71.

DeAngelis, D. L. (1992) *Dynamics of Nutrient Cycling and Food Webs,* London, Chapman & Hall.

DeAngelis, D. L. & Waterhouse, J. C. (1987) Equilibrium and nonequilibrium concepts in ecological models. *Ecological Monographs,* 57, 1–21.

DeAngelis, D. L., Gardner, R. H., Mankin, J. B., Post, W. M., & Carney, J. H. (1978) Energy flow and the number of trophic levels in ecological communities. *Nature,* 273, 406–7.

Debruyn, A. M. H., McCann, K. S., Moore, J. C., & Strong, D. R. (2007) An energetic framework for trophic control. In N. Rooney, K.S. McCann, D.L.G. Noakes eds. *From Energetics to Ecosystems for Trophic Control: The Dynamics and Structure of Ecological Systems.* Dordrecht, The Netherlands, Springer.

Diamond, J. M. (1975) Assembly of Species Communities. In M. L. Cody & J. M. Diamond, eds. *Ecology and Evolution of Communities.* Cambridge, Massachusetts, Belknap Press.

Dodson, S. I., Arnott, S. E., & Cottingham, K. L. (2000) The relationship in lake communities between primary productivity and species richness. *Ecology,* 81, 2662–79.

Doles, J. L. (2000) A survey of soil biota in the arctic tundra and their role in mediating terrestrial nutrient cycling. MSc thesis, Department of Biological Sciences, University of Northern Colorado, Greeley, Colorado.

Dunbar, M. J. (1954) Arctic and subarctic marine ecology: immediate problems. *Arctic,* 7, No. 3 and 4.

Dunne, J. A., Williams, R. J., & Martinez, M. D. (2002) Network structure and biodiversity loss in food webs: robustness increases with connectance. *Ecology Letters,* 5, 558–67.

Ehrlich, P. R. & Birch, L. C. (1967) The "balance of nature" and "population control". *American Naturalist,* 65, 86–8.

Ehrlich, P. R. & Holm, R. W. (1962) Patterns and populations. *Science*, 137, 652–7.
Eisenberg, J. F. (1981) *The Mammalian Radiations, an Analysis of Trends in Evolution, Adaptation, and Behavior*, Chicago, University of Chicago Press.
Elliott, E. T. & Coleman, D. C. (1988) Let the soil work for us. *Ecological Bulletins* 39, 23–32.
Ellis, J. E., Coughenour, M. B., & Swift, D. M. (1993) Climate variability, ecosystem stability, and the implications for range and livestock development. In R. H. Behnke Jr, I. Scoones, & C. Kerven, eds. *Range Ecology at Disequilibrium: New Models of Natural Variability and Pastoral Adaptation in African Savannas*. London, Overseas Development Institute.
Ellis, J. E. & Swift, D. M. (1988) Stability of African pastoral ecosystems: alternate paradigms and implications for development. *Journal of Range Management*, 41, 450–9.
Elton, C. (1927) *Animal Ecology*, New York, Macmillan.
Elton, C. (1958) *The Ecology of Invasions by Plants and Animals*, London, Methuen.
Engleberg, J. & Boyarski, L. L. (1979) The non-cybernetic nature of ecosystems. *American Naturalist*, 114, 317–24.
Estes, J. A. & Palmisano, J. F. (1974) Sea otters: their role in structuring nearshore communities. *Science*, 185, 1058–60.
Estes, J. A., Terborgh, J., Brashares, J. S., Power, M. E., Berger, J., Bond, W. J., Carpenter, S. R., Essington, T. E., Holt, R. D., Jackson, J. B. C., Marquis, R. J., Oksanen, L., Oksanen, T., Paine, R. T., Pikitch, E. K., Ripple, W. J., Sandin, S. A., Scheffer, M., Schoener, T. W., Shurin, J. B., Sinclair, A. R. E., Soule, M. E., Virtanen, R. & Wardle, D. A. (2011) Trophic downgrading of planet Earth. *Science*, 333, 301–6.
Evans, K. L., Greenwood, J. J. D., & Gaston, K. J. (2005) Dissecting the species–energy relationship. *Proceedings of the Royal Society of London B*, 272, 2155–63.
Fenchel, T. (1980) Suspension feeding in ciliated protozoa: functional response and particle size selection. *Microbial Ecology*, 6, 1–10.
Fisher, J. A. D. (2005) Exploring ecology's attic: overlooked ideas on intertidal food webs. *Bulletin of the Ecological Society of America*, 86, 145–51.
Ford, D. C., Lundberg, J., Palmer, A. N., Palmer, M. V., Dreybrodt, W., & Schwarcz, H. P. (1993) Uranium-series dating of the draining of an aquifer: The example of Wind Cave, Black Hills, South Dakota. *Geological Society of America Bulletin*, 105, 241–50.
Fonseca, A. R., Carvalho, C. F., & Souza, B. (2000) Functional response of *Chrysoperla externa* (Hagen) (Neuroptera: Chrysopidae) fed on *Schizaphis graminum* (Rondani) (Hemiptera: Aphididae) *Anais da Sociedade Entomológica do Brasil*, 29, 309–17.
Fortuna, M. A., Gómez-Rodríguez, C., & Bascompte, J. (2006) Spatial network structure and amphibian persistence in stochastic environments. *Proceedings of the Royal Society of London B*, 273, 1429–34.
Frank, K. A. (1995) Identifying cohesive subgroups. *Social Networks*, 17, 27–56.
Frank, K. A. (1996) Mapping interactions within and between cohesive subgroups. *Social Networks*, 18, 93–119.
Frank, K. A. & Yasumoto, J. (1996) Embedding subgroups in the sociogram: linking theory and image. *Connections* 19, 43–57.
Fridley, J. D. (2011) Invasibility, of Communities and Ecosystems. In D. Simberloff, & M. Rejmánek, eds. *Encyclopedia of Biological Invasions*. Berkeley and Los Angeles, University of California Press.
Fryer, G. (1959) The trophic interrelationships and ecology of some littoral communities of Lake Nyasa with especial reference to the fishes, and a discussion of the evolution of a group of rock-frequenting Cichlidae. *Proceedings of the Zoological Society of London*, 132, 153–281.

Gardner, M. R. & Ashby, W. R. (1970) Connectance of large dynamic (cybenetic) systems: critical values for stabillity. *Nature,* 228, 784–5.

Gardner, R. H., Cale, W. G., & O'Neill, R. V. (1982) Robust analysis of aggregation error. *Ecology,* 63, 771–9.

Gilpin, M. E. (1975) *Group Selection in Predator–prey Communities,* Princeton, NJ, Princeton University Press.

Gilpin, M. E. & Rosenzweig, M. L. (1972) Enriched predator–prey systems: theoretical stability. *Science,* 177, 902–04.

Gleason, H. A. (1926) The individualistic concept of the plant association. *Bulletin of the Torrey Botanical Club,* 53, 7–26.

Gleason, H. A. (1939) The individualistic concept of the plant association. *American Midland Naturalist,* 21, 92–110.

Goh, B. S. (1977) Global stability in many-species systems. *American Naturalist,* 142, 379–411.

Golley, F. B. (1993) *A History of the Ecosystem Concept in Ecology: More than the Sum of the Parts,* New Haven, Connecticut, Yale University Press.

Gough, L., Moore, J. C., Shaver, G. R., Simpson, R. T., & Johnson D. R. (2012) Above- and below ground responses of arctic tundra ecosystems to altered soil nutrients and mammalian herbivory. Ecology (in press).

Gould, S. J. (1982) Darwinism and the expansion of evolutionary theory. *Science* 216, 380–287.

Greenberg, S. M., Legaspi, B. C., & Jones, W. A. (2001) Comparison of functional response and mutual interference between two aphelinid parasitoids of *Bemisia argentifolii* (Homoptera: Aleyrodidae). *Journal of Entomological Science,* 36, 1–8.

Griffiths, B. S., Ritz, K., Bardgett, R. D., Cook, R., Christensen, S., Ekelund, F., Sørensen, S. J., Bååth, E., Bloem, J., de Ruiter, P. C., Dolfing, J., & Nicolardot, B. (2000) Ecosystem response of pasture soil communities to fumigation-induced microbial diversity reductions: an examination of the biodiversity–ecosystem function relationship. *Oikos,* 90, 279–94.

Gross, M. R. & Repka, J. (1998) Stability with inheritance in the conditional strategy. *Journal of Theoretical Biology,* 192, 445–53.

Gulbrandsen, J. (1991) Functional response of Atlantic halibut larvae related to prey density and distribution. *Aquaculture,* 94, 89–98.

Gyllenberg, H. G. & Eklund, E. (1974) Bacteria. In C. H. Dickinson & G. J. F. Pugh, eds. *Biology of Plant Litter Decomposition,* New York, Academic Press.

Hairston, N. G., Smith, F. E., & Slobodkin, L. B. (1960) Community structure, population control, and competition. *American Naturalist,* 94, 421–5.

Hairston, N. G. Jr. & Hairston, N.g. Sr. (1993) Cause–effect relationships in energy flow, trophic structure, and interspecific interactions. *American Naturalist,* 142, 379–411.

Harris, S. (1977) Science Cartoons Plus. Available at: http://www.sciencecartoonsplus.com/gallery/math/index.php# [accessed 14 November 2011].

Harrison, G. W. (1979) Global stability of food chains. *American Naturalist,* 114, 455–7.

Harrison, J. L. (1962) The distribution of feeding habits among animals in a tropical rain forest. *Journal of Animal Ecology,* 31, 53–63.

Hart, A. M. & Klumpp, D. W. (1996) Response of herbivorous fishes to crown-of-thorns starfish *Acanthaster planci* outbreaks. I. Substratum analysis and feeding ecology of *Acanthurus nigrofuscus* and *Scarus frenatus.* Marine ecology progress series, 132, 11–19.

Hassell, M. P. (1979) Nonrandom search in predator-prey models. *Fortschritte der Zoologie,* 25, 311–30.

Hassell, M. P. & May, R. M. (1974) Aggregation of predators and insect parasites and its effect on stability. *Journal of Animal Ecology,* 43, 567–94.

Hastings, A. (2001) Transient dynamics and persistence of ecological systems. *Ecology Letters,* 4, 215–20.

Hastings, A. (2004) Transients: the key to long-term ecological understanding? *Trends in Ecology and Evolution,* 19, 39–45.

Hastings, A. (2010) Timescales, dynamics, and ecological understanding. *Ecology,* 91, 3471–80.

Hastings, A. & Powell, T. (1991) Chaos in a three-species food chain. *Ecology,* 72, 896–903.

Hastings, A. & Wysham, D. B. (2010) Regime shifts in ecological systems can occur with no warning. *Ecology Letters,* 13, 464–72.

Hawkins, B. A. & Cornell, H. V. (1994) Maximum parasitism rates and successful biological control. *Science,* 266, 1886–90.

Heatwole, H. & Levins, R. (1972) Trophic structure stability and faunal change during recolonization. *Ecology,* 53, 531–4.

Hendrix, P. F., Parmelee, R. W., Crossley, D. A. J., Coleman, D. C., Odum, E. P., & Groffman, P. M. (1986) Detritus food webs in conventional and no-tillage agroecosystems. *BioScience,* 36, 374–80.

Hiatt, R. W. & Strasburg, D. W. (1960) Ecological relationships of the fish fauna on coral reefs of the Marshall Islands. *Ecological Monographs,* 30, 65–127.

Hofbauer, J. & Sigmund, K. (1989) On the stabilizing effect of predators and competitors on ecological communities. *Journal of Mathematical Biology,* 27, 537–48.

Hohberg, K. & Traunspurger, W. (2005) Predator–prey interaction in soil food web: functional response, size-dependent foraging efficiency, and the influence of soil texture. *Biology and Fertility of Soils,* 41, 419–27.

Holland, J. H. (1995) *Hidden Order: How Adaptation Builds Complexity,* New York, Helix Books (Addison Wesley).

Holling, C. S. (1959) The components of predation as revealed by a study of small mammal predation on the European pine sawfly. *The Canadian Entomologist,* 91, 293–320.

Holling, C. S. (1965) The functional response of predators to prey density and its role in mimicry and population regulation. *Memoirs of the Entomological Society of Canada,* 45, 5–60.

Holling, C. S. (1966) The strategy of building models of complex ecological systems. In K. E. F. Watt, ed. *Systems Analysis in Ecology.* New York, Academic Press.

Holling, C. S. (1968) The tactics of a predator. In T. R. E. Southwood, ed. *Insect Abundance.* Oxford, Blackwell Scientific Publications.

Holt, R. D. (2002) Food webs in space: on the interplay of dynamic instability and spatial processes. *Ecological Research,* 17, 261–73.

Holyoak, M. (2000) Habitat subdivision causes changes in food web structure. *Ecology Letters,* 3, 509–15.

Horrocks, R. D. & Szukalski, B. W. (2002) Using geographic information systems to develop a cave potential map for Wind Cave, South Dakota. *Journal of Cave and Karst Studies* 64, 63–70.

Horton, H. F. (2005) Food webs along a natural productivity gradient within the rooms and passages of Wind Cave, South Dakota. MSc thesis, Department of Biological Sciences, University of Northern Colorado, Greeley, Colorado.

Hunt, H. W., Coleman, D. C., Ingham, E. R., Ingham, R. E., Elliott, E. T., Moore, J. C., Rose, S. L., Reid, C. P. P., & Morley, C. R. (1987) The detrital food web in a shortgrass prairie. *Biology and Fertility of Soils,* 3, 57–68.

Hunt, H. W., Ingham, E. R., Coleman, D. C., Elliott, E. T., & Reid, C. P. P. (1988) Nitrogen limitation of production and decomposition in prairie, mountain meadow, and pine forest. *Ecology* 69, 1009–16.

Hutchinson, G. E. (1959) Homage to Santa Rosalia or why are there so many kinds of animals? *American Naturalist*, 93, 145–59.

Huxel, G. R. & McCann, K. (1998) Food web stability: the influence of trophic flows across habitats. *American Naturalist*, 152, 460–9.

Ingham, R. E., Trofymow, J. A., Ingham, E. R., & Coleman, D. C. (1985) Interactions of bacteria, fungi, and their nematode grazers: effects on nutrient cycling and plant growth. *Ecological Monographs*, 55, 119–40.

Ives, A. R. & Carpenter, S. R. (2007) Stability and diversity of ecosystems. *Science*, 317, 58–62.

Ives, A. R., Carpenter, S. R., & Dennis, B. (1999) Community interaction webs and zooplankton responses to planktivory manipulations. *Ecology*, 80, 1405–21.

Jalali, M., Tirry, L., & De Clercq, P. (2010) Effect of temperature on the functional response of *Adalia bipunctata* to *Myzus persicae*. *BioControl*, 55, 261–9.

Jansson, A. M. (1967) The food web of the *Cladophora* – belt fauna. *Helgoländer wissenshaftliche Meeresuntersunchungen*, 15, 571–88.

Jefferies, R. L. (1999) Herbivores, nutrients and trophic cascades in terrestrial environments. In H. Olff, V. K. Brown, & R. H. Drent, eds. *Herbivores: Between Plants and Predators*. Oxford, Blackwell Science.

Jefferies, R. L., Rockwell, R. F., & Abraham, K. F. (2004) Agricultural food subsidies, migratory connectivity and large-scale disturbance in arctic coastal systems: a case study. *Integrative and Comparative Biology*, 44, 130–9.

Jeschke, J. M. & Hohberg, K. (2008) Predicting and testing functional responses: An example from a tardigrade-nematode system. *Basic and Applied Ecology*, 9, 145–51.

Jesser, R. D. (1998) The effects of productivity on species diversity and trophic structure of detritus-based food webs within sediments of wind cave, South Dakota. MSc thesis, Department of Biological Sciences, University of Northern Colorado, Greeley, Colorado.

Johnson, D., Campbell, C. D., Lee, J. A., Callahan, T. V., & Gwynn-Jones, D. (2002) Arctic micro microorganisms respond more to elevated UV-B radiation than CO_2. *Nature*, 416, 82–3.

Johnston, R. F. (1956) Predation by short-eared owls on a *Salicornia* salt marsh. *The Wilson Bulletin*, 68, 91–102.

Jones, J. R. E. (1949) A further ecological study of calcareous streams in the "Black Mountain" district of South Wales. *Journal of Animal Ecology*, 18, 142–59.

Jonsson, T., Cohen, J. E., & Carpenter, S. R. (2005) Food webs, body-size, and species abundance in ecological community description. *Advances in Ecological Research*, 36, 1–84.

Keitt, T. H. (1997) Stability and complexity on a lattice: coexistence of species in an individual-based food web model. *Ecological Modelling*, 102, 243–58.

Keitt, T. H. & Johnson, A. R. (1995) Spatial heterogeneity and anomalous kinetics: emergent patterns in diffusion-limited predatory-prey interaction. *Journal of Theoretical Biology*, 172, 127–39.

Kitching, J. A. & Ebling, F. J. (1967) Ecological studies at Lough Ine. *Advances in Ecological Research*, 4, 197–291.

Kleiman, D. G. & Eisenberg, J. F. (1973) Comparisons of canid and felid social systems from and evolutionary perspective. *Animal Behavior*, 21, 637–59.

Knight, R. L. & Swaney, D. P. (1981) In defense of ecosystems. *American Naturalist*, 117, 991–2.
Knox, G. A. (1970) Antarctic marine ecosystems. In M. W. Holdgate, ed. *Antarctic Ecology*. New York, Academic Press.
Koen-Alonso, M. (2007) A process-oriented approach to the multispecies functional response. In N. Rooney, K. S. McCann, D. L. G. Noakes, & P. Yodzis, eds. *From Energetics to Ecosystems: the Dynamics and Structure of Ecological Systems*. Dordrecht, The Netherlands, Springer.
Koestler, A. (1967) *The Ghost in the Machine*, New York, Macmillan.
Kokkoris, G. D., Jansen, V. A. A., Loreau, M., & Troumbis, A. Y. (2002) Variability in interaction strength and implications for biodiversity. *Journal of Animal Ecology*, 71, 362–71.
Krause, A. E., Frank, K. A., Mason, D. M., Ulanowicz, R. E., & Taylor, W. W. (2003) Compartments revealed in food-web structure. *Nature*, 426, 282–5.
Kremer, J. N. & Nixon, S. W. (1978) *A Coastal Marine Ecosytem: Simulation and Analysis*, Berlin, Springer.
Krug, A. Z., Jablonski, D., & Valentine, J. W. (2007) Contrarian clade confirms the ubiquity of spatial origination patterns in the production of latitudinal diversity gradients. *Proceedings of the National Academy of Sciences USA*, 104, 18129–34.
Kuhn, T. S. (1970) *The Structure of Scientific Revolutions*, Chicago, University of Chicago Press.
Lane, P. (1986) Symmetry, change, perturbation, and observing mode in natural communities. *Ecology*, 67, 223–39.
Laska, M. S. & Wootton, J. T. (1998) Theoretical concepts and empirical approaches to measuring interaction strength. *Ecology*, 79, 461–76.
Lauenroth, W. K. & Milchunas, D. G. (1991) The shortgrass steppe. In R. T. Coupland, ed. *In Natural Grasslands: Introduction and Western Hemisphere*, Amsterdam, Elsevier.
Law, R. & Morton, R. D. (1996) Permanence and the assembly of ecological communities. *Ecology*, 77, 762–75.
Leonard, A. B. (1989) Functional response in *Antedon mediterranea* (Lamarck) (Echinodermata: Crinoidea): the interaction of prey concentration and current velocity on a passive suspension-feeder. *Journal of Experimental Marine Biology and Ecology*, 127, 81–103.
Leopold, A. (1949) *A Sand County Almanac*, New York, Oxford University Press.
Levin, S. A. (1998) Ecosystems and the biosphere as complex adaptive systems. *Ecosystems*, 1, 431–6.
Levin, S. A. (1999) *Fragile Dominion: Complexity and the Commons*, Cambridge, Massachusetts, Perseus Publishing.
Levin, S. A. (2002) Complex adaptive systems: exploring the known, the unknown and the unknowable. *Bulletin of the American Mathematical Society*, 40, 3–19.
Levine, J. M. & D'antonio, C. M. (1999) Elton revisited: a review of evidence linking diversity and invasibility. *Oikos*, 87, 15–26.
Levins, R. (1968) *Evolution in Changing Environments*, Princeton, Princeton University Press.
Lindeman, R. L. (1942) The trophic-dynamic aspect of ecology. *Ecology*, 23, 399–418.
Loreau, M., Mouquet, N., & Holt, R. D. (2003) Meta-ecosystems: a theoretical framework for a spatial ecosystem ecology. *Ecology Letters*, 6, 673–9.

Lotka, A. J. (1922) Contribution to the Energetics of Evolution. *Proceedings of the National Academy of Sciences USA*, 8, 147–51.
Luck, R. F., Lenteren, J. C. V., Twine, P. H., Kuenen, L., & Unruh, T. (1979) Prey or host searching behavior that leads to a sigmoid functional response in invertebrate predators and parasitoids. *Researches in Population Ecology*, 20, 257–64.
Ludwig, D., Walker, B., & Holling, C. S. (1997) Sustainability, stability, and resilience. *Conservation Ecology (Online)*, 1, 7.
MacArthur, R. (1955) Fluctuations of animal populations, and a measure of community stability. *Ecology*, 36, 533–6.
MacArthur, R. H. (1972) Strong, or weak, interactions? *Transactions of the Connecticut Academy of Arts and Sciences*, 44, 177–88.
MacArthur, R. H. & Wilson, E. O. (1963) An equilibrium theory of insular zoogeography. *Evolution*, 17, 373–87.
MacArthur, R. H. & Wilson, E. O. (1967) *The Theory of Island Biogeography*, Princeton, NJ, Princeton University Press.
MacGinitie, G. E. (1935) Ecological Aspects of a California Marine Estuary. *American Midland Naturalist*, 16, 629–765.
Mackintosh, N. A. (1964) A survey of Antarctic biology up to 1945. In R. Carrick, H. Holdgate, & J. Prevost, eds. *Biologie Antarctique*. Paris, Hermann.
Mack, M. C., Schuur, E. A. G., Bret-Harte, M. S., Shaver, G. R., & Chappin III, F. S. (2004) Ecosystem carbon storage in arctic tundra reduced by long-term nutrient fertilization. *Nature*, 431, 440–3.
Margalef, R. (1961) Communication of structure in planktonic populations. *Limnology and Oceanography*, 6, 124–8.
Margalef, R. (1968) *Perspectives in Ecological Theory*, Chicago, University of Chicago Press.
Marples, T. G. (1966) A radionuclide tracer study of arthropod food chains in a *Spartina* salt marsh ecosystem. *Ecology*, 47, 270–7.
Márquez, L., Caballos, M., & Domingues, P. (2007) Functional response of early stages of the cuttlefish *Sepia officinalis* preying on the mysid *Mesopodopsis slabberi*. *Marine Biology Research*, 3, 462–7.
Marshal, J. P. & Boutin, S. (1999) Power analysis of wolf–moose functional responses. *Journal of Wildlife Management*, 63, 396–402.
Martinez, N. D. (1992) Constant connectance in community food webs. *American Naturalist*, 139, 1208–18.
Mather, J. A. (1980) Some aspects of food intake in *Octopus joubini* Robson. *Veliger*, 22, 286–90.
May, R. M. (1972) Will a large complex system be stable? *Nature*, 238, 413–14.
May, R. M. (1973) *Stability and Complexity in Model Ecosystems*, Princeton, NJ, Princeton University Press.
May, R. M. (1976) Simple mathematical models with very complicated dynamics. *Nature*, 261, 459–67.
May, R. M. (1977) Thresholds and breakpoints in ecosystems with a multiplicity of stable states. *Nature*, 269, 471–7.
May, R. M. & Hassell, M. P. (1981) The dynamics of multiparasitoid-host interactions. *American Naturalist*, 117, 234–61.

McCabe, J. T. (1990) *Cattle Bring Us To Our Enemies: Turkana Ecology, Politics, and Raiding in a Disequilibrium System*, Ann Arbor, MI, University of Michigan Press.
McCann, K. (2000) The diversity–stability debate. *Nature*, 405, 228–33.
McCann, K., Hastings, A., & Huxel, G. R. (1998) Weak trophic interactions and the balance of nature. *Nature*, 395, 794–8.
McCann, K. & Yodzis, P. (1994) Biological conditions for chaos in a three-species food chain. *Ecology*, 75, 561–4.
McCann, K. S., Rasmussen, J. R., & Umbanhowar, J. (2005) The dynamics of spatially coupled food webs. *Ecology Letters*, 8, 513–23.
McCoull, C. J., Swain, R., & Barnes, R. W. (1998) Effect of temperature on the functional response and components of attack rate in *Naucoris congrex* Stål (Hemiptera: Naucoridae). *Australian Journal of Entomology*, 37, 323–7.
McGill, W. B., Hunt, H. W., Woodmansee, R. G., & Reuss, J. O. (1981) PHEONIX, a model of the dynamics of carbon and nitrogen in grassland soils. *Ecological Bulletin (Stockholm)*, 33, 49–115.
McKane, A. J. & Drossel, B. D. (2005) Modeling evolving food webs. In P. de Ruiter, V. Wolters, & J. C. Moore, eds. *Dynamic Food Webs*. Amsterdam, Academic Press.
McKenzie, L. W. (1960) Matrices with dominant diagonals and economic theory. In K. J. Arrow, ed. *Mathematical Methods in the Social Sciences*. Stanford, California, Stanford University Press.
McMurtie, R. E. (1975) Determinants of stability of large randomly connected systems. *Journal of Theoretical Biology*, 50, 1–11.
McNab, B. K. (1963) Bioenergetics and the determination of home range size. *American Naturalist*, 97, 133–40.
McNab, B. K. (1980) Food habits, energetics, and the population biology of mammals. *American Naturalist*, 116, 106–24.
McNaughton, S. J. & Coughenour, M. B. (1981) The cybernetic nature of ecosystems. *American Naturalist*, 117, 985–90.
Melville, K. (2003) Modeling the functional response of *Folsomia candida* (Willem) and its fungal prey. MSc thesis, Department of Biological Sciences, University of Northern Colorado, Greeley, Colorado.
Menge, B. A. & Sutherland, J. P. (1976) Species Diversity Gradients: Synthesis of the Roles of Predation, Competition, and Temporal Heterogeneity. *American Naturalist*, 110, 351–69.
Milchunas, D. G. & Lauenroth, W. K. (1992) Carbon dynamics and estimates of primary production by harvest, C14 dilution, and C14 turnover. *Ecology*, 73, 593–607.
Milchunas, D. G. & Lauenroth, W. K. (2001) Belowground primary production by carbon isotope decay and long-term root biomass dynamics. *Ecosystems*, 4, 139–50.
Milne, H. & Dunnet, G. M. (1972) Standing crop, productivity and trophic relations of the fauna of the Ythan estuary. In R. S. K. Barnes & J. Green, eds. *The Estuarine Environment*. London, Applied Science Publications.
Minshall, G. W. (1967) Role of allochthonous detritus in the trophic structure of a woodland springbrook community. Ecology, 48, 139–49.
Mittelbach, G. G. & Persson, L. (1998) The ontogeny of piscivory and its ecological consequences. *Canadian Journal of Fisheries and Aquatic Sciences*, 55, 1454–65.

Mittelbach, G. G., Steiner, C. F., Scheiner, S. M., Gross, K. L., Reynolds, H. L., Waide, R. B., Willig, M. R., Dodson, S. I., & Gough, L. (2001) What is the observed relationship between species richness and productivity? *Ecology*, 82, 2381–96.

Möbius, K. (1877) *Die Auster und Austernwirtschaft.*, Berlin, Wiegandt, Hempel und Parey.

Mohr, S. & Adrian, R. (2000) Functional responses of the rotifers *Brachionus calyciflorus* and *Barchonus rubens* feeding on armored and unarmored ciliates. *Limnology and Oceanography*, 45, 1175–80.

Montoya, J. M., Rodríguez, M. A., & Hawkins, B. A. (2003) Food web complexity and higher-level ecosystem services. *Ecology Letters*, 6, 614–22.

Moore, J. C. (1986) Micro-mesofauna dynamics and functions in dryland wheat-fallow agroecosystems. PhD dissertation, Department of Zoology and Entomology, Colorado State University, Colorado.

Moore, J. C. & de Ruiter, P. C. (1991) Temporal and spatial heterogeneity of trophic interactions within below-ground food webs. *Agriculture, Ecosystems and Environment*, 34, 371–9.

Moore, J. C. & de Ruiter, P. C. (1997) Compartmentalization of resource utilization within soil ecosystems. In A. C. G. A. V. K. Brown, ed. *Multitrophic Interactions in Terrestrial Systems*. London, Blackwell Science.

Moore, J. C. & de Ruiter, P. C. (2000) Invertebrates in detrital food webs along gradients of productivity. In D. C. Coleman and P. F. Hendrix eds. *Invertebrates as Webmasters in Ecosystems*. Oxford, UK, CABI Publishing.

Moore, J. C. & Hunt, H. W. (1988) Resource compartmentation and the stability of real ecosystems. *Nature*, 333, 261–3.

Moore, J. C., Berlow, E. L., Coleman, D. C., de Ruiter, P., Dong, Q., Hastings, A., Johnson, N. C., McCann, K. S., Melville, K., Morin, P. J., Nadelhoffer, K., Rosemond, A. D., Post, D. M., Sabo, J. L., Scow, K. M., Vanni, M. J., & Wall, D. (2004) Detritus, trophic dynamics and biodiversity. *Ecology Letters*, 7, 584–600.

Moore, J. C., de Ruiter, P. C., & Hunt, H. W. (1993) Influence of productivity on the stability of real and model ecosystems. *Science*, 261, 906–8.

Moore, J. C., de Ruiter, P. C., Hunt, H. W., Coleman, D. C., & Freckman, D. W. (1996) Microcosms and soil ecology: critical linkages between field studies and modelling food webs. *Ecology*, 77, 694–705.

Moore, J. C., McCann, K., & Ruiter, P. C. D. (2005) Modeling trophic pathways, nutrient cycling, and dynamic stability in soils. *Pedobiologia*, 49, 499–510.

Moore, J. C., McCann, K., Setälä, H., & de Ruiter, P. (2003) Top-down is bottom-up: does predation in the rhizosphere regulate aboveground dynamics? *Ecology*, 84, 846–57.

Moore, J. C., Walter, D. E., & Hunt, H. W. (1988) Arthropod regulation of micro- and mesobiota in below-ground detrital food webs. *Annual Review of Entomology*, 33, 419–39.

Moore, J. C., Walter, D. E., & Hunt, H. W. (1989) Habitat compartmentation and environmental correlates of food chain length. *Science*, 243, 238–40.

Müller, C. B., Adriaanse, I. C. T., Belshaw, R., & Godfray, H. C. J. (1999) The structure of an aphid–parasitoid community. *Journal of Animal Ecology*, 68, 346–70.

Murdoch, W. W. (1966) Community structure, population control, and competition: a critique. *American Naturalist*, 100, 219–26.

Muschiol, D., Markovi, M., Threis, I., & Traunspurger, W. (2008) Predatory copepods can control nematode populations: a functional-response experiment with *Eucyclops subterraneus* and bacterivorous nematodes. *Fundamental and Applied Limnology/Archiv für Hydrobiologie,* 172, 317–24.

Naeem, S., Thompson, L. J., Lawler, S. P., Lawton, J. H., & Woodfin, R. M. (1994) Declining biodiversity can alter the performance of ecosystems. *Nature,* 368, 734–7.

Nandini, S. & Sarma, S. S. S. (1999) Effect of starvation time on the prey capture behaviour, functional response and population growth of *Asplanchna sieboldi* (Rotifera). *Freshwater Biology,* 42, 121–30.

Nelmes, A. J. (1974) Evaluation of the feeding behavior of *Prionchulus punctatus* (Cobb), a nematode predator. *Journal of Animal Ecology,* 43, 553–65.

Neubert, M. G. & Caswell, H. (1997) Alternatives to resilience for measuring the responses of ecological systems to perturbations. *Ecology,* 78, 653–65.

Neutel, A. M. (2001) Stability of complex food webs: pyramids of biomass, interaction strenghts, and weight of trophic loops. Ph.D. dissertation, Department of Environmental Sciences, Utrecht University, Netherlands.

Neutel, A. M., Heesterbeek, J. A. P., & de Ruiter, P. C. (2002) Stability in real food webs: weak links in long loops. *Science,* 296, 1120–3.

Neutel, A. M., Heesterbeek, J. A. P., Van De Koppel, J., Hoenderboom, G., Vos, A., Kaldeway, C., Berendse, F., & de Ruiter, P. C. (2007) Reconciling complexity with stability in naturally assembling food webs. *Nature,* 449, 599–602.

Neutel, A.-m., Roerdink, J. B. T. M., & de Ruiter, P. C. (1994) Global stability of two-level detritus-decomposer food chains. *Journal of Theoretical Biology,* 171, 351–3.

Newman, J. R. (1956) *The World of Mathematics,* New York, Simon and Schuster.

Nielsen, Ó. K. (1999) Gyrfalcon predation on ptarmigan: numerical and functional responses. *Journal of Animal Ecology,* 68, 1034–50.

Niering, W. A. (1963) Terrestrial Ecology of Kapingamarangi Atoll, Caroline Islands. *Ecological Monographs,* 33, 131–60.

Nixon, S. W. & Oviatt, C. A. (1973) Ecology of a New England Salt Marsh. *Ecological Monographs,* 43, 463–98.

Norberg, J., Swaney, D. P., Dushoff, J., Lin, J., Casagrandi, R., & Levin, S. A. (2001) Phenotypic diversity and ecosystem functioning in changing environments: A theoretical framework. *Proceedings of the National Academy of Sciences USA,* 98, 11376–98.

Noy-meir, I. (1973) Desert ecosystems. I. Environment and producers. *Annual Review of Ecology & Systematics,* 4, 25–52.

O'Neill, R. V. (1969) Indirect estimation of energy fluxes in animal food webs. *Journal of Theoretical Biology,* 22, 284–90.

O'Neill, R. V. (2001) Is it time to bury the ecosystem concept? (With full military honors, of course!). *Ecology,* 82, 3275–84.

O'Neill, R. V., Deangelis, D. L., Waide, J. B., & Allen, T. F. H. (1986) *A Hierarchical Concept of the Ecosystem,* Princeton, NJ, Princeton University Press.

O'Neill, R. V., Krummel, J. R., Gardner, R. H., Sugihara, G., Jackson, B., Deangelis, D. L., Milne, B. T., Turner, M. G., Zygmunt, B., Christensen, S. W., Dale, V. H., & Graham, R. L. (1988) Indices of landscape pattern. *Landscape Ecology,* 1, 153–62.

Odum, E. P. (1953) *Fundamentals of Ecology,* Philidelphia, Saunders.

Odum, E. P. (1962) Relationships between structure and function in the ecosystem. *Japanese Journal of Ecology,* 12, 108–18.

Odum, E. P. (1963) *Ecology*, New York, Holt, Rinehart, and Winston.
Odum, E. P. (1969) The strategy of ecosystem development. *Science*, 164, 262–79.
Oechel, W. C., Vourlitis, G. L., Hastings, S. J., Zulueta, R. C., Hinzman, L., & Kane, D. (2000) Acclimation of ecosystem CO_2 exchange in the Alaskan Arctic in response to decadal climate warming. *Nature*, 406, 978–81.
Oksanen, L., Fretwell, S. D., Arruda, J., & Niemelä, P. (1981) Exploitation ecosystems in gradients of primary productivity. *American Naturalist*, 118, 240–61.
Paine, R. T. (1966) Food web complexity and species diversity. *American Naturalist*, 100, 65–75.
Paine, R. T. (1980) Food webs: linkage, interaction strength and community infrastructure. *Journal of Animal Ecology*, 49, 667–85.
Paine, R. T. (1988) Food webs: road maps of interactions or grist for theoretical development? *Ecology*, 69, 1648–54.
Paine, R. T. (1992) Food-web analysis through field measurements of per capita interaction strength. *Nature*, 355, 73–5.
Palmer, A. N. & Palmer, M. V. (2000) Speleogenesis of the Black Hills Maze Caves, South Dakota, U.S.A. In A. B. Klimchouk, D. C. Ford, A. N. Palmer, and W. Dreybrodt, eds. *Speleogenesis: Evolution of karst aquifers*. Huntsville, AL, National Speleological Society.
Patten, B. C. & Finn, J. T. (1979) Systems approach to continental shelf ecosystems. In E. Halfon, ed. *Theoretical Systems Ecology*, New York, Academic Press.
Paustian, K., Andrén, O., Clarholm, M., Hansson, A.-c., Johansson, G., Largerlöf, J., Lindberg, T., Petterson, R., & Sohlenius, B. (1990) Carbon and nitrogen budgets of four agro-ecosystems with annual and perennial crops, with and without N fertilization. *Journal of Applied Ecology*, 27, 60–84.
Paviour-smith, K. (1956) The biotic community of a salt marsh in New Zealand. *Transactions of the Royal Society of New Zealand*, 83, 525–54.
Peters, R. H. (1983) *The Ecological Implications of Body Size*, Cambridge, Cambridge University Press.
Petipa, T. S., Pavlova, E. V., & Mironov, G. N. (1970) The food web structure, utilization transport of energy by trophic levels in the plankton communities. In J. H. Steele, ed. *Marine Food Chains*, Edinburgh, Oliver and Boyd.
Pfister, C. A. (1995) Estimating competition coefficients from census data: a test with field manipulations of tidepool fishes. *American Naturalist*, 146, 271–91.
Phillips, D. A., Fox, T. C., Ferris, H., & Moore, J. C. (2006) Increases in atmospheric CO_2 and the soil food web. In J. Nösberger, S. P. Long, R. J. Norby, M. Stitt, G.r. Hendrey, & H. Blum eds. *Managed Ecosystems and CO_2 Case Studies, Processes and Perspectives*, Heidelberg, Springer-Verlag.
Phillipson, J. (1981) Bioenergetic options and phylogeny. In C. R. Townsend & P. Calow, eds. *Physiological Ecology: an Evolutionary Approach to Resource Use*. Oxford, Blackwell Science.
Pichlova, R. & Vijverberg, J. (2001) A laboratory study of the functional response of Leptodora *kindtii* to some cladoceran species and copepod nauplii. *Archiv für Hydorbiologie*, 150, 529–44.
Pimm, S. L. (1979a) Complexity and stability: another look at MacArthur's original hypothesis. *Oikos*, 33, 351–7.
Pimm, S. L. (1979b) The structure of food webs. *Theoretical Population Biology*, 16, 144–58.
Pimm, S. L. (1982) *Food webs*, London, Chapman & Hall.

Pimm, S. L. (1991) *The Balance of Nature? Ecological Issues in the Conservation of Species and Communities,* Chicago, University of Chicago Press.

Pimm, S. L. & Lawton, J. H. (1977) Number of trophic levels in ecological communites. *Nature,* 268, 329–31.

Pimm, S. L. & Lawton, J. H. (1978) On feeding on more than one trophic level. *Nature,* 275, 542–4.

Pimm, S. L. & Lawton, J. H. (1980) Are food webs divided into compartments? *Journal of Animal Ecology,* 49, 879–98.

Polis, G. A. & Strong, D. R. (1996) Food web complexity and community dynamics. *American Naturalist,* 147, 813–46.

Post, D. M. (2002) The long and short of food chain length. *Tree,* 17, 269–77.

Post, D. M., Conners, M. E., & Goldberg, D. S. (2000) Prey preference by a top predator and the stability of linked food chains. *Ecology,* 81, 8–14.

Preisser, E. L. (2003) Field evidence for a rapidly cascading underground food web. *Ecology,* 84, 869–74.

Qazim, S. Z. (1970) Some problems related to the food chains in a tropical estuary. In J. H. Steele, ed. *Marine Food Chains.* Edinburgh, Oliver and Boyd.

Real, L. A. (1977) The kinetics of functional response. *American Naturalist,* 111, 289–300.

Redpath, S. M. & Thirgood, S. J. (1999) Numerical and functional responses in generalist predators: hen harriers and peregrines on Scottish grouse moors. *Journal of Animal Ecology,* 68, 879–92.

Roberts, A. (1974) The stability of a feasible random ecosystem. *Nature,* 251, 607–8.

Rohde, K. (1992) Latitudinal gradients in species diversity: the search for the primary cause. *Oikos,* 65, 514–27.

Rooney, N., McCann, K., Gellner, G., & Moore, J. C. (2006) Structural asymmetry and the stability of diverse food webs. *Nature,* 442, 265–9.

Rooney, N., McCann, K., & Moore, J. C. (2008) A landscape theory for food web architecture. *Ecology Letters,* 11, 867–81.

Root, R. B. (1967) The niche exploitation pattern of the blue-gray gnatcatcher. *Ecological Monographs,* 37, 317–50.

Rosenzweig, M. L. (1971) Paradox of enrichment: destabilization of exploitation ecosystems in ecological time. *Science,* 171, 385–7.

Rosenzweig, M. L. (1995) *Species Diversity in Space and Time,* Cambridge, Cambridge University Press.

Rosenzweig, M. L. & Abramsky, Z. (1993) How are diversity and productivity related? In R. E. R. A. D. Schulter, ed. *Species Diversity in Ecological Communities: Historical and Geographical Perspectives.* Chicago, University of Chicago Press.

Rosenzweig, M. L. & Macarthur, R. H. (1963) Graphical representation and stability conditions of predator–prey interactions. *American Naturalist,* 97, 209–23.

Rozdilsky, I. D., Stone, L., & Solow, A. (2004) The effects of interaction compartments on stability for competitive systems. *Journal of Theoretical Biology,* 227, 277–82.

Ruesink, J. L. (1998) Variation in per capita interaction strength: thresholds due to nonlinear dynamics and nonequilibrium conditions. *Proceedings of the National Academy of Sciences USA,* 95, 6843–7.

Sala, E. & Graham, M. H. (2002) Community-wide distribution of predator-prey interaction strength in kelp forests. *Proceedings of the National Academy of Sciences USA,* 99, 3678–83.

Salthe, S. H. (1985) *Evolving Hierarchical Systems: Their Structure and Representation*, New York, Columbia University Press.
Sarbu, S. M., Kane, T. C., & Kinkle, B. K. (1996) A chemoautrophically based cave ecosystem. *Science*, 272, 1953–5.
Schaller, F. (1968) *Soil Animals*, Ann Arbor, Michigan, University of Michigan Press.
Schmitz, O. J. (1997) Press perturbations and the predictability of ecological interactions in a food web. *Ecology*, 78, 55–69.
Schoener, T. W. (1974) Resource partitioning in ecological communities. *Science*, 185, 27–39.
Schröter, D., Wolters, V., & de Ruiter, P. C. (2003) C and N mineralisation in the decomposer food webs of a European forest transect. *Oikos*, 102, 294–308.
Shaver, G. R. & Chapin, F. S. (1991) Production: Biomass relationships and element cycling in contrasting Arctic vegetation types. *Ecological Monographs*, 61, 1–31.
Shaver, G. R., Giblin, A. E., Nadelhoffer, K. J., Thieler, K. K., Downs, M. R., Laundre, J. A., & Rastetter, E. B. (2006) Carbon turnover in Alaskan tundra soils: effects of organic matter quality, temperature, moisture and fertilizer. *Journal of Ecology*, 94, 740–53.
Shelford, V. E. (1931) Some concepts of bioecology. *Ecology*, 12, 455–67.
Shelford, V. E. (1935) The major communities. *Ecological Monographs*, 5, 252–92.
Simberloff, D. S. & Wilson, E. O. (1969) Experimental zoogeography of islands. The colonization of empty islands. *Ecology*, 50, 278–96.
Simberloff, D. S. & Wilson, E. O. (1970) Experimental zoogeography of islands: a two year record of colonization. *Ecology*, 51, 934–7.
Simon, H. A. (1962) The architecture of complexity. *Proceedings of the American Philosophical Society*, 106, 467–82.
Simpson, R. T., Frey, S. D., Six, J., & Thiet, R. K. (2004) Preferential accumulation of microbial carbon in aggregate structures of no-tillage soils. *Soil Science Society of America Journal*, 68, 1249–55.
Six, J., Elliott, E. T., & Paustian, K. (1999) Aggregate and soil organic matter dynamics under conventional and no-tillage systems. *Soil Science Society of America Journal*, 63, 1350–8.
Skalski, G. T. & Gilliam, J. F. (2001) Functional responses with predator interference: viable alternatives to the Holling Type II model. *Ecology*, 82, 3083–92.
Snider, R. J., Shaddy, J. H., & Butcher, J. W. (1969) Culture techniques for rearing soil arthropods. *The Michigan Entomologist*, 1, 357–362.
Solé, R. V. & Montoya, J. M. (2001) Complexity and fragility in ecological networks. *Proceedings of the Royal Society of London B*, 268, 2039–45.
Sterner, R. W., Bajpai, A., & Adams, T. (1997) The enigma of food chain length: absence of theoretical evidence for dynamic constraints. *Ecology*, 78, 2258–62.
Stieglitz, M., McKane, R. B., & Klausmeier, C. A. (2006) A simple model for analyzing climatic effects on terrestrial carbon and nitrogen dynamics: an arctic case study. *Global Biogeochemical Cycles*, 20, GB3016.
Stinner, B. R., Crossley Jr., D. A., Odum, E. P., & Todd, R. L. (1984) Nutrient budgets and internal cycling of N, P, K, Ca, and Mg in conventional tillage, no-tillage, and old field ecosystems on the Georgia Piedmont. *Ecology*, 65, 354–69.
Strong, D. R. (1992) Are trophic cascades all wet? Differentiation and donor-control in speciose ecosystems. *Ecology*, 73, 747–54.
Summerhayes, V. S. & Elton, C. S. (1923) Contributions to the ecology of Spitzbergen and Bear Island. *Journal of Ecology*, 11, 214–86.

Sundell, J., Norrdahl, K., Korpimäki, E., & Hanski, I. (2000) Functional response of the least weasel, *Mustela nivalis nivalis*. *Oikos*, 90, 501–508.
Swift, M. J., Vandermeer, J., Ramakrishnan, P. S., Anderson, J. M., Ong, C. K., & Hawkins, B. A. (1996) Biodiversity and agroecosystem function. In J. H. Cushman, H. A. Mooney, E. Medina, O. E. Sala, & E. D. Schulze, eds. *Functional Roles of Biodiversity: a Global Perspective*. Chichester, UK, John Wiley & Sons.
Tahil, A. S. & Juinio-menez, M. A. (1999) Natural diet, feeding periodicity and functional response to food density of the abalone, *Haliotis asinina* L., (Gastropoda). *Aquaculture Research*, 30, 95–107.
Tansley, A. G. (1935) The use and abuse of vegetational concepts and terms. *American Naturalist*, 16, 284–307.
Taylor, D. L. & Collie, J. S. (2003) Effect of temperature on the functional response and foraging behavior of the sand shrimp *Crangon septemspinosa* preying on juvenile winter flounder *Pseudopleuronectes americanus*. *Marine Ecology Progress Series*, 263, 217–34.
Taylor, W. D. (1978) Responses of ciliate Protozoa to the abundance of their bacterial prey. *Microbial Ecology*, 4, 207–14.
Teal, J. M. (1962) Energy flow in the salt marsh ecosystem of Georgia. *Ecology*, 43, 614–24.
Teng, J. & McCann, K. S. (2004) Dynamics of compartmented and reticulate food webs in relation to energetic flows. *American Naturalist*, 164, 85–100.
Thrall, P. H. & Burdon, J. J. (2003) Evolution of Virulence in a Plant Host-Pathogen Metapopulation. *Science*, 299, 1735–7.
Tilman, D. (1982) *Resource Competition and Community Structure*, Princeton, NJ, Princeton University Press.
Tilman, D. (1994) Competition and biodiversity in spatially structured habitats. *Ecology*, 75, 2–16.
Tilman, D., Wedin, D., & Knops, J. (1996) Productivity and sustainability influenced by biodiversity in grassland ecosystems. *Nature*, 379, 718–20.
Tittensor, D. P., Mora, C., Jetz, W., Lotze, H. K., Ricard, D., Berghe, E. V., & Worm, B. (2010) Global patterns and predictors of marine biodiversity across taxa. *Nature*, 466, 1098–101.
Ulanowicz, R. E. (1986) *Growth and Development: Ecosystems Phenomenology*. New York, Springer-Verlag.
Urban, D. & Keitt, T. (2001) Landscape connectivity: a graph–theoretic perspective. *Ecology*, 82, 1205–18.
van de Koppel, J., Huisman, J., van der Wall, R., & Olff, H. (1996) Patterns of herbivory along a productivity gradient: an empirical and theoretical investigation. *Ecology*, 77, 736–45.
Varley, G. C. (1970) The concept of energy aaplied to a woodland community. In A. Watson, ed. *Animal Populations in Relation to their Food Resources*. Oxford, Blackwell Science.
Verdy, A. & Caswell, H. (2008) Sensitivity analysis of reactive ecological dynamics. *Bulletin of Mathematical Biology*, 70, 1634–59.
Verhoef, H. A. & Brussaard, L. (1990) Decomposition and nitrogen mineralization in natural and agro-ecosystems: the contribution of soil animals. *Biogeochemistry*, 11, 175–211.
Viertel, B. (1992) Functional response of suspension feeding anuran larvae to different particle sizes at low concentrations (Amphibia). *Hydrobiologia*, 234, 151–73.
Volterra, V. (1927) Variazioni e fluttazioni del numero d'individui in specie animali conviventi. *Venezia: Regio Comitato Talassografico Italiano*, Mem. CXXXI.
Wade, M. J. & Goodnight, C. J. (1991) Wright's shifting balance theory: an experimental study. *Science*, 253, 1015–18.

Waide, R. B., Willig, M. R., Steiner, C. F., Mittelbach, G., Gough, L., Dodson, S. I., Juday, G. P., & Parmenter, R. (1999) The relationship between productivity and species richness. *Annual Review of Ecology and Systematics*, 30, 257–300.

Wall, D. H. & Moore, J. C. (1999) Interactions underground. Soil biodiversity, mutualism, and ecosystem processes. *BioScience*, 49, 109–17.

Walsh, G. E. (1967) An ecological study of a Hawaiian mangrove swamp. In G. H. Lauff, ed. *Estuaries—Publication number 83*. Washington, DC, American Association for the Advancement of Science.

Wander, M. M. & Bidart, M. G. (2000) Tillage practice influences on the physical protection, bioavailability and composition of particulate organic matter. *Biology and Fertility of Soils*, 32, 360–7.

Wardle, D. A., Bardgett, R. D., Klironomos, J. N., Setala, H., Van Der Putten, W. H., & Wall, D. H. (2004) Ecological linkages between aboveground and belowground biota. *Science*, 304, 1629–33.

Warren, P. H. (1990) Variation in food-web structure: the determinants of connectance. *American Naturalist*, 136, 689–700.

Webb, C. T. (2007) What is the role of ecology in understanding ecosystem resilience? *BioScience*, 57, 470–1.

Webb, C. T. & Levin, S. A. (2005) Cross-system perspectives on the ecology and evolution of resilience. In E. Jen, ed. *Robust Design: a Repertoire of Biological, Ecological, and Engineering Case Studies*, Oxford, Oxford University Press.

Weintraub, M. N. & Schimel, J. P. (2005) Nitrogen cycling and the spread of shrubs control changes in the carbon balance of arctic tundra ecosystems. *Bioscience*, 55, 408–15.

Whicker, A. & Detling, J. K. (1988) Ecological consequences of prairie dog disturbances. *BioScience*, 38, 778–85.

Whittacker, R. H. (1975) *Communities and Ecosystems*, 2nd Ed., New York, Macmillan.

Wiener, N. (1948) *Cybernetics*, New York, Wiley.

Wilson, E. O. (1969) The species equilibrium. *Brookhaven Symposia in Biology*, 22, 38–47.

Wolters, V., ed. (1997) Functional implications of biodiversity in soil. *Ecosystems Research Report*, 24, 45–8.

Woodwell, G. M. (1967) Toxic substances and ecological cycles. *Scientific American*, 216, 24–31.

Wootton, J. T. (1994) Predicting direct and indirect effects: an integrated approach using experiments and path analysis. *Ecology*, 75, 151–65.

Yodzis, P. (1980) The connectance of real ecosystems. *Nature*, 284, 544–5.

Yodzis, P. (1981) The stability of real ecosystems. *Nature*, 289, 674–6.

Yodzis, P. (1988) The indeterminacy of ecological interactions as perceived through perturbation experiments. *Ecology*, 69, 508–12.

Yodzis, P. & Innes, S. (1992) Body-size and consumer-resource dynamics. *American Naturalist*, 139, 1151–73.

Zaret, T. M. & Paine, R. T. (1973) Species introduction in a Tropical Lake. *Science*, 182, 449–55.

Zwart, K. B., Burgers, S. L. G. E., Bloem, J., Bouwman, L. A., Brussaard, L., Lebbink, G., Didden, W. A. M., Marinissen, J. C. Y., Vreeken-buijs, M. J., & de Ruiter, P. C. (1994) Population dynamics in the belowground food webs in two different agricultural systems. *Agriculture, Ecosystems, and Environment*, 51, 187–98.

Index

Page numbers given in bold indicate where a subject is defined

Allochthonous resources, sources, inputs 24, **27**–9, **35**, 46–9, 101, 143–4, 165, 206, 297; *see also* donor-controlled; detritus
Apparent complexity 22, 234, 303, **305**–8
Autochthonous resources, sources, inputs 24, **27**–9, 35, 46–9, 101, 143, 206; *see also* detritus
Asymptotically ambiguous states 285, **303**–7

Biomass 8, 10, 20–2, 27, **30**
  distribution of 79–89, 132–142
  pyramid of 8, 21, **79**–89, 93, 121, 123, 131–4, 139–40, 147–52, 165, 169–71, 243, 252–3, 268, 295
Body size 34, 60, 69, 75, 109–10, 155–7, 169, 177, 229, 243, 285, 307
Bottom-up control 4, 7

Central Plains Experimental Range (CPER) food web, *see* soil food webs
Chaos 125, 157, **161**–65, 170, 215, 248, 306
Community matrix, *see* Jacobian Matrix, *A*
Community perspective **5**–7, 21
Compartments 22, 91–93, **125**, 171–207, 213, 224, 227, 233, 252, 260, 283, 289, 297–300, 305–308; *see also* energy channel
Complexity 3, 11, 15, 22, 37, 64–8, 123–5, 172–5, 206, 231–4, 245, 252–83, 292, 303–8
Connectance, *C* 61, **64**–8, 71, 173–80, 195–7, 232, 239–45, 256, 264, 265, 278, 292
Connectedness, *see* food web
Conservation of matter 9, 27, 75, 220
Cybernetic 8, **9**, 19

Death rate, *d* 20, 24, 29, **30**, 47, 50, 56, 73, 75, 77, 78, 91, 92, 98–103, 111, 112, 128, 130, 138, 143–5, 151, 161, 164, 167, 201, 218–22, 227, 247, 249, 267, 285, 287, 289, 295
Decomposition 1, 59, 70, 199, 270–1

Detritus 7, 8, 10, **13**, 16, 17, 18, 20, 54–71, 72–93, 98, 124, 167, 169, 170, 173, 175, 242, 266, 269, 274, 275, 276, 277, 279, 281, 282, 297, 306; *see also* allocthonous inputs; autochthonous inputs
  Compartments 173, 174, 180–4, 186–90, 195, 196, 199; *see also* energy channel
  Donor control 160, 206–7
  Feedback 18, 57, 173
  Input rate, R_D **29**, 30, 46–51, 101, 143–6, 152–5, 161, 214, 220, 224, 225, 232, 238, 252, 253, 254, 260, 299
  Interaction strengths 100–21
  Quality (labile and resistant) 35, 55–61, **59**, 181, 198, 200, 276
Detritus-based models 23, 24, **27**, 29, 44, 50–3, 125, 143–7, 160, 169, 170, 297; *see also* food chain, detritus-based food chain
  Equilibrium of 47
  Resilience of, return-time of 50–2, 144–7, 152–5, 169, 206, 213, 299
  Stability of 44, 48–50
Diagonal dominance 44; *see also* Routh-Hurwitz criterion
  Quasi-diagonal dominance **39**–41, 138, 161
Disturbance 39, **41**, 138; *see also* equilibrium, deviation from
Donor control 7, 16, 41, **46**–50, 53, 69, 160–1, 165, 196, 206–7, 292; *see also* food chain, detritus based-food chains; detritus-based models
Dynamic architectures **22**, 232, 235–9, 250, 252

Ecosystem perspective 5, **7**–11, 21, 175
Eigenvalues 15, **37**–8, 45, 49, 146, 203, 216, 245–7
Energetic efficiency, *e* **33**–5, 50, 104, 152, 155, 158, 167, 222, 294, 295, 297

Energetic efficiency, e (*cont.*)
  Assimilation efficiency, a 29, 33, **34**, 35, 47, 58, 76, 77, 98, 101, 128, 149, 188, 287, 296
  Production efficiency, p 29, 33–**5**, 47, 58, 76, 77, 98, 101, 128, 149, 188, 287
Energetic organization **3**, 103, 105, 123–5, 128–131, 133, 140, 141, 229, 233, 238, 243, 244, 250, 252, 257, 283, 298
Energy budgets 35, 78
Energy channels **125**, 133, 181–207, 225–6, 271, 282, 309
Enrichment 21, 123, 124, 134, 140, **141**–71, 204–6, 213–8, 294, 295, 297
  CO_2 enrichment 199
  Paradox of enrichment 147, **159**–65, 204, 205, 215–6
Environmental heterogeneity 227–8, 283; *see also* compartments; energy channels
Equilibrium 36–41, 45–50, 52, 67, 95–102, 128, 139, 144–9, 157–170, 198, 201, 204, 215, 216, 239, 240, 245–9, 280, 284, 285, 289–92, 296, 301, 304–9; *see also* steady-state
  Deviation from 141, **142**, 147, 155
Exploitation ecosystem hypothesis (EEH) **37**, 39, 44, 50, 213, 290, 291

Feasibility **45**, 50, 51, 125, 126, 143–156, 167, 168, 227, 229, 247, 252, 292, 297
  Dynamics, $F_{[\lambda]}$ 123–5, 147–56, 159, 214–29, 280–92, 296–309
  Productivity, $F_{[NPP]}$ 18, **45**, 123–6, 147–55, 214–23, 249, 280–3, 298–300, 309
  Spatial networks 298–300
  Species diversity 213–23
  System, $F_{[sys]}$ 123–6, 214–29, 226, 230–3, 280–92, 296–309
Feedback 9, 10, 11, 18, 32, 39, 40, 57, 70, 232, 250, 276, 300
  Loops 40, 173
  Self-feedback 39
Feeding rates 20, 30–3, 42, 75–92, 99–102, 105–9, 111, 113–21, 127, 129, 131, 132, 134, 135, 138, 139, 167, 286–8; *see also* functional response
Food chain 2, 3, 4, 8, 39, 40, 44–7, 50–2, 61, 65, 71, 132, 141–70, 175, 176, 180, 195, 201–5, 213–27, 245, 248, 250, 252–7, 264, 265, 268, 278, 280, 283, 292, 297, 298, 301, 304
  Detritus-based food chains 21, **28**, 44, 50–2, 143–6, 152–5, 169, 206, 213, 214, 292
  Food chain length – maximum (FCL$_{Max}$) 23, **61**, 64–6, 71, 245, 256, 264, 265, 278

Food chain length – mean (FCL$_{Mean}$) 23, **61**, 64–6, 71, 245, 256, 264, 265, 278
Primary producer-based food chains **28**, 44, 51, 143, 145, 146, 149, 152–7, 214, 221, 223
Food web 3, 7, 8, **11**–20
  Community food web 11–**3**, 54, 55, 180, 187, 192, 195
  Connectedness web 13–**4**, 21–4, 38, 43–4, 54–71, 72, 75, 79, 92, 94, 103, 104, 109 113, 128, 129, 172, 177, 180, 207, 235, 243–50, 252, 256, 258, 265, 270, 272, 276, 278, 285, 305, 306
  Energy flow/flux web 8, 13, 14, 23, 54, 196, 225, 285
  Food web architecture 21, 22, 93, 142, 173, 202, 235–8, 240, 243, 248 296, 307
  Food web structure 42, 70, 93, 103, 135, 137, 225, 231, 243, 250, 253, 264, 266, 282, 283, 284, 285, 288, 289, 296, 305, 306, 308, 309
  Functional food web 14, **15**, 25, 75, 80, 93–121, 127, 178
  Sink food web 12, 13, 14
  Soil, *see* soil food webs
  Source food web 11–**3**
Functional groups 16, 17, **19**, 20, 21, 24, 54–71, 72–5, 79–93, 94, 97–105, 110–21, 123, 127, 128, 134, 141, 146, 172, 183, 199, 242, 250, 256, 265, 271, 278, 285, 286
Functional response 29, **30**–3, 36, 49, 58, 75, 111, 164, 166, 170, 201, 202, 213, 215, 216, 221, 295, 302
  Estimating interaction strengths 94–9, 101
  Holling type I 30–1, 33, 36, 50, 52, 98, 99, 109, 121, 143, 157–9
  Holling type II 30–**2**, 33, 49, 157–9, 161–6, 200, 201, 295, 302
  Holling type III 30, **32**, 33, 110, 157, 201, 225

Home range 24, 56, 69, 109, 110, 126, 155, 156, 192, 213, 228, 229, 307
Horseshoe Bend food web, *see* soil food webs

Immobilization **25**, 249; *see also* mineralization
Inorganic nitrogen 55–**9**, 198
Interaction matrix, *see* Jacobian matrix, A
Interaction strength **37**–**8**, 123, 124, 201, 202, 205, 232, 245–8, 267, 283, 284, 287, 288; *see also* per capita effects
  Detritus interactions 160–1
  Diversity and complexity 64–8, 173–6, 207
  Energy flux 21, 103–11
  Estimation of 15, 24, 94–102
  Feasibility, *see* feasibility

Loops 42, 43; *see also* trophic interaction loops
Stable patterns 22, 103, 113–21, 127–40
Internal cycling 27, 48, 49, 51, 170; *see also* autochthonous resources, sources, inputs
Intrinsic rate of increase, *r*, *see* productivity and enrichment
Island Biogeography 7, 9, 22, **239**–42, 298
  Colonization rate 6, 7, 237, 239–42, 247, 250, 260, 280, 283, 284
  Extinction rate 7, 158, 213, 218, 220, 226, 239–42, 247, 283, 290, 303
  Species turn-over 187, 188, 202, 207

Jacobian matrix, *A* **37**, 38, 40, 53, 93, 95, 98–103, 110, 126, 128–31, 187, 188, 201, 207, 231, 250, 264, 280, 284, 287, 289, 290, 291, 308
  Detritus 100–2
  Diagonal elements 38–47, 67, 102, **103**, 128, 130, 131, 137, 138, 161, 218, 280, 292
  Dimensions of **54**, 64, 98, 103, 231, 235, 238, 242, 244, 248, 249
  Elements **98**, 104, 121, 218
  Examples of **37**, 40, 43, 45, 47, 100, 161
  Loops 135–8, 264; *see also* trophic interaction loops
  Off-diagonal elements 39, 41, 64, 67, **99**–103, 138, 178, 292
  Patterns 38–41, 174
  Sign structure 38, 235

Kjettslinge food web, *see* soil food webs

Life history
  Strategies 55, 195, 215, 220, 230, 285
  Traits (attributes, characteristics, parameters) 10, 15, 18–22, 24, 30, 56, 57, 71, 187, 188, 196, 197, 207, 215, 220, 222, 229, 231, 247, 248, 283, 285, 300, 305, 307
Lotka-Volterra 7, 11, 16, 27, 36, 45, 49, 95, 98, 121, 141, 143, 147, 148, 214, 248, 292, 302, 304, 305
Lotka-Volterra cascade model (LVCM) 292, 294
Lovinkhoeve food web, *see* soil food webs

Mass balance 25, 75, **77**, 93, 99, 144; *see also* conservation of matter
Michaelis-Menton equation **32**; *see also* functional response
Microbes 190
  Bacteria and fungi (saprophytic and mycorrhizal – VAM) 8, 17, 48, 69, 73, 74, 136, 151, 167, 180–4, 187–92, 198–200, 256–63, 265–8, 269–78, 281, 282
  Bacterial and fungal pathways 180–4, 187–92, 198–200, 277, 281; *see also* energy channel
  Examples of bacteria and fungi in food webs 55–64, 73, 74, 79–91, 100–20, 136, 256–63, 265–8, 269–78
Mineralization **25**, 75–9, 249, 270, 285
  Carbon mineralization 72, 91, 92, 198; *see also* CO_2 emission
  CO_2 emissions (respiration) 24, 35, 56, 57, 72, 78, 198, 200
  Nitrogen mineralization 72, 79, 80, 90–3, 112, 198, 200, 287, 288
  Rates of 72, 75–9, 80, 90–3, 111, 112, 198, 286–8
Mortality
  Due to natural death 49, 76, 77, 111, 265, 289
  Due to predation 31–2, 76, 77, 102, 111–3, 143, 265, 287, 289

Omnivory 45, **68**–70

Paradox of enrichment, *see* enrichment
Per capita effects **37**, 95, 101, 135, 141; *see also* interaction strength; Jacobian matrix, *A*
Permanence 233, **303**–5, 308
Perpetual/persistent transient state 301, 303
Physiological parameters 72, 73, 247, 248
Primary producer-based models 23, 24, **27**, 28, 44, 46, 50–3, 125, 143–7, 159, 169; *see also* primary-producer-based food chains
Productivity 22, **51**, 69, 124–6, 232–4; *see also* enrichment
  Body size and homerange 155–7
  Ecosystem estimates 165–9
  Energetic efficiencies 152–5
  Feasibility 144–6, 214–8
  Productivity gradient 22, 141–71, 208–29, 252–83
  Resilience 144–7
  Species richness 208, 218–23
  Trophic structure and dynamics 147–52, 157–65; *see also* paradox of enrichment; exploitation ecosystem hypothesis

Reactivity **290**–7
Resilience, *see* return time, *RT*; transient dynamics
Return time, *RT* **50**–2, 144–7, 152, 155, 161, 169–70, 206, 213–7, 227, 290
Routh-Hurwitz criterion **38**–41

Schiermonnikoog food web, *see* soil food webs
Soil fauna 17, 55, 80, 93, 241, 260, 279, 285
  Detritivores 17, 24, 40, 46, 48, 50, 69, 98, 101, 143, 181, 206
  Examples of soil fauna in food webs 55–64, 73, 74, 76, 79–91, 100–20, 136, 256–63, 265–8, 269–78
  Functional response of 58
Soil food webs 13, 16, 25, 55, 61, 64–8, 71
  Central Plains Experimental Range (CPER) food web 17, 21–5, 54–60, 66, 69–74, 79–89, 90, 98–106, 110–4, 120, 129–31, 135–8, 165, 181–3, 186, 191–2, 267
  Horseshoe Bend food web 25, 63–6, 74, 83–5, 89–92, 107–9, 117–20
  Kjettslinge food web 25, 63–6, 73–4, 83–5, 89–91, 108–9, 119–20
  Lovinkhoeve food web 25, 62, 64, 66, 74, 83, 85, 89, 90, 91, 106–9, 112–20, 190, 191, 199, 285–8
  Schiermonnikoog food web 233, 253, 260–8, 280–2, 288
  Toolik Lake food web 64, 190, 199, 233, 253, 268–80
  Wind Cave food web 48, 232, 253–63, 280, 281
Spatial
  Asynchrony 303
  Averaging 37, 71, 178, 234, 307; *see also* temporal averaging
  Heterogeneity 289, 306; *see also* temporal heterogeneity
  Modularity 298–301
  Networks 297, 298–301, 307, 309; *see also* temporal networks
  Scale 172, 187, 205, 213, 226, 232, 234, 236, 285, 289, 305, 307, 308; *see also* temporal scale
  Systems 296
Species richness, S 10, 22, **61**, 64, 67, 125, 173, 174, 176, 177, 179, 196, 208–29, 239–41, 243, 271, 296
Stability 3, 9–11, 15, 16, 19–23, 27, 33, **36**–44, 46–54, 60, 61–8, 94–7, 102, 103, 113–21, 123–6, 206, 207, 231–4, 307, 309; *see also* permanence; transient dynamics
  Asymptotic 231–3, 244, 284–9, 292, 294, 303, 304, 306, 307, 309
  Diversity, complexity, and compartments 173–5, 200–4, 208–29
  Donor control 206–7; *see also* detritus-based models; food chain, detritus-based food chain
  Energetic organization 127–40
  Enrichment 141–71, 260–8, 280–3
  Local stability **37**, 38, 44, 157, 227, 245–8, 285

Lypanov function **44**, 46, 49, 305; *see also* stability, global
  Global **44**, 46
  Qualitative 38–9, 146
Steady-state, *see* equilibrium
  Densities or biomass 15, 21, **45**, 50, 76, 77, 79, 91, 92, 144, 149, 150, 167, 198, 213, 287, 299
  Dynamics **45**, 77, 91, 131, 132, 134, 213, 231, 232, 285–9
  Flux rates 132

Temporal
  Asynchronony/synchrony 186, 303
  Averaging 37, 70–1, 178, 187, 234, 307
  Heterogeneity 22, 289, 306; *see also* spatial heterogeneity
  Networks 194, 309; *see also* spatial networks
  Scale 58, 59, 70, 73, 172, 179, 205, 213, 226, 232, 234, 236, 285, 289, 305, 307, 308; *see also* spatial scale
Thermodynamics, 1st and 2nd laws 9, 25, 27, 287; *see also* conservation of matter
Toolik Lake food web, *see* soil food webs
Top-down control 3, **7**
Transient dynamics 22, 216–9, 229, 233, 285, 289–96, 301, 306, 307, 308; *see also* return time
Transient state, perpetual/persistent 301, 303
Transitions 68, 125, 148–51, 161, 164, 170, 203, 215–7, 244–50, 276, 279, 280–2, 290
  in dynamic states 125, 147, 150–2, 155, 157, 161, 163, 164, 165, 170, 174, 204, 215–8, 248, 273, 280–3, 290
  in trophic structure 125, 149–50, 157, 165, 215, 280–3, 290
Trophic cascade 3, 4, 7, **142**, 150, 151, 238, 257
Trophic interaction loops 16, **42**–3, 127, 135–40, 147, 151, 252, 280
  Critical loop **268**, 280; *see also* maximum loop weight
  Loop length **42**–3, 135–7, 139
  Loop weight 16, **42**–3, 135–40, 147, 151, 171, 233, 252, 264–8, 280, 282, 288, 289
  Maximum loop weight **42**, 137–40, 264–7, 280, 282
Trophic level 8, **44**, 45, 55, 68, 69, 79, 175, 228, 230, 233, 252–4, 280–3, 294, 305
  Biomass 105, 121, 132–4, 165, 171, 191, 205, 243, 268, 295
  Energetic efficiency 152, 229
  Food chain length **61**, 124, 141–52, 170, 171

Functional response 110
Interaction strength 105, 131, 132
Spatial distribution 225, 260
Trophic interaction loops 138, 171, 268
Trophic structure **22**, 53, 139, 224, 225, 229
 Body size 155–7
 Case studies of 235–83
 Distribution of biomass 139, 147–52
 Enrichment and stability 141–71
 Productivity and dynamics 213–8
 Transient dynamics 292, 295, 301
Tuesday Lake food web 239, 243–5

VAM, *see* microbes

Wind Cave food web, *see* soil food webs

DATE DUE

PRINTED IN U.S.A.